Amino Acids in Higher Plants

Editor

Jayant Khillare

scitus
academics

Amino Acids in Higher Plants

Edited by **Jayant Khillare**

ISBN: 978-1-68117-042-8

Library of Congress Control Number: 2015935437

www.scitusacademics.com
Box No. 4766, 616 Corporate Way,
Suite 2, Valley Cottage,
NY 10989

Printed in United States of America on Acid Free Paper ∞

Contents

Preface

Amino acid biochemistry in plants has been a major topic research. There are many aspects of amino acid metabolism in plants that has to be understood. The biosynthesis and metabolic regulation of synthesis, of several protein amino acid still pose questions such as histidine synthesis, the regulation of methionine synthesis and proline metabolism which is dealt with in a limited way. Amino acid derived secondary metabolites are still rich field of biochemists and recent progress in understanding metabolisms is encouraging. Plants convert inorganic nitrogen into amino acids, the building blocks for proteins. They also make a wide range of other nitrogen compounds to help protect themselves from pests and diseases. An understanding of these compounds can therefore help in devising better crop protection and production methods. This volume contains essays by scientists who have studied aspects of plant nitrogen nutrition and amino acid biosynthesis. There are chapters on protein amino acids, nonprotein amino acids, betaines, glutathione, polyamines, and other secondary metabolites derived from amino acids. The results of these studies will be of interest to students and professionals in biochemistry and botany.

Editor

1

Plastid-Localized Amino Acid Biosynthetic Pathways of Plantae Are Predominantly Composed of Non-Cyanobacterial Enzymes

Adrian Reyes-Prietoa, [1,3] and Ahmed Moustafa[2,3]

[1]Canadian Institute for Advanced Research and Department of Biology, University of New Brunswick, Fredericton, Canada

[2]Department of Biology and Biotechnology Graduate Program, American University in Cairo, Egypt.

ABSTRACT

Studies of photosynthetic eukaryotes have revealed that the evolution of plastids from cyanobacteria involved the recruitment of non-cyanobacterial proteins. Our phylogenetic survey of .100 Arabidopsis nuclear-encoded plastid enzymes involved in amino acid biosynthesis identified only 21 unambiguous cyanobacterial-derived proteins. Some of the several non-cyanobacterial plastid enzymes have a shared phylogenetic origin in the three Plantae lineages. We hypothesize that during the evolution of plastids some enzymes encoded in the host nuclear genome were mistargeted into the plastid. Then, the activity of those foreign enzymes was sustained by both the plastid metabolites and interactions with the native cyanobacterial enzymes. Some of the novel enzymatic activities were favored by selective compartmentation of additional complementary enzymes. The mosaic phylogenetic composition of the plastid amino acid biosynthetic pathways and the reduced number of plastid-encoded proteins of non-cyanobacterial origin suggest that enzyme recruitment underlies the recompartmentation of metabolic routes during the evolution of plastids.

Primary plastids of plants and algae are the evolutionary outcome of an endosymbiotic association between eukaryotes and cyanobacteria[1]. The establishment of the permanent photosynthetic endosymbionts involved critical evolutionary scenarios, such as cyanobacteria surviving the digestive process[2], the establishment of mechanisms for metabolite exchange between both partners[3, 4], the evolution of system for the transport of cytoplasmic-translated proteins into the endosymbiotic cells[5], and the loss or transfer of genes from the endosymbiont into the host nuclear genome[6]. These latter mechanisms contributed to a significant reduction in the plastid gene-coding capacity. Typical plastid genomes encode only circa 10% of the plastid proteome[7]. Recent surveys of diverse plant and algal nuclear genomes have identified dozens to hundreds of plastid-targeted proteins of non-cyanobacterial origin[8] and cases of plastid-localized pathways composed of enzymes of diverse phylogenetic origin. These pathways include the Calvin Cycle[9, 10] and the shikimate biosynthetic route[11].

In addition to their photosynthetic capabilities, key biochemical pathways such as the *de novo*synthesis of fatty acids[12], isoprenoid synthesis[13], and critical steps of nitrogen assimilation[14], including several pathways for amino acid biosynthesis (Fig. 1) occur in plastids of angiosperms. Nitrogen assimilation (NA) in plastids begins with nitrite (NO_2^-) uptake and its subsequent reduction to ammonia (NH_3) by the enzyme nitrite reductase. Ammonia and glutamate are converted into glutamine by the plastid glutamine synthetase (GS) and then the amido group of the glutamine is transferred to a molecule of 2-oxoglutarate by the glutamate synthetase (GOGAT), producing a net gain of one glutamate molecule. The concerted activity of these two plastid enzymes constitutes the "GS/GOGAT cycle", a pivotal step in the biosynthesis of diverse amino acids and other metabolites (Fig. 1). Numerous enzymes participating in the biosynthesis of chorismate[11], histidine[15], aromatic[16], branched-chain[17], and aspartate-derived[18] amino acids are localized in plastids of angiosperms and green algae[19]. The biosynthesis of methionine[20] and cysteine[21] apparently occurs, but not exclusively, in plastids as well. Enzymes involved in the biosynthesis of proline[22] and arginine[23] have been also identified in plant plastids. Thus, diverse experimental and *in silico* evidences strongly suggest that a number of enzymes catalyzing critical reactions for the biosynthesis of several amino acids are localized in angiosperm plastids[14].

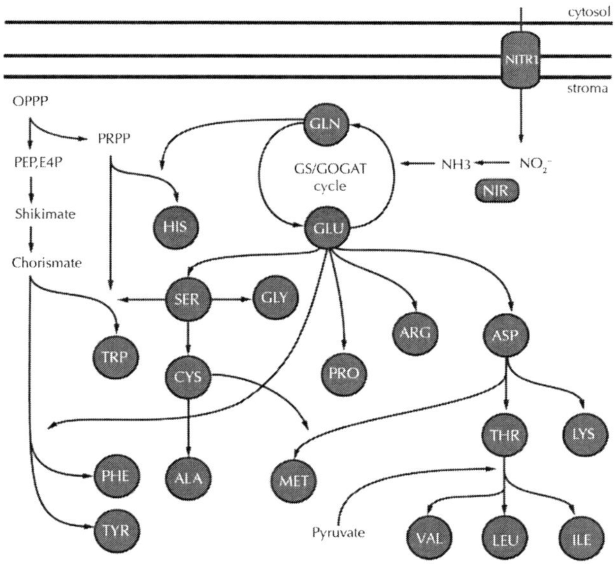

Figure 1: Schematic metabolic map of the amino acid biosynthetic pathways localized in the *Arabidopsis thaliana* plastids.

Our central aims are to evaluate if the dissimilar phylogenetic history described for some*Arabidopsis* plastid biochemical routes prevails in the plastid amino acid biosynthesis (AAB) pathways, and, importantly, if this phylogenetic mosaicism is shared among the three different Plantae (*sensu* Cavalier-Smith[24]) lineages: Viridiplantae (plants and green algae), Rhodophyta (red algae), and Glaucophyta. Considering the ancient (circa 1.5 billion years ago[25, 26]) evolutionary divergence between the three Plantae lineages, we would expect some dissimilarities in the subcellular localization of certain enzymatic reaction or even entire metabolic pathways (i.e., metabolic compartmentation[27]) as consequence of more than one billion years of independent evolution. However, if the Plantae plastids have a single origin through primary endosymbiosis, we expect as well to identify a number of shared non-cyanobacterial enzymes that represent common recruitments during the early evolution of the plastid proteome. We used the well-characterized and curated biochemical and genomic knowledgebase from *Arabidopsis* as a reference to investigate the evolution of the plastid-localized AAB pathways.

RESULTS

Even though we inferred ML phylogenies of the 158 *Arabidopsis thaliana* proteins involved in AAB, our central analysis was focused on a subset of 103 proteins that we identified as plastid-localized products. As expected, not all ML trees inferred from single-locus alignments are easily interpretable. To alleviate irregular taxa distribution and stochastic errors inherent to molecular phylogenetic estimations, such as short sequences, compositional biases, use of oversimplified substitution models and incomplete lineage sorting, we focused our analysis on 92 ML trees, including 62 plastid-localized proteins (Table 1), where at least two Plantae lineages branch together in the same clade, regardless of whether they form a monophyletic group or are intermingled with algal lineages harboring plastids of secondary origin. In order to contrast each ML tree against a *null hypothesis*, we defined our reference "type tree" as the topological pattern where the *Arabidopsis* (Plantae in general) plastid proteins branch with cyanobacterial homologs reflecting an ancestral endosymbiotic origin of the encoding gene. After visual inspection of the 62 trees of plastid-localized, we identified only 21 proteins (corresponding to 13 different enzymatic activities) of unambiguous cyanobacterial origin. Additionally, we identified 8 trees where the cyanobacterial provenance of the plastid-enzyme cannot be entirely discerned. In contrast, we identified 33 trees presenting topologies of apparent incongruence with the cyanobacterial ancestry of the corresponding enzyme (Table 1). The non-cyanobacterial origin of these 33 proteins was additionally supported in 15 phylogenetic estimations that recovered cyanobacterial homologs in the same tree but branching in different clades and distant from *Arabidopsis* and other Plantae rejecting the possibility of cyanobacterial origin of those enzymes in Plantae. We identified as well four cases of cyanobacterial-derived enzymes likely localized in the cytoplasm of the plant cells.

Table 1: Phylogenetic origin of 62 plastid-localized enzymes involved in amino acid biosynthesis

Amino acid biosynthesis pathway	Plastid-localized isoenzymes	Cyanobacterial	Non-cyanobacterial	ML bootstrap (%)
Glutamine				
Glutamine synthetase	1		1	100
Glutamate				
Ferredoxin-dependent Glutamate synthase (GOGAT)	2	2		>63
Tryptophan				
Anthranilate synthase alpha subunit	2		2	>70
Anthranilate synthase beta subunit	2	2		<50
Anthranilate phosphoribosyl transferase	1		1	<50
Phosphoribosyl anthranilate isomerase	3		3	95
Tryptophan synthase apha subunit	2	2		>75
Tryptophan synthase beta subunit	2	2		100
Phenylalanine/Tyrosine				
Arogenate dehydratase/ prephenate dehydratase	6		6	<50
Arogenate dehydrogenase	2		2	<50
Histidine				
ATP phosphoribosyltransferase	2		2	<50
BBMII isomerase	1		1	<50
Imidazoleglycerol-phosphate dehydratase	2	2		>94

Histidinol-phosphate aminotransferase	2		2	>95
Histidinol phosphate phosphatase	1		1	<50
Aspartate-derived				
Aspartate kinase	1		1	<50
Lysine				
L-diaminopimelate aminotransferase	1		1	<50
Diaminopimelate epimerase	1	1		58
Diaminopimelate decarboxylase	2		2	>94
Threonine				
Threonine synthase	1		1	<50
Methionine				
Cystathionine gamma-synthase	1		1	100
Isoleucine and Valine				
Acetolactate synthase large subunit	1		1	99
Acetolactate synthase small subunit	1	1		95
Ketolacid reductoisomerase	1		1	<50
Leucine				
Isopropylmalate isomerase	4		4	<50
Isopropylmalate dehydrogenase	4	4		>75
Arginine				
N-Acetylglutamate synthase	1		1	100
N-Acetylglutamate kinase	1	1		69
N-Acetylglutamate-5-P reductase	1		1	91
N 2 - Acetylornithine:glutamate acetyltransferase	1	1		99
Carbamoyl-P synthetase Large subunit	1	1		100
Ornithine carbamoyltransferase	1		1	<50
P II nitrogen sensing protein GLB I	1	1		<50
Serine				
3-phosphoglycerate dehydrogenase	2		2	<50
Glycine				

Serine (glycine) hydroxymethyltransferase	2		2	65
Serine (alanine):glyoxylate aminotransferase	1		1	94
Alanine				
Cysteine desulfurase	1	1		<50
Total	62	21	33+8	

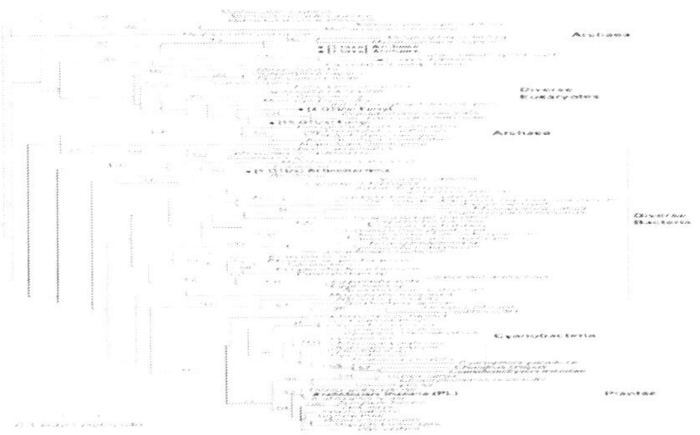

Figure 2: Maximum Likelihood phylogenetic tree of the imidazoleglycerol-phosphate dehydratase (At4g14910.1). The numbers at the nodes indicate RaxML bootstrap values (only values >50% are shown). Branch lengths are proportional to the number of substitutions per site (see the scale bars). Colored boxes highlight Plantae lineages (glaucophytes, red algae and viridiplants). PL, plastid localized protein.

The Enzymes of the GS/GOGAT Cycle

The plastid glutamine synthetase (GS) in *Arabidopsis*, viridiplants in general, and red algae has a putative host origin[28]. The *Cyanophora* homolog was not recovered in our plastid GS tree, but it was retrieved in the several ML trees estimated when *Arabidopsis* non-plastid GS isoenzymes were used as queries. The only GS encoded in the genome of the red alga *Cyanidioschyzon merolae* is probably cytosolic-localized[29]; however, the GS subcellular localization in glaucophyte algae is unknown. In contrast to the convoluted evolution of the plastidic GS, the two *Arabidopsis* plastid Fd-GOGAT are of cyanobacterial provenance (Table 1). The Fd-GOGAT encoding gene is still present in plastid genomes of red algae. The plastid NADH-dependent GOGAT, more abundant in non-photosynthetic tissues, is of a putative host origin. Given that GS is likely cytosolic in *Cyanidioschyzon* and the Fd-GOGAT is plastid-localized, the existence of a plastid GS/GOGAT cycle in plastids of red algae is unlikely[30]. These results and previous evidence indicate that the plastid-localization of the GS/GOGAT cycle is a particularity of streptophytes and that those two enzymatic players have distinct ancestral origins.

Aromatic Amino Acids and Histidine Biosynthesis

Our analyses indicate that the plastid routes for the biosynthesis of the aromatic amino acids tryptophan, tyrosine, and phenylalanine (Fig. 1) comprise only 4 out of 11 (i.e., < 40%) enzymatic components of cyanobacterial origin (see Table 1). An interesting case is the plastid-localized alpha subunits of the anthranilate synthase (ASA), which have non-cyanobacterial origin in Plantae lineages, branching in well-supported clades with Planctomycetes bacteria homologs (Fig. 3a and Table 1; >70% bootstrap support, BS). In contrast, the two different plastid-localized ASA beta subunits seem to be cyanobacterial-derived in Viridiplantae and Glaucophyta (Fig. 3b; no BS); however, the branching position of the red algal sequences is not well resolved. The ASA beta subunit is still plastid-encoded in both glaucophytes and red algae. These results suggest that the plastidic ASA is, at least in viridiplants and glaucophytes, an oligomeric complex constituted by enzyme subunits of disparate phylogenetic origins. The overall result illustrates the mosaic composition of the Plantae plastid tryptophan biosynthesis pathway, similarly as described before in the secondary plastids of diatoms and other stramenopiles[31]. Several of the enzymes (e.g., chorismate mutase, prephenate aminotransferase, arogenate dehydratase/prephenate dehydratases, and arogenate dehydrogenases) of non-cyanobacterial origin involved in tyrosine and phenylalanine biosynthesis have no apparent homologs in cyanobacterial genomes.

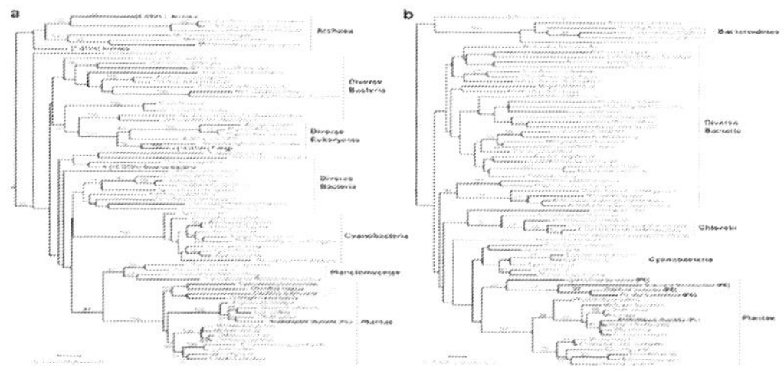

Figure 3: Phylogenetic analysis of the plastid anthranilate synthase complex. Maximum Likelihood phylogenetic tree of the (a) alpha subunit, or component I (At5g05730.1) and (b) beta subunit, or component II (At1g25220.1) of the anthranilate synthase. The numbers at the nodes indicate RaxML bootstrap values (only values >50% are shown). Branch lengths are proportional to the number of substitutions per site (see the scale bars). Colored boxes highlight Plantae lineages PL, plastid localized protein; PE, plastid encoded protein.

The plastidic route for histidine biosynthesis includes only one enzyme (out of 11 enzymes) of unambiguous cyanobacterial provenance in all Plantae: the imidazoleglycerol-phosphate dehydratase. The ML trees of the ATP phosphoribosyltransferase and histidinol phosphate phosphatase are inconclusive and a cyanobacterial origin for these enzymes cannot be conclusively disproved. Some enzymes in the histidine biosynthetic pathway were likely recruited for plastid roles from bacterial sources other than the cyanobacterial ancestor of the plastid. Our results demonstrate that the plastid pathway for the synthesis of histidine has a mosaic phylogenetic constitution as well and depicts a complex evolutionary history that involves independent losses and gains of non-cyanobacterial enzymes in the different Plantae lineages. For example, the histidinol-phosphate aminotransferase (Fig. 4) has a shared non-cyanobacterial origin in the tree Plantae lineages, indicating this enzyme was recruited for plastid functions by the Plantae ancestor from Chloroflexi bacteria (≥95% BS). Other cases indicate that some non-cyanobacterial plastid enzymes are unique evolutionary innovations in viridiplants, such as the host-derived histidinol dehydrogenase.

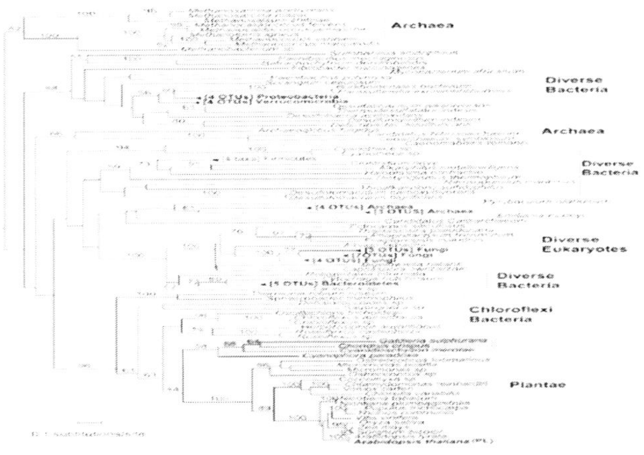

Figure 4: Phylogenetic analysis of the plastid histidinol-phosphate aminotransferase (At1g71920.1). The numbers at the nodes indicate RaxML bootstrap values (only values >50% are shown). Branch lengths are proportional to the number of substitutions per site (see the scale bars). Colored boxes highlight Plantae lineages. PL, plastid localized protein.

Aspartate-Derived and Branched-Chain Amino Acid Biosynthesis

The phylogenetic scrutiny of the plastid-localized enzymes involved in the biosynthesis of aspartate, lysine, threonine, methionine, and isoleucine (22 enzymatic activities represented by 41 nuclear-encoded proteins) revealed only three proteins of unambiguous cyanobacterial origin involved in the biosynthesis of these amino acids. The three cyanobacterial-derived enzymes are the dihydrodipicolinate reductase (only present in streptophytes), the diaminopimelate epimerase and the small subunit of the acetolactate synthase (Fig. 5b; 95% BS). The lysine biosynthesis sub-route encompasses three enzymes of non-cyanobacterial origin in viridiplants (dihydrodipicolinate synthase) and Plantae lineages (diaminopimelate aminotransferase and two diaminopimelate decarboxylases), respectively (Table 1). Both the biosynthesis of threonine from aspartate-semialdehyde and methionine comprises enzymes of non-cyanobacterial provenances. Remarkably, the phylogenetic tree of the large subunit of acetolactate synthase (ALS; Fig. 5a) illustrates the non-cyanobacterial origin of this enzyme in viridiplants and *Cyanophora*, but its cyanobacterial derivation in the case of the plastid-encoded red algal homolog. Considering that the ALS small subunit branches in the same clade with cyanobacterial homologs (Fig. 5b), the overall ALS results (Fig. 5) suggest that the ALS in viridiplants and *Cyanophora* is another example of a plastid oligomeric complex constituted by proteins of disparate ancestral origins. The other five enzymes (three reactions) participating in isoleucine biosynthesis have non-cyanobacterial origins (Table 1). The plastid biosynthesis of the branched-chain (BCH) amino acids leucine and valine from pyruvate involves the activity as of ALS as well and the non-cyanobacterial enzymes ketol-acid reductoisomerase, dihydroxy acid dehydratase (only present in viridiplants) and several BCH aminotransferases. Finally, several isopropylmalate dehydrogenases involved in leucine biosynthesis represent the only cases of cyanobacterial-derived enzymes in this pathway (Table 1).

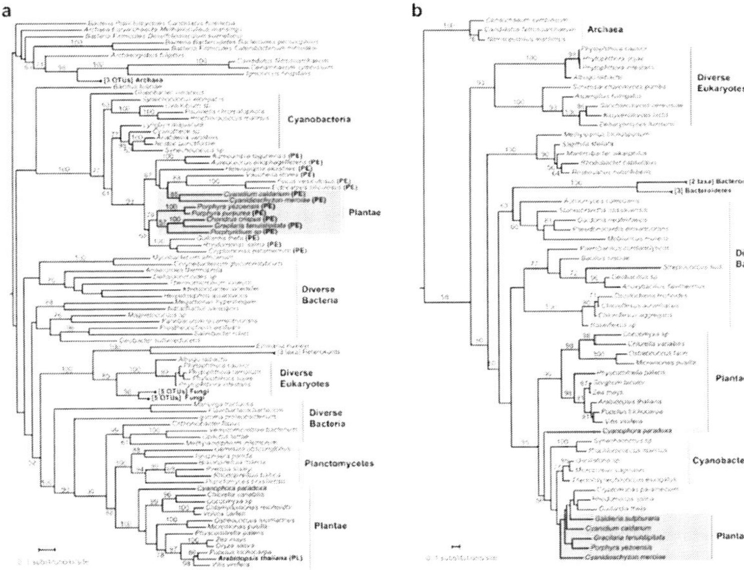

Figure 5: Phylogenetic trees of the acetolactate synthase. Large (At3g48560.1) and (b) small (At2g31810) subunits. The numbers at the nodes indicate RaxML bootstrap values (only values >50% are shown). Branch lengths are proportional to the number of substitutions per site (see the scale bars). PL, plastid localized protein; PE, plastid encoded protein.

Arginine and Proline Biosynthesis

Several enzymes that participate in arginine biosynthesis from glutamate and ornithine, have been localized in plastids[32, 33]. The overall phylogenetic analysis of the 12 enzymes involved in arginine biosynthesis revealed that the N-acetylglutamate kinase, N2-acetylornithine-glutamate acetyltransferase, and the large subunit of the carbamoyl phosphate synthetase (CPS; Fig. S110) are the only three proteins of unambiguous cyanobacterial origin. The phylogenetic tree of the small subunit of the CPS indicates the non-cyanobacterial origin of this protein in Viridiplantae but cyanobacterial-derived and plastid-encoded in red algae. These results suggest CPS is another possible case of a plastid oligomer constituted by subunits of different origins, similar to the ASA (Fig. 3) and ALS cases (Fig. 5). Even though the plastid occurrence of the arginine pathway awaits further experimental verification, our results show a mosaic composition for this pathway with only a minor contribution of cyanobacterial-derived components. Plastid proline biosynthesis from glutamate involves the activity of the non-cyanobacterial enzymes pyrroline-5-carboxylate synthase (only present in viridiplants) and the pyrroline-5-carboxylate reductase.

Serine, Glycine, Cysteine, and Alanine Biosynthesis

The three plastidic enzymes that participate in serine biosynthesis from 3-phosphoglycerate have non-cyanobacterial origin. The phylogeny of the plastid enzyme serine:glyoxylate aminotransferase (SGAT; Fig. 6), which is part of the route for glycine biosynthesis from glyoxylate, suggests a unique non-cyanobacterial origin in the three Plantae lineages (94% BS). This result suggests that the Plantae common ancestor recruited SGAT for plastid functions from other bacterial sources. The plastid-localized glutamate:glyoxylate aminotransferase, involved in glycine biosynthesis from glyoxylate, is of a non-cyanobacterial origin as well. Thus, the two plastid-localized alternative routes for glycine biosynthesis evolved from enzymes of different origins. The plastid enzymes serine O-acetyltransferase and cysteine synthase/O-acetylserine lyase, which catalyze the conversion of serine to cysteine, are other cases of non-cyanobacterial proteins recruited for plastid functions. Finally, the plastid cysteine desulfurase, which catalyzes the alanine synthesis from cysteine, is a protein of cyanobacterial origin (albeit with no significant bootstrap support), which is present only in viridiplants and *Cyanophora* (Table 1).

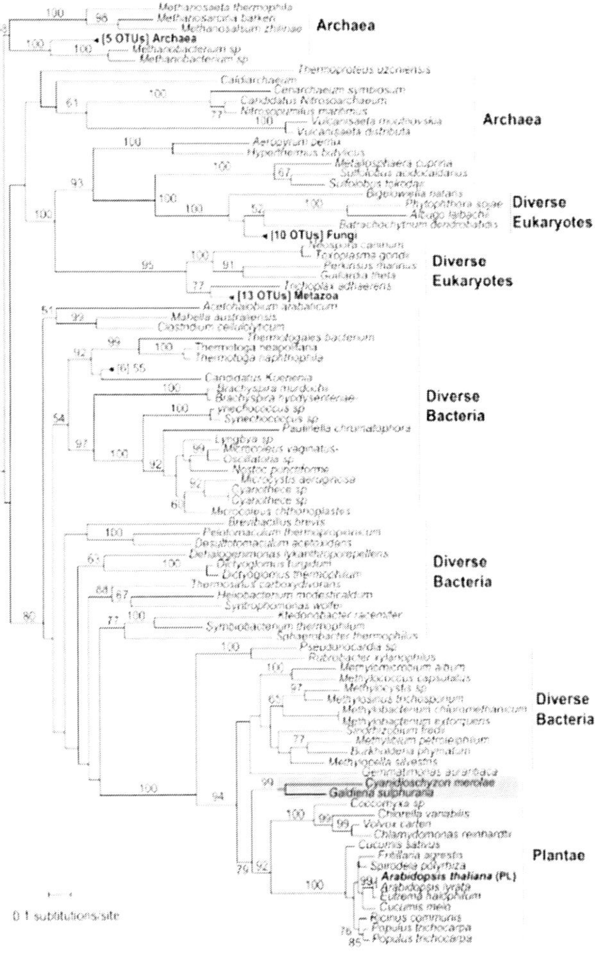

Figure 6: Phylogenetic trees of the plastid-localized serine: glyoxylate aminotransferase (At2g13360.2). The numbers at the nodes indicate RaxML bootstrap values (only values >50% are shown). Branch lengths are proportional to the number of substitutions per site (see the scale bars). PL, plastid localized protein.

DISCUSSION

Our phylogenetic survey shows that circa two-thirds (41/62 proteins and 26/39 enzymatic activities) of the *Arabidopsis* (and streptophytes in general) plastid enzymes involved in AAB evolved from non-cyanobacterial sources. We suggest that the phylogenetic mosaicism of the plastid AAB pathways is, in part, an outcome of ancient recruitment of non-cyanobacterial enzymes for plastid functions throughout the establishment of the endosymbiotic relationships that gave rise to the photosynthetic organelle. A key question is whether the phylogenetic mosaic composition of the AAB plastid pathways in *Arabidopsis*, and streptophytes in general, is an ancestral trait shared with other Plantae lineages. Our results reveal that several of the streptophytes plastid non-cyanobacterial enzymes branch in moderate to well-supported clades (>80% BS) together with green algae, red algae, or glaucophytes. Other ML trees with no significant support are consistent with this branching pattern as well. Therefore, the most parsimonious scenario to explain the shared non-cyanobacterial ancestry of the Plantae plastid enzymes is that several alien enzymes were "re-compartmentalized" by the host protein sorting system into the novel photosynthetic organelle early during the evolution of the Plantae ancestor[34, 35]. The common non-cyanobacterial ancestry of several enzymes involved in AAB illustrates that, in addition to their photosynthetic role, plastids became critical compartments for the "assembly" of the high energy consuming routes of nitrogen assimilation.

Even though enzyme re-compartmentation is a likely explanation for the phylogenetic mosaicism of the plastid AAB routes, in principle, we cannot discard that the plastid ancestor acquired the genes encoding non-cyanobacterial enzymes via horizontal gene transfer (HGT) prior to engulfment by the eukaryotic host. HGT has been largely recognized as a major force in prokaryote genome evolution and we assume this pervasive force had some impact on the evolution of the genome of the plastid ancestor. It is possible to speculate that a fraction of genes recruited via HGT by the plastid ancestor was subsequently transferred into the host nuclear genome via endosymbiotic gene transfer (EGT[6]). However, it is indispensable to consider, based upon the magnitude of ancestral HGT observed in plastid genomes, whether this scenario is sufficient to explain the extensive phylogenetic mosaicism evident in plastid AAB pathways. The genes of the RuBisCO operon (*rbcL, rbcS*)[36], the gene *cbbX*[37, 38] from red alga, and the seven genes involved in menaquinone/phylloquinone biosynthesis (*menF, menD, menC, menB, menE, menH* and *menA*)[39]of cyanidiales are the few well-documented cases of Plantae plastid genes of likely ancient (i.e., before plastid origins) non-cyanobacterial origin. However, our blastp search of the proteins encoded in seven Plantae plastid genomes versus all bacterial sequences in GenBank suggests that most plastid protein-encoding genes are of cyanobacterial origin (> 90% in *Cyanophora paradoxa, Pyropia yezoensis* and 3 diverse viridiplants, and > 70% in extremophilic cyanidiales). Thus, the low number of plastid genes of possible non-cyanobacterial origin (< 10%) in mesophilic Plantae makes it unlikely that the pre-endosymbiosis HGT scenario best explains the extensive phylogenetic mosaicism of the plastid AAB pathways (i.e., 66% of non-cyanobacterial enzymes). There is no reason to assume that HGT has more extensively affected genes encoding enzymes involved in the diverse pathways for AAB than other gene sets before the plastid establishment. As indirect comparative reference, it is important to note that the genome of the cyanobacterial-derived organelle in the filose amoeba*Paulinella chromatophora* FK01 contains only 33 genes acquired via HGT before the endosymbiosis that gave rise to the chromatophore (i.e., 4% of the chromatophore genes are non-cyanobacterial)[40]. This evidence provides an independent estimation of the relatively low number of alien genes present in other organelle genome of cyanobacterial origin. In summary, these results indicate that the non-cyanobacterial phylogenetic (66% of the proteins) signal observed in the composition of the plastid-localized AAB pathways is higher than that estimated for the gene repertoires in cyanobacterial-derived organelles (4–10%) of independent origins. This comparison suggests that post-endosymbiosis enzyme recruitment and re-compartmentation are the most likely explanations for our findings.

A key aspect of the plastid proteome evolution is the elucidation of the evolutionary forces that might have driven the re-compartmentation of the several enzymes involved in AAB into the organelle. Intracellular compartmentation by the presence of diverse membranous organelles generates a non-homogenous distribution of soluble compounds and enzymes critical for the organization and regulation of the eukaryotic cell metabolism[41]. Thus, differential metabolite concentrations permit efficient enzymatic activities and effective regulation of metabolic routes. Most noticeable cases of metabolic intracellular compartmentation involve the localization of entire biochemical pathways in particular organelles, such as the fatty acid beta-oxidation pathway, the tricarboxylic acid cycle and oxidative phosphorylation in the mitochondrion, the glycolysis in the cytosol, and the Calvin cycle, and terpenoid biosynthesis in the plastid. There are cases of re-compartmentation of complete pathways to different organelles such as the glycolysis relocation inside peroxisomes in trypanosomatids[42]. As envisaged by Ginger *et al.*, physical retargeting of entire metabolic pathways to new cellular compartments implies establishment and retuning of the regulatory mechanisms[42]. We hypothesize that re-compartmentation, or even *de novo* assembly, of AAB pathways into the photosynthetic organelle comprised both adaptive and non-selective changes derived from the stochastic traffic (i.e., mistargeting) of cytosolic and mitochondrial enzymes into the plastids. In principle, it is plausible to suppose that the physical relocation of non-cyanobacterial enzymes into the photosynthetic organelle occurred by incidental mistargeting via the TIC/TOC protein import machinery. Subsequently, we hypothesize that the catalytic activities of some of the mistargeted enzymes were transiently and randomly coupled to the plastid metabolite pools, biochemical intermediaries, effectors and the activity of native enzymes of the photosynthetic organelle. However, how were entire non-cyanobacterial pathways compartmentalized and assembled in the new photosynthetic organelle? If we assume that several proteins were occasionally mistargeted to subcellular locations different from the "correct" compartment, then it is plausible to predict that some enzymes catalyzing reactions associated with anabolic pathways were stochastically active in "wrong" organelles ("minor mistargeting model"[27]). Moreover, a certain degree of intermingling between the mistargeted proteins and the endosymbiont enzymes catalyzing similar, preceding or subsequent (e.g., fortuitous substrate channeling; use of common or non-specific substrates) reactions possibly sustained a steady state of novel catalyzed reactions inside the photosynthetic organelle. It is reasonable to expect that transitory enzymatic activities, sustained by recurrent mistargeting of non-cyanobacterial enzymes, produced slightly diverse plastid phenotypes[27]. These new plastid-localized enzymatic activities were initially neutral. However, the evolution of the obligate host-organelle interdependence in the Plantae ancestor opened a unique window for selective forces acting over the alien enzymatic activities and their incidental interactions in plastids. Thus, the advantageous re-compartmentation of some foreign enzymes into the organelle and the inexorable tendency of endosymbiont genomes to be reduced overtime established an ideal scenario for both replacement of some endosymbiont enzymes and subsequent selection of other "complementary" mistargeted enzymes. Overall, the phylogenetic results reflect a process of protein recruitment underlying the re-compartmentation of the AAB routes during the evolution of the plastid proteome.

Nevertheless, we still need to explain the ulterior evolution of novel regulatory mechanism and metabolic control exerted by the light quality and intensity and the redox intermediaries of the photosynthesis (e.g., phytochromes, ferredoxin, and NADPH)[14] over the plastid amino acid production. Plastid nitrogen assimilation is tightly coupled with the flow of carbon compounds (e.g., products of photorespiration, glycolysis, and the tricarboxylic acid cycle) into the organelle, which are an essential source of carbon skeletons for amino acid biosynthesis[14]. Under this scenario, we predict that the active transport (i.e., availability) and the concentration of carbon compounds (e.g., organic acids) and ammonia were also critical conditions for the successful re-compartmentation of several AAB pathways in the plastid. Consistent with this scenario, it is well known that nitrogen assimilation in plastids depends mostly on photosynthetic energy: in plant photosynthetic cells, 80% of the redox intermediaries required for nitrogen assimilation are regenerated by photochemically-reduced ferredoxin[14]. The activities of GS and Fd-GOGAT are up-regulated by light- and sugar-signaling in plant plastids[43]. Additionally, GS activity decreases when the photosynthetic rate is low[44] and the GOGAT activity is mediated by light via phytochrome[45]. Overall, different plastid routes such as the GS/GOGAT cycle, the chorismate biosynthesis, key steps of histidine biosynthesis, and synthesis of the aspartate-family require ATP and NADPH[14], which are actively produced in the plastid during the photosynthesis light-dependent reactions. The plastid pathways for the synthesis of branched-chain amino acids from pyruvate are light-regulated as well[46]. The ATP/AMP and NADPH/ $NADP^+$ ratios inside the photosynthetic organelle provided a favorable microenvironment to randomly sustain significant catalytic activities of mistargeted enzymes. At particular plastid internal substrates concentrations, some transitory enzymatic activities were sufficient to generate slightly different biochemical phenotypes. Thus, we consider that the rate of production of ATP and NADPH during light-dependent reactions of photosynthesis was an important selective element that favored the assembly of energy-demanding and redox-regulated AAB routes during the evolution of the plastid. In summary, diverse metabolic data suggest that the plastid redox-energy balance favored the re-compartmentation of these enzymes into the photosynthetic organelle.

Our results demonstrate that in addition to the quintessential role of plastids for the emergence of the eukaryotic photoautotrophic lifestyle (which relies mostly on an ancestral protein core of cyanobacterial origin), these photosynthetic organelles resulted in target compartments for the gradual assembly and re-compartmentation of entire biosynthetic pathways by recruiting non-cyanobacterial enzymes. Our results reveal that several of these non-cyanobacterial plastid enzymes are encoded by genes likely acquired over time by the host via HGT from diverse prokaryotic sources[47, 48]. Well-known examples of this evolutionary process include essential plastid proteins such as the ATP/ADP translocator[49] and enzymes involved in starch metabolism[50] that likely originated from ancestral parasitic Chlamydiae-like bacteria. These scenarios suggest that transitory energy parasites were important genetic donors for the establishment of the plastid as well[50, 51]. The fact that many of these non-cyanobacterial plastid enzymes are shared between green, red, and glaucophyte algae suggests that at least some part of this foreign enzymatic collection was anciently assembled in their common ancestor. Although, our comparative search and phylogenetic evidence[52] do not support the possibility that the Plantae non-cyanobacterial plastid enzymes of the AAB pathways were originally present in the genome of the plastid ancestor, and later transferred to the nucleus of the host through EGT, it is still possible that extant cyanobacterial genomes are radically different from those ancient cyanobacteria that gave rise to the Plantae plastid more than a billion years in the past.

In the recent years, it has been demonstrated that the incorporation of bacterial genes has been an important factor in the evolution of primary plastids[8, 51]. It has been estimated that circa 40% of the plastid proteins shared between red algae and Viridiplantae originated from the host repertoire and/or diverse bacteria[8]. The detailed evolutionary history of the plastid NA network, which we present here, emphasizes the role of the metabolic re-compartmentation and evolutionary innovation for the assembly of several amino acid biosynthetic pathways and their relevance in the integration of the plastid protein repertoire. The overall implication is that these processes demonstrate the ancestral and critical participation of the cyanobacterial-derived compartment in the evolution of the whole plant cell metabolism beyond its primary distinctive contribution as photoautotrophic partner.

METHODS

Identification of Enzymes Involved in Plastid Amino Acid Biosynthesis

We used information of the Plant Metabolic Network database (PMN; www.plantcyc.org, May-August, 2010) to identify 158 reported, or predicted, *Arabidopsis thaliana* proteins involved in AAB. The corresponding amino acid sequences were retrieved from the *Arabidopsis* Information Resource database (TAIR; http://www.arabidopsis.org) for further phylogenetic analyses. The final protein set included several cases of paralogous sequences encoded in the *Arabidopsis thaliana*genome.

Protein Subcellular Localization

We analyzed information from previous experimental investigations that report the subcellular localization of most of the 158 nuclear-encoded *Arabidopsis* proteins involved in AAB. Additionally, we used three different computational protocols (Predotar, TargetP, and WoLF PSORT)[53, 54, 55] to predict the subcellular localization of the 158 *Arabidopsis*proteins. We also used ChloroP 1.1[56] to investigate the presence of transit peptides in the predicted plastid-targeted proteins.

Phylogenetic Analysis of *Arabidopsis* Enzymes Involved in Amino Acid Biosynthesis

Each of the 158 retrieved *Arabidopsis* proteins was used as individual query to identify homolog sequences from our local protein database comprising the RefSeq GenBank data and conceptual translations of transcriptomic datasets from diverse eukaryotes, which do not have genome sequence available yet[57]. We established a blastp cutoff E-value ≤ 1e-10 to identify homolog proteins. In order to partially reduce sampling biases given the taxonomic composition of the database and to maximize the taxon coverage, we constrained the incorporation of the BLAST hits into the multiple alignment with an increasing maximum number on entries for each major taxonomic rank, starting from one sequence per species up to a maximum of 200 sequences per "Kingdom" according to the NCBI Taxonomy definitions (http://www.ncbi.nlm.nih.gov/taxonomy)[57]. The homologues protein sequences were aligned with MUSCLE[58]. Resulting multiple protein alignments were manually verified and edited using Se-Al 2.0 (http://tree.bio.ed.ac.uk/software/seal/). Partial sequences with more than 50% of missing amino acid residues were discarded from further analyses. We estimated maximum likelihood (ML) trees from each multiple alignment using the protein substitution model LG+F+**Γ** implemented in RAxML HPC-MPI 7.2.8 with 100 bootstrap replicates[59]. Visual inspection of the estimated unrooted ML trees was carried out with Archaeopteryx 0.997 beta version (http://www.phylosoft.org/archaeopteryx/). During the visual inspection of the unrooted ML trees, we used the Archaeopteryx *Root/Reroot* option to manually select at least three alternative rooting nodes (i.e., outgroup selection) in each tree to reduce potential misidentification of taxa branching patterns (i.e., monophyletic groups) as result of possible rooting artifacts. Our tree visual inspection aimed to distinguish between plastid enzymes of unambiguous cyanobacterial origin and those plastid proteins branching with homologs from taxonomic groups different from cyanobacteria. We distinguished plastid proteins of cyanobacteria provenance when the *Arabidopsis* (viridiplants in general) sequences branch in the same clade with cyanobacterial homologs and either any member of the Plantae group and/or algae with plastids of secondary origin (e.g., "chromists", chlorarachniophytes, and euglenids). In contrast, we defined plastid proteins of non-cyanobacterial origin in those cases as the *Arabidopsis* protein branches with other Plantae, algal homologs, and lineages other than cyanobacteria. ML trees in Newick format and refined multiple protein alignments are available upon request.

Searching for Non-Cyanobacterial Genes in Plastids Genomes

In order to identity plastid genes acquired via horizontal gene transference (HGT), we carried out a blastp reciprocal search (cutoff E-value ≤ 1e-15) of all proteins encoded in the plastid genomes of seven diverse Plantae (*Arabidopsis thaliana* [85 protein models], *Mesostigma viride* [105],*Chlamydomonas reinhartii* [69], *Cyanophora paradoxa* [149], *Pyropia yezoensis* [209],*Cyanidioschyzon merolae* [203], and *Cyanidium caldarium* [209]) against all prokaryotic sequences present in GenBank (as of June 2012). We used the taxonomic profile of the top-ten BLASTP hits as proxy to identify the phylogenetic origin of each plastid-encoded query. We arbitrarily defined a plastid-encoded protein of cyanobacterial origin when at least six out of ten top hits (i.e., ≥60%) were cyanobacterial (different species) homologs. In contrast, those plastid queries with less than five cyanobacteria within the top-ten hits (i.e., <50%) were considered non-cyanobacterial. Considering that horizontal incorporation of protein-coding genes into plastids genomes is highly constrained[52], this blastp similarity approach provides a conservative estimation of the number of alien genes present in the genome of the plastid ancestor.

ACKNOWLEDGEMENTS

The authors acknowledge Andreas P.M. Weber and Shawn R. MacLellan for their comments on the manuscript. ARP further acknowledges the Integrated Microbiology Program of the Canadian Institute for Advanced Research. The present work was supported by the Natural Sciences and Engineering Research Council of Canada (project 402421-2011), the Canada Foundation for Innovation (project 28276), and the New Brunswick Innovation Foundation (project RIF2012-006) awarded to ARP.

REFERENCES

1. Reyes-Prieto, A., Weber, A. P. & Bhattacharya, D. The origin and establishment of the plastid in algae and plants. *Annu Rev Genet* 41, 147–168 (2007).
2. Sagan, L. On the origin of mitosing cells. *J. Theor. Biol.* 14, 225–IN6 (1967).
3. Bhattacharya, D., Archibald, J. M., Weber, A. P. & Reyes-Prieto, A. How do endosymbionts become organelles? Understanding early events in plastid evolution. *Bioessays* 29,1239–1246 (2007).
4. Weber, A. P. M., Linka, M. & Bhattacharya, D. Single, ancient origin of a plastid metabolite translocator family in Plantae from an endomembrane-derived ancestor. *Eukaryot. Cell* 5,609–12 (2006).
5. Cavalier-Smith, T. The origins of plastids. *Biol. J. Linn. Soc.* 17, 289–306 (1982).
6. Martin, W. & Herrmann, R. Gene transfer from organelles to the nucleus: how much, what happens, and Why? *Plant. Physiol.* 118, 9–17 (1998).
7. Leister, D. Chloroplast research in the genomic age. *Trends Genet* 19, 47–56 (2003).
8. Suzuki, K. & Miyagishima, S. Y. Eukaryotic and eubacterial contributions to the establishment of plastid proteome estimated by large-scale phylogenetic analyses. *Mol Biol Evol* 27, 581–590 (2010).
9. Martin, W. & Schnarrenberger, C. The evolution of the Calvin cycle from prokaryotic to eukaryotic chromosomes: a case study of functional redundancy in ancient pathways through endosymbiosis. *Curr. Gen.* 32, 1–18 (1997).
10. Reyes-Prieto, A. & Bhattacharya, D. Phylogeny of Calvin cycle enzymes supports Plantae monophyly. *Mol Phylogenet Evol* 45, 384–391 (2007).
11. Richards, T. A. *et al.* Evolutionary origins of the eukaryotic shikimate pathway: gene fusions, horizontal gene transfer, and endosymbiotic replacements. *Eukaryot. Cell* 5, 1517–31(2006).

12. Ohlrogge, J. B., Kuhn, D. N. & Stumpf, P. K. Subcellular localization of acyl carrier protein in leaf protoplasts of Spinacia oleracea. *Proc Natl Acad Sci U S A* 76, 1194–1198 (1979).
13. Lichtenthaler, H. K., Schwender, J., Disch, A. & Rohmer, M. Biosynthesis of isoprenoids in higher plant chloroplasts proceeds via a mevalonate-independent pathway. *FEBS Lett* 400,271–274 (1997).
14. Lancien, M. Lea, P. J. & Azevedo, R. A. Amino Acid Synthesis in Plastids in Wise. R. R. & Hoober, J. K. *The Structure and Function of Plastids,* Springer. 355–385 (2006).
15. Stepansky, A. & Leustek, T. Histidine biosynthesis in plants. *Amino Acids* 30, 127–142(2006).
16. Rippert, P., Puyaubert, J., Grisollet, D., Derrier, L. & Matringe, M. Tyrosine and phenylalanine are synthesized within the plastids in *Arabidopsis*. *Plant Physiol* 149,1251–1260 (2009).
17. Diebold, R., Schuster, J., Daschner, K. & Binder, S. The branched-chain amino acid transaminase gene family in *Arabidopsis* encodes plastid and mitochondrial proteins. *Plant Physiol* 129, 540–550 (2002).
18. Jander, G. & Joshi, V. Aspartate-Derived Amino Acid Biosynthesis in *Arabidopsis thaliana.The Arabidopsis Book* 7, 1–15 (2008).
19. Vallon, O. & Spalding, M. H. Amino acid metabolism in Stern, D. & Harris, E. *The Chlamydomonas Sourcebook 2nd Edition, Volume 2: Organellar and Metabolic Processes, Elsevier.* 115–158 (2009).
20. Ravanel, S. *et al.* Methionine metabolism in plants: chloroplasts are autonomous for de novo methionine synthesis and can import S-adenosylmethionine from the cytosol. *J Biol Chem*279, 22548–22557 (2004).
21. Schmutz, D. & Brunold, C. Intercellular Localization of Assimilatory Sulfate Reduction in Leaves of *Zea mays* and *Triticum aestivum*. *Plant Physiol* 74, 866–870 (1984).
22. Rayapati, P. J., Stewart, C. R. & Hack, E. Pyrroline-5-Carboxylate Reductase Is in Pea (*Pisum sativum* L.) Leaf Chloroplasts. *Plant Physiol* 91, 581–586 (1989).
23. Peltier, J. B. *et al.* The oligomeric stromal proteome of *Arabidopsis thaliana* chloroplasts. *Mol Cell Proteomics* 5, 114–133 (2006).

24. Cavalier-Smith, T. Eukaryote kingdoms: seven or nine? *Biosystems* 14, 461–481 (1981).

25. Yoon, H. S., Hackett, J. D., Ciniglia, C., Pinto, G. & Bhattacharya, D. A molecular timeline for the origin of photosynthetic eukaryotes. *Mol. Biol. Evol.* 21, 809–18 (2004).

26. Hedges, S. B., Blair, J. E., Venturi, M. L. & Shoe, J. L. A molecular timescale of eukaryote evolution and the rise of complex multicellular life. *BMC Evol. Biol.* 4, 2 (2004).

27. Martin, W. Evolutionary origins of metabolic compartmentalization in eukaryotes. *Philos Trans R Soc Lond B Biol Sci* 365, 847–855 (2010).

28. Ghoshroy, S., Binder, M., Tartar, A. & Robertson, D. L. Molecular evolution of glutamine synthetase II: Phylogenetic evidence of a non-endosymbiotic gene transfer event early in plant evolution. *BMC Evol Biol* 10, 198 (2010).

29. Terashita, M., Maruyama, S. & Tanaka, K. Cytoplasmic localization of the single glutamine synthetase in a unicellular red alga, *Cyanidioschyzon merolae* 10D. *Biosci. Botech. Bioch.*70, 2313–5 (2006).

30. Imamura, S. *et al.* Nitrate assimilatory genes and their transcriptional regulation in a unicellular red alga *Cyanidioschyzon merolae*: genetic evidence for nitrite reduction by a sulfite reductase-like enzyme. *Plant Cell Physiol* 51, 707–717 (2010).

31. Jiroutova, K., Horak, A., Bowler, C. & Obornik, M. Tryptophan biosynthesis in stramenopiles: eukaryotic winners in the diatom complex chloroplast. *J Mol Evol* 65, 496–511 (2007).

32. Ferrario-Méry, S., Besin, E., Pichon, O., Meyer, C. & Hodges, M. The regulatory PII protein controls arginine biosynthesis in *Arabidopsis*. *FEBS letters* 580, 2015–20 (2006).

33. Slocum, R. D. Genes, enzymes and regulation of arginine biosynthesis in plants. *Plant Physiol Biochem* 43, 729–745 (2005).

34. Rodríguez-Ezpeleta, N. *et al.* Monophyly of primary photosynthetic eukaryotes: green plants, red algae, and glaucophytes. *Curr. Biol.* 15, 1325–30 (2005).

35. Price, D. C. *et al. Cyanophora paradoxa* Genome Elucidates Origin of Photosynthesis in Algae and Plants. *Science* 335, 843–847 (2012).

36. Delwiche, C. F. & Palmer, J. D. Rampant horizontal transfer and duplication of rubisco genes in eubacteria and plastids. *Mol. Biol. Evol.* 13, 873–82 (1996).
37. Maier, U. G., Fraunholz, M., Zauner, S., Penny, S. & Douglas, S. A nucleomorph-encoded CbbX and the phylogeny of RuBisCo regulators. *Mol. Biol. Evol.* 17, 576–83 (2000).
38. Fujita, K., Ehira, S., Tanaka, K., Asai, K. & Ohta, N. Molecular phylogeny and evolution of the plastid and nuclear encoded cbbX genes in the unicellular red alga *Cyanidioschyzon merolae*. *Genes Genet. Syst.* 83, 127–33 (2008).
39. Gross, J., Meurer, J. & Bhattacharya, D. Evidence of a chimeric genome in the cyanobacterial ancestor of plastids. *BMC Evol. Biol.* 8, 117 (2008).
40. Reyes-Prieto, A. *et al.* Differential gene retention in plastids of common recent origin. *Mol. Biol. Evol.* 27, 1530–7 (2010).
41. Bowsher, C. G. & Tobin, A. K. Compartmentation of metabolism within mitochondria and plastids. *J Exp Bot* 52, 513–527 (2001).
42. Ginger, M. L., McFadden, G. I. & Michels, P. A. Rewiring and regulation of cross-compartmentalized metabolism in protists. *Philos Trans R Soc Lond B Biol Sci* 365, 831–845(2010).
43. Thum, K. E., Shasha, D. E., Lejay, L. V. & Coruzzi, G. M. Light- and carbon-signaling pathways. Modeling circuits of interactions. *Plant Physiol* 132, 440–452 (2003).
44. Woodall, J., Boxall, J. G., Forde, B. G. & Pearson, J. Changing perspectives in plant nitrogen metabolism: the central role of glutamine synthetase. *Science Progress* 1–26(1996).
45. Becker, T. W., Nef-Campa, C., Zehnacker, C. & Hirel, B. Implication of the phytochrome light regulation of the tomato gene(s) encoding ferredoxin dependent glutamate synthase. *Plant. Physiol. Bioch.* 725–729 (1993).
46. Schulze-Siebert, D., Heineke, D., Scharf, H. & Schultz, G. Pyruvate-Derived Amino Acids in Spinach Chloroplasts: Synthesis and Regulation during Photosynthetic Carbon Metabolism.*Plant Physiol* 76, 465–471 (1984).
47. Doolittle, W. F. You are what you eat: a gene transfer ratchet could account for bacterial genes in eukaryotic nuclear genomes. *Trends Genet* 14, 307–311 (1998).

48. Watkins, R. F. & Gray, M. W. The frequency of eubacterium-to-eukaryote lateral gene transfers shows significant cross-taxa variation within amoebozoa. *J Mol Evol* 63, 801–814(2006).

49. Linka, N. *et al.* Phylogenetic relationships of non-mitochondrial nucleotide transport proteins in bacteria and eukaryotes. *Gene* 306, 27–35 (2003).

50. Ball, S., Colleoni, C., Cenci, U., Raj, J. N. & Tirtiaux, C. The evolution of glycogen and starch metabolism in eukaryotes gives molecular clues to understand the establishment of plastid endosymbiosis. *J Exp Bot* 62, 1775–1801 (2011).

51. Moustafa, A., Reyes-Prieto, A. & Bhattacharya, D. Chlamydiae has contributed at least 55 genes to Plantae with predominantly plastid functions. *PLoS One* 3, e2205 (2008).

52. Rice, D. W. & Palmer, J. D. An exceptional horizontal gene transfer in plastids: gene replacement by a distant bacterial paralog and evidence that haptophyte and cryptophyte plastids are sisters. *BMC Biol.* 4, 31 (2006).

53. Small, I., Peeters, N., Legeai, F. & Lurin, C. Predotar: A tool for rapidly screening proteomes for N-terminal targeting sequences. *Proteomics* 4, 1581–1590 (2004).

54. Emanuelsson, O. Brunak, S. Von Heijne, G. & Nielsen, H. Locating proteins in the cell using TargetP, SignalP and related tools. *Nat Protoc* 2, 953–971 (2007).

55. Horton, P. *et al.* WoLF PSORT: protein localization predictor. *Nucleic Acids Res* 35, W585–7(2007).

56. Emanuelsson, O., Nielsen, H. & von Heijne, G. ChloroP, a neural network-based method for predicting chloroplast transit peptides and their cleavage sites. *Protein Sci.* 8, 978–84(1999).

57. Moustafa, A., Bhattacharya, D. & Allen, A. E. iTree: A high-throughput phylogenomic pipeline. *5th Cairo International Biomedical Engineering Conference* 103–107 (2010).

58. Edgar, R. C. MUSCLE: multiple sequence alignment with high accuracy and high throughput. *Nucleic Acids Res* 32, 1792–1797 (2004).

59. Stamatakis, A. RAxML-VI-HPC: maximum likelihood-based phylogenetic analyses with thousands of taxa and mixed models. *Bioinformatics* 22, 2688–2690 (2006).

Chapter 2

Amino Acids Biosynthesis and Nitrogen Assimilation Pathways: A Great Genomic Deletion During Eukaryotes Evolution

RLM Guedes[1], F Prosdocimi[2,3], GR Fernandes[1,2], LK Moura[2], HAL Ribeiro[1], and JM Ortega[1]

[1]Departamento de Bioquímica e Imunologia, Instituto de Ciências Biológicas, Universidade Federal de Minas Gerais, Belo Horizonte, 31270-901, MG, Brazil

[2]Programa de Pós-Graduação em Ciências Genômicas e Biotecnologia, Universidade Católica de Brasilia, Brasilia, 70790-160, DF, Brazil

[3]Instituto de Bioquímica Médica, Universidade Federal do Rio de Janeiro, Rio de Janeiro, 21941-902, RJ, Brazil

ABSTRACT

Background

Besides being building blocks for proteins, amino acids are also key metabolic intermediates in living cells. Surprisingly a variety of organisms are incapable of synthesizing some of them, thus named Essential Amino Acids (EAAs). How certain ancestral organisms successfully competed for survival after losing key genes involved in amino acids anabolism remains an open question. Comparative genomics searches on current protein databases including sequences from both complete and incomplete genomes among diverse taxonomic groups help us to understand amino acids auxotrophy distribution.

Results

Here, we applied a methodology based on clustering of homologous genes to seed sequences from autotrophic organisms *Saccharomyces cerevisiae* (yeast) and *Arabidopsis thaliana* (plant). Thus we depict evidences of presence/absence of EAA biosynthetic and nitrogen assimilation enzymes at phyla level. Results show broad loss of the phenotype of EAAs biosynthesis in several groups of eukaryotes, followed by multiple secondary gene losses. A subsequent inability for nitrogen assimilation is observed in derived metazoans.

Conclusions

A Great Deletion model is proposed here as a broad phenomenon generating the phenotype of amino acids essentiality followed, in metazoans, by organic nitrogen dependency. This phenomenon is probably associated to a relaxed selective pressure conferred by heterotrophy and, taking advantage of available homologous clustering tools, a complete and updated picture of it is provided.

BACKGROUND

Creation and analysis of groups of orthologous genes have been widely used for gene function prediction, evolutionary and divergence time studies [1]. Moreover, orthology is also a valuable source for evolutionary comprehension of pathways through phylogenetic analysis. In respect to a central issue on cellular metabolism, the order of appearance for universal cellular metabolisms was estimated by Cunchillos and Lecointre [2, 3], with amino acid catabolism and anabolism being respectively the first and second pathways to appear, even earlier than glycolysis and gluconeogenesis. The amino acids biosynthesis, rather than linear and universal series of reactions with homologues occurring in different organisms, sometimes relies on alternative pathways, as shown by Hernández-Montes et al. [4]. Moreover, gene loss and pathway depletion, important events in genome evolution, can be inferred from the orthologous groups through comparative genomics. Today, a vast amount of information is provided by intensive genome sequencing, and the efforts of grouping homologous genes had reached great standards.

Amino acid anabolism is responsible for about 20% of the energy that cells spend on protein synthesis [5, 6]. The nutritional requirements of essential amino acids and nitrogen are of striking importance and they have been estimated as ~22mg/kg of EAAs and 3mg/kg of N in human body [7, 8]. More recent approaches for dietary requirement calculations, using amino acid oxidation as an indicator, reveal that the requirement is over fivefold what the classical approaches indicated, and the requirement has now been determined for each of the nine human EAAs [9]. It is of general understanding that plant, as well as fungi, synthesize all amino acids required for protein synthesis and that evolutionary processes culminated in human inability to synthesize nine amino acids (histidine, phenylalanine, tryptophan, valine, isoleucine, leucine, lysine, methionine and threonine), thus called essential amino acids (EAAs), which must be obtained through diet. Amino acids also constitute our source of organic nitrogen. There have been few attempts to understand why some amino acids have become essential. However, genome deletion events have happened in the past and many organisms have lost a number of important enzymes necessary for *de novo* biosynthetic pathways. Hitherto, the pattern of loss versus retention for amino acids biosynthetic pathways was analyzed for a few protists and metazoans by Payne and Loomis [10]. They verified that the set of essential amino acids is the same in animals and protists. Curiously, most of the retained amino acids are intermediates in secondary pathways like purine ring biosynthesis and nitrogen metabolism.

An overview for the presence/absence of the enzymes which compose the amino acid biosynthetic pathways, among distinct phyla in the tree of life, could be accomplished with (i) rich protein databases such as the UniProt Knowledgebase (UniProtKB) [11] comprising over 10 million full-length sequences and (ii) the current initiatives to group these proteins by evolutionary relatedness - called homologues - such as COG-Cluster of Orthologous Groups [12] and KEGG Orthology [13]. Unfortunately these initiatives consider only proteins derived from complete genomes and thus a large amount of information is currently lost, with over 6 million remaining full-length proteins that belong to organisms with still incomplete genomes.

Here, we applied a methodology that takes into account all available protein information to depict, at phyla level, the EAA biosynthetic and nitrogen assimilation enzymes scenarios to inspect how and when amino acid auxotrophy has first appeared along evolution.

A Great Genomic Deletion model is proposed to explain the phenotypic inability to synthesize amino acids that appears independently in distinct phylogenetically distant clades of eukaryotes. Such events should be followed by subsequent steps of gene loss due to relaxed selective pressure in already incomplete pathways, leading to an eventual loss of all genes for a particular biosynthesis pathway in some clades. Accordingly, in metazoans but Cnidaria, dependence on organic nitrogen accompanies the evolution of heterotrophy, thus organisms become dependent even on NEAA for supplying their nitrogen requirements.

RESULTS

Clustering Homologues of Amino Acid Biosynthetic Enzymes

To determine the distribution of amino acid biosynthetic enzymes, a homologue clustering process was developed to allow the use of both complete and incomplete genomes [14, 15]. The procedure starts with Seed Linkage software [14] that clusters cognate proteins from multiple organisms beginning with a single seed sequence through connectivity saturation with it. Since basal eukaryotes such as plants and fungi are autotrophic, sequences coding for all the enzymes used in the biosynthesis of EAAs from the plant *Arabidopsis thaliana* and the fungus *Saccharomyces cerevisiae*were manually inspected using KEGG Pathway and used as seeds to search for homologues. Moreover, our group has been developing a procedure to enrich secondary databases such as COG[12] and KEGG Orthology (to be published) with UniRef50 clusters [16] available from UniProt, therefore allowing the inclusion of data from incompletely sequenced genomes. Since these searches recruit sequences from diverse clades, which may or may not contain organisms with completely sequenced genomes, we represented this information in Figure 1 as: (a) black filled circles for phyla containing complete genomes; (b) grey filled circles comprise clades with at least one draft genome available, but no complete genome, and (c) empty circles represent phyla with no complete nor draft genomes. Protein fragments are not included in the search for homologues because they may represent partial sequenced full length proteins at mRNA level or incompletely modeled from genome. Moreover since some full length proteins might have not been captured in databases due to high sequence divergence, a second search round used UniProt to query all clustered sequences. This step also captures partial sequences (entries labeled as fragments in UniProt) which were approved by the coverage filtering applied (see Methods for details). These additional significant hits are represented by triangles in Figure 1. Furthermore, enzymes required for the biosynthesis of the indicated amino acids are ordered in the anabolic pathway from left to right. All pathways refer to EAAs biosynthesis except serine and glycine (the rightmost ones) used as experimental controls. Serine is represented with two alternative pathways observed in human and other eukaryotes: S (1), from 3P-D-glycerate; and S (2), from pyruvate. Glycine is also represented by two pathways: G (1) and G (2), both coming from serine; and G (3), coming from threonine. As expected, serine and glycine biosynthesis were found to be potentially proficient in almost all phyla. This control supports the searching mechanism and attest for the efficacy of methods applied. A few exceptions were observed and deserve comments: (i) Serine biosynthetic pathways was found to be absent in Rhodophyta, although the complete genome of *Cyanidioschyzon merolae* is available. We manually inspected this result with regular BLAST searches and did not find additional evidence, although a translation of partial CDS was obtained for glycine biosynthetic enzyme G1 (Figure 1, triangle); (ii) Serine biosynthesis seems absent in Apicomplexa as well, a clade comprising two *Plasmodium* complete genomes lacking enzymes S1 and S4; (iii) Considering the animals, besides being able to find serine biosynthetic enzymes, we fail to support the NEAA character of glycine for Mollusca. However, evidences could be obtained for ancient organisms such as Placozoa and Porifera. For the Microsporidia *E. cuniculi*, an obligatory intracellular parasitic fungus with complete genome, it has been reported that "the repertoire for the biosynthesis of amino acids is restricted to asparagines synthetase and serine hydroxymethyltransferase genes", then serine was known as an EAA [17]. Thus, absence of evidence may not guarantee the absence of the gene. However, out of 28 phyla, discarding both the four clades with no genome project or in progress (open circles) and the ones with complete genome (filled symbols), we could not provide evidence of glycine biosynthesis for two phyla (Fornicata and Mollusca). However evidence for serine has been provided in all of them.

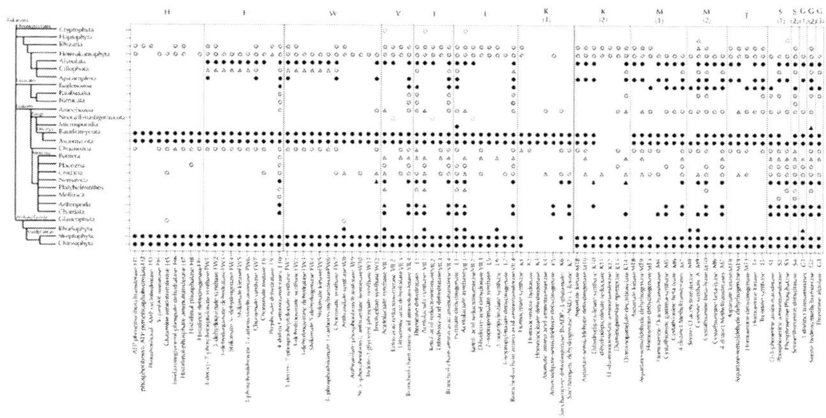

Figure 1: Essential amino acid anabolic pathways. Schematic representation for presence/absence of anabolic enzymes for nine essential amino acids and the non-essential amino acids serine and glycine. Eukaryotic taxonomic tree displayed at phyla level. Circles represent detection of complete proteins and triangles detection of complete and fragmented proteins. Black: phyla containing complete genomes; Grey: at most organisms with draft genomes; White: phyla with no complete or draft genomes. *Saccharomyces cerevisiae* (Ascomycota) and*Arabidopsis thaliana* (Streptophyta) were used as seeds. The 4 distinct aminotransferases in phenylalanine pathway are: (i) aspartate aminotransferase (ii) histidinol-phosphate aminotransferase (iii) aromatic amino acid aminotransferase (iv) tyrosine aminotransferase. The 4 distinct methyltransferases in methionine pathway are: (i) 5-methyltetrahydropteroyltriglutamate--homocysteine methyltransferase (ii) homocysteine S-methyltransferase (iii) betaine-homocysteine methyltransferase (iv) 5-methyltetrahydrofolate--homocysteine methyltransferase. The 3 distinct transaminases in glycine pathway are: alanine-glyoxylate transaminase, serine-glyoxylate transaminase and serine-pyruvate transaminase.

Data presented in Figure 1 clearly depicts the presence of complete biosynthetic pathways for EAAs in both plants (Chlorophyta and Streptophyta) and fungi (Ascomycota and Basidiomycota), as stated above. In previous work we hypothesized that a great event of genome deletion on which many of the intermediate enzymes for biosynthetic pathways for amino acids have vanished, ended up affecting the usage of EAAs in chordate proteomes [18,19]. In 2006, Payne and Loomis [10] using pFam protein signatures reported that protists and animals share essentiality for the nine amino acids. Here we provide a broader analysis covering all genomes available today and trying to map how and when the Great Genomic Deletion has happened. Evidence was found suggesting that this loss of capability to synthesize EAAs is conspicuous at the base of metazoan evolution, simultaneously affecting the complete set of EAAs. The phenomenon is characterized as an initial phenotypic deficiency, observed in Choanozoa, followed by multiple secondary gene losses. Accordingly, some enzymes found in Chordata such as K14, M4 and M9 are missing in Arthropoda. Remarkably, some components such as VIL1 and M7 are maintained in most metazoan clades, despite of pathway loss.

Actually, a Great Deletion causing concurrent phenotypic loss of amino acid biosynthesis capability affects both metazoan and non-metazoan eukaryotes. Several clades containing complete genomes (black filled symbols) such as Rhodophyta, Euglenozoa and Apicomplexa, show similar EAAs pattern. Moreover, some evidence is provided suggesting the absence of complete pathways in the non-Dikarya Fungi Microsporidia and Neocallimastigomycota. This gives support to separate events of Great Genomic Deletion for the origin of EAAs auxotrophy in at least three other branches. Similarly to Choanozoa, clades such as Heterokontophyta and Rhizaria present various enzymes and some complete pathways. Evidences of complete pathways for all EAAs but histidine (H) were obtained in Heterokontophyta. Valine (V), isoleucine (I), lysine (K) and threonine (T) are potentially synthesized in Rhizaria as well as methionine (M) in Euglenozoa and Amoebozoa. However it is possible that other EAAs may also be synthesized in some of these clades. The anabolic capabilities suggested by the current data might be underestimated because we have only draft genomes available for most of these organisms. The Choanozoa clade contains only draft genomes. Though we observed more enzymes than in metazoan clades, a final picture of Choanozoan phenylalanine biosynthesis, for example, might require completion of genome sequencing. Further gene loss occurs during metazoan evolution; however, for Placozoa, Porifera and Cnidaria, the Great Genomic Deletion seems to be well established. Since the first available sponge genome is still an ongoing project and its proteins are not yet deposited in UniProt, we manually inspected the deduced proteome using regular BLAST alignments (see Methods) and evidenced auxotrophy for all nine EAAs. The same simple approach was applied to all phyla (Figure 1, triangles). Other clades that do not present any enzymes were omitted from Figure 1, such as Apusozoa and Jakobida.

Lysine Biosynthesis

Inspection of Figure 1 depicts a remarkable difference on lysine (K) biosynthesis pathways present in fungi and plants. Since the occurrence of an α-aminoadipate (AAA) pathway K(1) in Fungi [20] as opposite to a diaminopimelate (DAP) pathway K(2) known to be present in plants, algae and bacteria[21,22] has already been reported, we set up to depict the complete scenario for K biosynthesis including prokaryotes (Figure 2). A third pathway K(3) preferentially used by Archaea but also reported to exist in bacterial groups [23] was also considered, therefore sequences from the *Pyrococcus horikoshii* archaea were also used as seed for homologue sequence clustering. Data supports the view that the K(2) pathway, found to be complete in plants, is often present in prokaryotic clades of bacteria and archaea, in agreement with previous findings [21,22]. Curiously, nine bacterial clades (Acidobacteria, Chlorobi, Deferribacteres, Deinococcus-Thermus, Fusobacteria, Chlamydiae, Synergistetes, Tenericutes and Thermotogae) -- all of which contain complete genomes -- do not present K12 enzyme, but there are three other alternative subsets of enzymes present in prokaryotes that could circumvent this step in lysine biosynthesis. Chlamydiae may represent an evidence of amino acid essentiality extended to prokaryotes, since diaminopimelate decarboxylase (K14) is absent and there are no known alternatives to this reaction. The set of enzymes responsible for the K (3) pathway, was found to occur in prokaryotes, and it is complete in the archaeal clades Crenarchaeota and Euryarcheota, as well as in the bacterial clades Chloroflexi and Proteobacteria, and probably in Actinobacteria and Bacteroidetes. Remarkably, the first four enzymes that constitute this pathway are coincident with the K (1) pathway (indicated by gray shading). The complete K (1) pathway occurs in Proteobacteria (and possibly in Actinobacteria, Bacteroidetes and Firmicutes, as evidenced by regular BLAST) and fungi. Thus, it is tempting to assume that a variant synthesis of K occurred in Archaea and, being modified in one of the four bacterial phyla above (with the addition of three enzymes: aminoadipate-semialdehyde dehydrogenase, saccharopine dehydrogenase NADP+ and saccharopine dehydrogenase NAD+), ended up constituting the fungi-occurring K biosynthetic pathway. The eukaryotic clades Rhizaria and Heterokontophyta, which present the K (2) pathway, appear to group with plants.

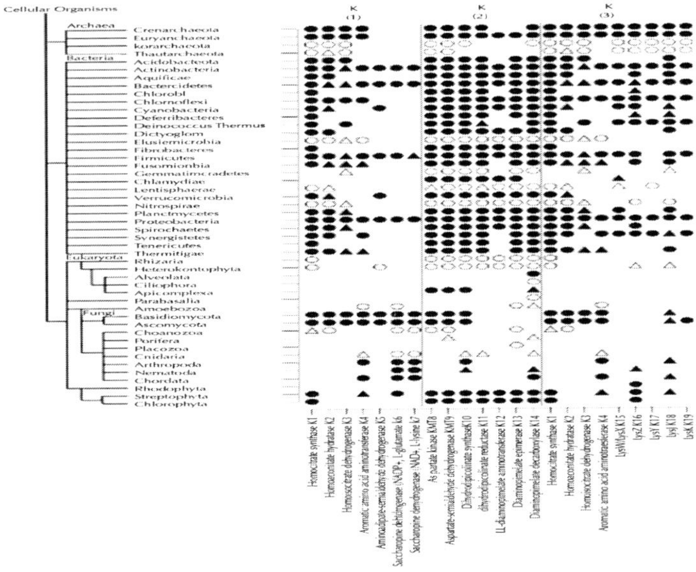

Figure 2: Lysine anabolic pathways. Schematic representation for presence/absence of enzymes involved in lysine biosynthesis. K(1) represents Fungi -aminoadipate (AAA) pathway; K(2) bacteria, plants, and algae diaminopimelate (DAP) pathway; K(3) archaea α-aminoadipate (AAA) variant pathway. Taxonomic tree displayed at phyla level. Circles represent detection of complete proteins and triangles detection of complete and fragmented proteins. Colors are as for Figure 1. *Saccharomyces cerevisiae*(Ascomycota), *Arabidopsis thaliana* (Streptophyta) and *Pyrococcus horikoshii* (Euryarchaeota) were used as seeds.

Nitrogen Auxotrophy

Consumption of amino acids is an important route for nitrogen assimilation in other biological compounds for heterotrophic organisms, such as those comprised by some of the clades shown in Figure 1 (e.g. Chordata). Assimilation of free ammonium in eukaryotes is done by a cytoplasmatic reaction catalyzed by glutamate dehydrogenase

(EC: 1.4.1.4) which incorporates ammonium into alpha-ketoglutarate yielding glutamate, using electrons from a reduced cytoplasmatic co-enzyme NADPH. Two isoforms are present in fungi and one in plants, the latter having the additional option to not only assimilate nitrogen, but also to fixate it, often with the association of nitrogen-fixating bacteria. Thus, to investigate if the Great Genomic Deletion of biosynthetic enzymes for EAAs co-occurred with the heterotrophy for nitrogen, we generated clusters of the assimilative isoforms (EC:1.4.1.4) and, as a control, the mitochondrial enzymes (EC:1.4.1.2) which tend to operate in the reverse direction, i.e. glutamate degradation, by oxidizing it and delivering ammonium, loading electrons in NAD+ co-enzyme. In yeast, the cytoplasmic assimilative isoforms are named *GDH1* and*GDH3*, and the catabolic (mitochondrial) is known as *GDH2*. *Arabidopsis thaliana* proteins were also used as seed together with the *Saccharomyces cerevisiae* sequences: one known as putative *GDH*which grouped with the fungi assimilative ones, and three catabolic *GDHs*, that grouped with the human mitochondrial *GLUD1*, though not with the yeast catabolic *GHD2*. Results are shown in Figure3A. The left column shows a cluster that groups assimilative isoforms with the two from yeast and the putative *GDH* from *A. thaliana*. The catabolic mitochondrial isoforms from yeast (central column) and plant (right column) formed two independent clusters. In metazoan organisms, an assimilative enzyme was found in the basal group Cnidaria, all others being dependent on amino acid consumption to build nitrogenated compounds such as DNA, Porifera included. Assimilative isoforms were also lacking in Choanozoa although complete genomes are unavailable. The same was observed for Placozoa. Comparing these results with those shown in Figure 1, it is remarkable that Choanozoa, while still registering many amino acid biosynthetic enzymes (37 out of 61, redundancy eliminated) shows a simultaneous deletion in both EAAs biosynthesis and nitrogen assimilation. It is also apparent that the Great Genomic Deletion attains its almost final broad distribution in Cnidaria, which may be the last metazoan clade still capable to assimilate nitrogen from free ammonium. Therefore a few biosynthetic enzymes remain, in this clade and other Metazoa, probably by connective functions in metabolism (e.g. EC: 1.2.1.31 aminoadipate-semialdehyde dehydrogenase K5 and EC: 1.5.1.7 saccharopine dehydrogenase K7 also participates in the lysine degradation pathway). We have also observed that mammalian *GDH* (*GLUD1*) presents a specialized allosteric control [24] which might have turned the enzyme toward glutamate catabolism rather than anabolism. Such control was first observed in Ciliophora [25] and it is thought to have been transferred by lateral gene transfer to the metazoan ancestor [26]. To confirm the grouping in three clusters of enzymes with so similar activities, Figure 3B shows a phylogenetic tree built with eukaryotic glutamate dehydrogenase sequences, which clustered the isoforms in total accordance with data shown in Figure 3A.

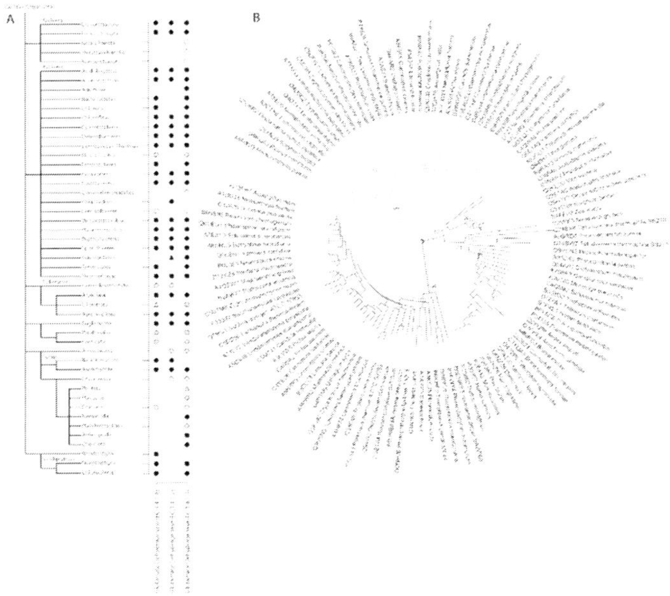

Figure 3: Glutamate dehydrogenases. Schematic representation for presence/absence of glutamate dehydrogenases. A: Left column: assimilative *GDH1* and *GDH3* from *Saccharomyces cerevisiae* and putative*GDH* from *Arabdopsis thaliana*; Central column: catabolic *GDH2* from *Saccharomyces cerevisiae*; Right column: catabolic *GDH1*, *GDH2* and *GDH3*from *Arabdopsis thaliana*. Taxonomic tree displayed at phyla level. Circles represent detection of complete proteins and triangles detection of complete and fragmented proteins. Colors are as for Figure 1. *Saccharomyces cerevisiae* (Ascomycota) and *Arabidopsis thaliana* (Streptophyta) were used as seeds. B: Phylogenetic tree with eukaryotic sequences from glutamate dehydrogenase isoforms. Green branches: EC1.4.1.4; Red branches: EC: 1.4.1.2; Blue branches: EC: 1.4.1.3.

The non-Metazoa eukaryotes with complete genomes, such as Alveolata, Apicomplexa and Euglenozoa, lack EAA biosynthetic enzymes (Figure 1) but keep the capability of nitrogen assimilation (Figure 3). Fornicata and Parabasalia, although represented only by draft genomes, have shown to contain the nitrogen assimilation enzyme even if they appear to be auxotrophic for all EAAs. Lacking detection of any isoform of glutamate dehydrogenase and with available draft genomes is Rhizaria (no complete genomes available), which still presents some EAA biosynthetic capability. It is possible that the dependency of organic nitrogen has been attained earlier in Rhizaria, although complete sequencing is required for a sound conclusion. In general, data support a tendency for nitrogen heterotrophy succeeding the amino acid essentiality. In Rhodophyta, a clade containing complete genomes sequenced, surprisingly no catabolic homologues were found; however a sequence that clusters with the assimilative isoforms has been found.

We also investigated nitrogen assimilation in prokaryotes. Homologues of assimilative enzymes are present and detected by our clustering procedure, but besides finding homologues of the catabolic seeds in bacterial clades, assimilative enzymes were not found in Aquificae, Chlamydiae and Synergistetes, all of them containing complete genomes available. This absence is consistent with the lysine auxotrophy suggested in Chlamydiae (Figure 2) and support the idea that EAA auxotrophy is associated with the lack of nitrogen assimilation even in the prokaryotic clades. It is hard to infer differential enzymatic activity in prokaryotes, since the annotated sequences available often report mixed use of coenzyme, either NADPH or NAD, although the homologous tools had grouped them distinctively. If the homology is related to function, it may indicate that these organisms also demand the consumption of NEAA to constitute a source of organic nitrogen. The presented scenario suggests that the loss of nitrogen assimilation forcing consumption of NEAA shortly succeeds the Great Genomic Deletion of EAA biosynthetic enzymes in metazoans. If this hypothesis is true, the Cnidaria would be an exception.

EAA Biosynthetic Enzymes Maintained

The remaining EAA biosynthetic enzymes in organisms that do not have the complete amino acid pathway (Figure 1) are more susceptible to evolutionary modifications. It is also possible that prologue sub functionalization occurred in the common ancestor of animals, fungi and plants, and thus the divergent copy has remained in detriment of the original gene. Considering both hypothesis we set up to analyze enzymes from EAA and functional NEAA pathways present in metazoans. Phylogenetic trees for acetolactate synthase (VIL1 code in Figure 1) and for a group of alanine-glyoxylate, serine-glyoxylate and serine-pyruvate transaminases (G1 code in Figure 1) are represented in Figure 4. As expected, the distance between the ancestors of the two prototrophic groups varies, plant (green circles) and fungi (yellow circles): 0.4 and 0.7, for VIL1 (Figure 4A) and G1 (Figure 4B), respectively. The distance from the ancestors of plant (green circles) to metazoans (red circles) are relatively higher for the remaining enzyme VIL1: 1.0 (as compared to 0.4 measured from plant to fungi, 2.5 fold) than for the NEAA biosynthetic enzyme G1: 0.7 (as compared to 0.7 measured from plant to fungi, 1.0 fold). Thus, the remaining EAA enzymes are experiencing higher divergence after the attainment of amino acids auxotrophy.

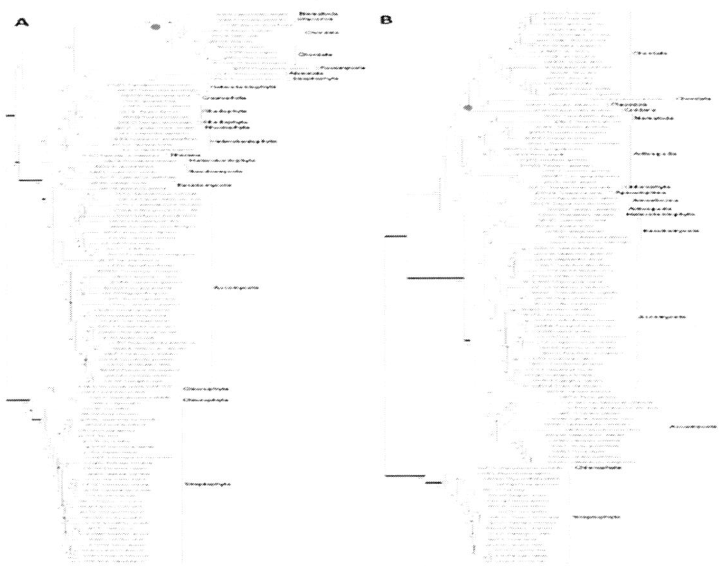

Figure 4: Phylogenetic analyses for EAA and NEAA enzymes. Phylogenetic trees for (A) acetolactate synthase (VIL1 code in Figure 1), an enzyme for EAA valine, isoleucine and leucine biosynthesis and (B) a group of alanine-

glyoxylate, serine-glyoxylate and serine-pyruvate transaminases (G1 code in Figure 1), a NEAA biosynthetic enzyme for glycine biosynthesis. The green, yellow and red circles are marking the plant (Streptophyta), fungi (Dikarya) and animals (Metazoa) branches, respectively. In (A), the distance (given by substitutions per site) from the green circle to the yellow and red circles are, respectively, 0.4 and 1.0. In (B), these values are, respectively, 0.7 and 0.7.

To support this observation, Figure 5 shows the ratios calculated for 12 enzymes. Only trees that show significant bootstraps for the branches of interest were considered. Enzyme codes in bars are described as in Figure 1. The Y axis at the right side corresponds to the distance measured from plant (Streptophyta) to the ancestor of fungi (Dikarya). This distance was assumed as a background distance to normalize the distances measured "from" plant (green bars) "to" the clades indicated in the X axis. The three enzymes on the right, S1, G1 and G2, belong to NEAA pathways, and the ratios are low. For the enzymes H5, FW7, F8, VIL1, VIL3, MT3 and M7, the ratio shown by green bars are conversely high, ranging from around 1.5 up to 7 fold. These preliminary data suggest that the additional evolutionary modifications have occurred in distinct levels in the enzymes maintained after the loss of biosynthetic capability. M (2) pathway appears as incomplete in Basidiomycota (Figure 1; M8 is absent), however MT3 enzyme used here is present in threonine pathway which is complete in this clade. K6 and K10 are involved in incomplete pathways, respectively, in plants and fungi. Accordingly, the distance measured from plant to fungi is high, and so is the drift between plant to Chordata (K6) or to Arthropoda (K10), therefore yielding balanced lower ratios. Since the ancestor of fungi and plants seems to be equally distant from both of these two groups, and the divergence between plant and Fungi/Metazoa group tends to a trifurcation (see Figure 4), the yellow bars (which represent the distance from fungi to the animal clades in the X axis divided by the background distance from plant to fungi) are similar to the ratios represented by the green bars, independently of how much modification has been occurred to the animal sequences (e.g. VIL1, MT3, G1). Furthermore, a detailed inspection of phylogenetic trees seems to indicate that subfunctionalized paralogues have appeared in basal clades such as Fungi, and those divergent paralogues remain in the more recent groups of organisms, while the copy that previously participated in the biosynthesis was actually deleted in animals. Note some Streptophyta and Ascomycota divergent paralogues (outparalogues) [27] grouped with animal sequences under 100% bootstrap (Figure 4A). Accordingly, similar divergent paralogues were observed for M7 enzyme (Ascomycota and Basidiomycota divergent paralogues grouped with animal sequences, 98% bootstrap,

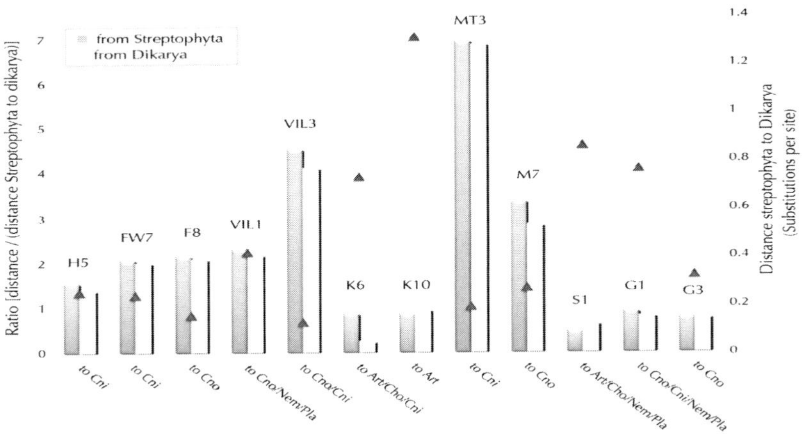

Figure 5: Relative distance of Metazoa enzymes from homologues of EAA and from NEAA biosynthetic enzymes present in plant and fungi. Phylogenetic trees were obtained for 12 enzymes, using all eukaryotic clustered proteins. Codes for enzymes are the same as in Figure1 and are shown over the bars. For normalization, a background distance from the plant phylum Streptophyta to the fungi subkingdom Dikarya was measured and represented by triangles (right Y axis). The distance "from" either Streptophyta (green bars) or Dikarya (yellow bars), "to" the branches that group the clades indicated below the bars, were measured and normalized by the distance Streptophyta/Dikarya, yielding the ratio represented by bars (left Y axis). Only the three enzymes on the right (S1, G1 and G2) participate of biosynthesis of NEAAs: serine (S1) and glycine (G1 and G2). K6 and K10 are enzymes that compose lysine biosynthetic pathways which are not complete, respectively, in Streptophyta or Dikarya (see Figure 1). Abbreviations: Art, Arthropoda; Cho, Choanozoa; Cni, Cnidaria; Nem, Nematoda; Pla, Placozoa.

DISCUSSION

The advance on genome sequencing and computational methods for clustering homologous proteins has been helping the scientific community to reevaluate several aspects of basic biology. Here we have applied clustering of protein sequences chosen from two clades of organisms that are known to be autotrophic for the biosynthesis of Essential Amino Acids (EAAs). Furthermore, we searched for the enzymes responsible for nitrogen assimilation, incorporating ammonium into glutamate. Lack of cytoplasmic glutamate dehydrogenase leads to a dependency of amino acids consumption as the source of organic nitrogen, i.e., the organism in a certain sense actually becomes auxotrophic to both EAAs and NEAAs (Non-Essential Amino Acids), in order to build other nitrogen-containing molecules.

The work presented here takes advantage of both the Seed Linkage software and a home-built UniProt Enriched KEGG Orthology database (UEKO) as source of information, to rapidly group homologues of fungi and plant amino acid sequences, respectively represented by *Saccharomyces cerevisiae* and *Arabidopsis thaliana*. KEGG Orthology contains to date more than 1 million sequences from nearly 1,000 genomes and it was enriched by a procedure developed by our group to attain 2,442,384 sequences from 25,024 organisms, constituting the UEKO database (UniRef50 enriched KEGG Orthology database, to be published elsewhere and further distributed). Counting the total recruited sequences reported in this work (31,392), the percentage of recruitment by (i) Seed Linkage, (ii) original KO or (iii) the enriched portion of KO (UEKO) was, respectively, 6%, 44% and 50%. Moreover, 26% of all detected enzymes for the phyla represented in Figures 1, 2 and 3 were exclusively detected by Seed Linkage software and/ or UEKO database. These numbers reinforce the relevance on the development of homologous searching capability, improving the ability of KEGG Orthology database to build a scenario for the biological processes of interest such as those presented here. Moreover, on top of the search for homologues represented by circles in the Figures, a complementary search using the 31,392 clustered sequences allowed the investigation of all UniProt sequences, including fragments (e.g. UniProt accession B7QGP4, VIL1 from Arthropoda) and some full length proteins not accessed by the initial search (e.g. UniProt accession D3AYE6, complete protein K14, from Amoebozoa; actually a more recent version of KO already incorporates this entry). It is important to notice that, in UniProt, the technical term fragment is applied to partial CDS sequences, a product of incompletely sequenced mRNA, as well as amino acid sequences modeled from the genome that lack initial methionine. Thus they might represent additional evidence of the enzyme presence rather than a reminiscent pseudogene. Stringent criteria (1×10^{-10} e-value, 50% identity and 50% subject coverage cutoffs) were adjusted with extensive manual inspection and additional evidences were included as triangles in the Figures. One evidence collected as triangle claimed our attention, since it came from a clade bearing the complete genome of the well annotated organism *Drosophila melanogaster* (Figure 1, enzyme VIL1, phylum Arthropoda). Manual inspection reveals that the evidence yielded by the additional search (represented by triangle) returned a hit from *Ixodes scapularis* (a genome under "assembly" status), but remarkably, the gene was found to be missing in the fly. Thus, this represents a recent gene loss within a non-functional pathway.

The main interest of this work was to depict the evolution of amino acids essentiality, or heterotrophy. Grouping organisms into phyla level allowed easy labeling of clades that comprise organisms with sequenced or draft genomes, as shown in Figures 1, 2 and 3, making it possible to infer deletion events distinctively in these clades. It is important to notice that many phyla contain complete genomes, which allowed us to figure out the deletion process with more certainty. However, the picturing of the entire scenario allowed the analysis to be extended to the branched clades, although this requires additional caution on interpretation. Even escaping the scope of this work, it suggests a demand for planned choice of genomes to be completely sequenced, since as clearly shown here we lack information from several phyla such as the ones represented with empty circles (e.g. Cryptophyta, Haptophyta, Neocallimastigomycota and Glaucophyta). Enzymes not found by our analysis requires further attention and search using more sensitive methods and detailed manual or even experimental analysis, to detect divergent sequences; in other words, the absence of evidence is not evidence of absence. However, the present work exemplifies a method that can be easily applied to other scenarios of gene/pathway loss.

The scenario of amino acid auxotrophy supports the hypothesis of a Great Genomic Deletion model of amino acid biosynthesis in association with heterotrophy. This phenomenon has probably occurred several times, particularly at the origin of metazoans. This deletion has been likely associated with endosymbiotic relationships or with the development of systems specialized in nutrient absorption. It seems that amino acid essentiality has been originated as a phenotypic loss of pathways early in Choanozoa, followed by multiple losses during metazoan evolution. Similar progresses of deletions occur closer to Heterokontophyta and Rhizaria, culminating in Apicomplexa. Rhodophyta and Microsporidia also attain the auxotrophy.

Moreover, remaining enzymes set apart from their original roles in amino acid biosynthetic metabolism seem to be more prone to evolutionary changes whilst enzymes present in complete pathways are more structurally conserved among distant phyla (Figures 4 and 5). Although a detailed investigation is needed, our preliminary analysis suggests that the copies which remained in metazoan genomes may have suffered subfunctionalization and sometimes this might have occurred in more ancestral organisms. Thus, in some sense, the orthologue enzyme might actually have been deleted in animals, and the divergent copy is the one remaining. These divergent copies are sometimes named outparalogues. We are currently investigating substitution rate ratios and promoter elements in these genes.

Subsequent deletion includes the enzymes implicated in nitrogen assimilation, which takes place just after the broad deletion of EAAs biosynthetic enzymes (since except metazoans, other eukaryotic clades lack biosynthetic pathways and contains a nitrogen assimilative enzyme), as observed in more derived metazoans, but not Cnidaria. Most Cnidaria are carnivorous, so one possibility is that Cnidaria may benefit from the assimilation of organic nitrogen under long periods of fasting, however this finding needs additional investigation. Thus, the simplest explanation, is that the loss of nitrogen assimilative enzymes are related to lower selective pressure associated with the origin of the most heterotrophic organisms, animals.

To our knowledge this is the first initiative to clarify the complete scenario using powerful homologous grouping approaches and the total repertoire of sequenced genomes.

CONCLUSIONS

The procedures described here provide a deeper analysis of amino acid and nitrogen heterotrophy among distinct taxa, extended to include the entire set of available proteins. They show that amino acid essentiality was a broad phenomenon in eukaryotes, followed by the subsequent nutritional requirement of organic nitrogen, in animals.

METHODS

Software and Databases

Seed Linkage clustering software [14] and detailed explanation of usability can be obtained athttp://www.biodados.icb.ufmg.br/eaa/ *webcite*. Seed Linkage requires BLAST (version used was 2.2.20), MySQL (version 5.0.77) [28] and PHP (version 5.1.6) [29].

The protein database is composed of UniProtKB entries (version used was 2010_09) available athttp://www.biodados.icb.ufmg.br/eaa/ *webcite*. Except where otherwise indicated, all fragmented proteins were removed from analyses by parsing the description line in FASTA files.

To enrich KEGG Orthology clusters with incomplete genome proteins UniRef50 Enriched KEGG Orthology (UEKO) was built with the procedure described by Fernandes *et al* [15]. A local MySQL database was used.

Procedure

Amino acid biosynthetic pathways were depicted with KEGG Pathway [30] manual inspection where UniProtKB identifiers for the enzymes used in this work could be retrieved for the model autotrophic organisms *Saccharomyces cerevisiae*, *Arabidopsis thaliana* and, for the archaeal lysine biosynthesis,*Pyrococcus horikoshii*. The procedure starts with the selected sequences used as seed for Seed Linkage search in UniProtKB. The homologous cluster is enriched by (i) entries in KEGG Orthology (KO) belonging to the same KO where the seed is found and (ii) UEKO entries for this same KO. All steps were conducted with MySQL consults and PERL v5.8.8 [31] scripts. To verify the recruitment, seed sequences were used in PSI-BLAST alignments with the recruited sequences, having the PSI-BLAST iterations stopped whenever the score obtained for the seed sequence itself decreases to below 50% of the initial score. Results of search for homologues are represented by circles in the Figures.

Simple BLASTp analysis (10^{-10} e-value cutoff) were also conducted with all UniProt proteins, comprising both UniProt complete and fragment entries, for each phylum against all clustered proteins in this project. Resulting output was filtered to remove alignments with less than both 50% identity and 50% subject coverage. Results of this analysis are represented by triangles in the Figures.

Taxonomy Information

All UniProtKB identifiers could be associated with an organism taxonomy ID with the file available atftp://ftp.uniprot.org/pub/databases/uniprot/current_release/knowledgebase/idmapping *webcite*.

Further association of organism taxonomy ID with phyla classification was achieved through a local database built with NCBI taxonomy information obtained at ftp://ftp.ncbi.nih.gov/pub/taxonomy*webcite*.

Genome statuses were obtained by NCBI Genome Project analysis at:http://www.ncbi.nlm.nih.gov/genomeprj *webcite*.

Phylogenetic Analyses

For phylogenetic analysis Prankster [32] was used for multiple sequence alignment and MEGA4 [33] to construct the phylogenetic tree using the neighbor-joining method [34] with 500 bootstrap replicates. Branch distances were obtained from phylogenetic trees, from the ancestors of Streptophyta, Dikarya and clades of metazoans. Only branches with significant bootstrap were used. With the distances, a ratio was calculated as below:

$$\text{Distance } F - T / \text{Distance } S - D$$

Where F (from) is either Streptophyta or Dikarya ancestor and T (to) is an animal ancestor (see Figure5, X axis); and S and D are the ancestors of Streptophyta and Dikarya, respectively. Phylogenetic trees used to compose Figure 5 can be accessed at our server athttp://www.biodados.icb.ufmg.br/eaa/.

AUTHORS' CONTRIBUTIONS

The work presented here was carried out in collaboration between all authors. FP and JMO defined the research theme. RLMG developed the clustering procedure, created the dataset and conducted the experiments. RLMG and GFR created the figures. RLMG, FP and LKM conducted phylogenetic analyses. GRF created the procedure of Uniref50 enrichment of KEGG Orthology database. HALR developed the PSI-BLAST validation method. JMO, FP and RLMG wrote the paper. All authors supervised and approved the final manuscript.

ACKNOWLEDGEMENTS

Authors thank Dr. Darren Natale from PIR (USA) and Elisa Donnard (LICR) for critically reviewing this manuscript, Henrique Velloso for helping with taxonomic data and Laryssa Santos Queiroz with pathway inspections. This work has been sponsored by the Brazilian Ministry of Education (CAPES) and Foundation for Research Support of Minas Gerais State (FAPEMIG).

This article has been published as part of *BMC Genomics* Volume 12 Supplement 4, 2011: Proceedings of the 6th International Conference of the Brazilian Association for Bioinformatics and Computational Biology (X-meeting 2010). The full contents of the supplement are available online at http://www.biomedcentral.com/1471-2164/12?issue=S4

REFERENCES

1. Blair J, Shah P, Hedges SB: Evolutionary sequence analysis of complete eukaryote genomes. BMC Bioinformatics 2005, 6:53.
2. Cunchillos C, Lecointre G: Early steps of metabolism evolution inferred by cladistic analysis of amino acid catabolic pathways. Comptes Rendus Biologies 2002, 325:119-129.
3. Cunchillos C, Lecointre G: Ordering events of biochemical evolution. Biochimie 2007, 89:555-573.
4. Hernández-Montes G, Díaz-Mejía JJ, Pérez-Rueda E, Segovia L: The hidden universal distribution of amino acid biosynthetic

networks: a genomic perspective on their origins and evolution. Genome Biol 2008, 9:R95.

5. Reeds PJ, Wahle KWJ, Haggarty P: Energy costs of protein and fatty acid synthesis. Proceedings of the Nutrition Society 1982, 41:155-159.
6. Aoyagi Y, Tasaki I, Okumura J, Muramatsu T: Energy cost of whole-body protein synthesis measured in vivo in chicks. Comp Biochem Physiol A Comp Physiol 1988, 91:765-768.
7. Millward DJ: Metabolic Demands for Amino Acids and the Human Dietary Requirement: Millward and Rivers (1988) Revisited. The Journal of Nutrition 1998, 128:2563S-2576S.
8. Millward DJ, Rivers JP: The nutritional role of indispensable amino acids and the metabolic basis for their requirements. Eur J Clin Nutr 1988, 42:367-393.
9. Elango R, Ball R, Pencharz P: Amino acid requirements in humans: with a special emphasis on the metabolic availability of amino acids. Amino Acids 2009, 37:19-27.
10. Payne SH, Loomis WF: Retention and Loss of Amino Acid Biosynthetic Pathways Based on Analysis of Whole-Genome Sequences. Eukaryotic Cell 2006, 5:272-276.
11. Consortium TU: The Universal Protein Resource (UniProt) in 2010. Nucleic Acids Research 2010, 38:D142-D148.
12. Tatusov R, Fedorova N, Jackson J, Jacobs A, Kiryutin B, Koonin E, Krylov D, Mazumder R, Mekhedov S, Nikolskaya A, et al: The COG database: an updated version includes eukaryotes. BMC Bioinformatics 2003, 4:41.
13. Kanehisa M, Goto S, Kawashima S, Okuno Y, Hattori M: The KEGG resource for deciphering the genome. Nucleic Acids Research 2004, 32:D277-D280.
14. Barbosa-Silva A, Satagopam V, Schneider R, Ortega JM: Clustering of cognate proteins among distinct proteomes derived from multiple links to a single seed sequence. BMC Bioinformatics 2008, 9:141.
15. Fernandes GR, Barbosa DVC, Prosdocimi F, Pena IA, Santana-Santos L, Coelho Junior O, Barbosa-Silva A, Velloso HM, Mudado MA, Natale DA, et al: A procedure to recruit members to enlarge protein family databases– the building of UECOG (UniRef-

Enriched COG Database) as a model. Genetics and molecular research GMR 2008, 7:910-924.

16. Suzek B, Huang H, McGarvey P, Mazumder R, Wu C: UniRef: comprehensive and non-redundant UniProt reference clusters. Bioinformatics 2007, 23:1282-1288.
17. Katinka MD, Duprat S, Cornillot E, Metenier G, Thomarat F, Prensier G, Barbe V, Peyretaillade E, Brottier P, Wincker P, et al: Genome sequence and gene compaction of the eukaryote parasite Encephalitozoon cuniculi. Nature 2001, 414:450-453.
18. Prosdocimi F, Mudado MA, Ortega JM: A set of amino acids found to occur more frequently in human and fly than in plant and yeast proteomes consists of non-essential amino acids. Computers in Biology and Medicine 2007, 37:159-165.
19. Santana-Santos L, Prosdocimi F, Ortega JM: Essential amino acid usage and evolutionary nutrigenomics of eukaryotes–insights into the differential usage of amino acids in protein domains and extra-domains. Genetics and molecular research GMR 2008, 7:839-852.
20. Miyazaki T, Miyazaki J, Yamane H, Nishiyama M: -Aminoadipate aminotransferase from an extremely thermophilic bacterium, Thermus thermophilus. Microbiology 2004, 150:2327-2334.
21. Velasco AM, Leguina JI, Lazcano A: Molecular Evolution of the Lysine Biosynthetic Pathways. Journal of Molecular Evolution 2002, 55:445-449.
22. Hudson AO, Bless C, Macedo P, Chatterjee SP, Singh BK, Gilvarg C, Leustek T: Biosynthesis of lysine in plants: evidence for a variant of the known bacterial pathways. Biochim Biophys Acta 2005, 1721:27-36.
23. Nishida H, Nishiyama M, Kobashi N, Kosuge T, Hoshino T, Yamane H: A Prokaryotic Gene Cluster Involved in Synthesis of Lysine through the Amino Adipate Pathway: A Key to the Evolution of Amino Acid Biosynthesis. Genome Research 1999, 9:1175-1183.
24. Smith TJ, Schmidt T, Fang J, Wu J, Siuzdak G, Stanley CA: The Structure of Apo Human Glutamate Dehydrogenase Details Subunit Communication and Allostery. Journal of Molecular Biology 2002, 318:765-777.

25. Allen A, Kwagh J, Fang J, Stanley CA, Smith TJ: Evolution of Glutamate Dehydrogenase Regulation of Insulin Homeostasis Is an Example of Molecular Exaptation†. Biochemistry 2004, 43:14431-14443.

26. Andersson J, Roger A: Evolution of glutamate dehydrogenase genes: evidence for lateral gene transfer within and between prokaryotes and eukaryotes. BMC Evolutionary Biology 2003, 3:14.

27. Sonnhammer ELL, Koonin EV: Orthology, paralogy and proposed classification for paralog subtypes. Trends in Genetics 2002, 18:619-620.

28. MySQL. [http://www.mysql.com].

29. PHP. [http://www.php.net].

30. Kanehisa M: KEGG: From genes to biochemical pathways. Bioinfomatics: Databases and Systems Kluwer Academic Publishers; 1999, 63-76.

31. Perl. [http://www.perl.org/].

32. Löytynoja A, Goldman N: An algorithm for progressive multiple alignment of sequences with insertions. Proceedings of the National Academy of Sciences of the United States of America 2005, 102:10557-10562.

33. Tamura K, Dudley J, Nei M, Kumar S: MEGA4: Molecular Evolutionary Genetics Analysis (MEGA) Software Version 4.0. Molecular Biology and Evolution 2007, 24:1596-1599.

34. Saitou N, Nei M: The neighbor-joining method: a new method for reconstructing phylogenetic trees. Molecular Biology and Evolution 1987, 4:406-425.

Induction of a bZIP Type Transcription Factor and Amino Acid Catabolism-Related Genes in Soybean Seedling in Response to Starvation Stress

Takashi Yuasa[1,2] Yuri Nagasawa[3], Katsumasa Osanai[4] Ayami Kaneko[4] Daichi Tajima[4] Nang Myint Phyu Sin Htwe[4] Yushi Ishibashi[1], and Mari Iwaya-Inoue[1]

[1]Department of Bioresource Sciences, Faculty of Agriculture, Kyushu University, Higashiku Hakozaki 6-10-1, Fukuoka 812-8581, Japan

[2]Department of Agricultural and Environmental Sciences, Faculty of Agriculture, University of Miyazaki, Gakuen Kihanadai Nishi 1-1, Miyazaki 889-2192, Japan

[3]School of Agriculture, Kyushu University, Higashiku Hakozaki 6-10-1, Fukuoka 812-8581, Japan

[4]Graduate School of Bioresource and Bioenvironmental Sciences, Kyushu University, Higashiku Hakozaki 6-10-1, Fukuoka 812-8581, Japan

ABSTRACT

To address roles of bZIP transcription factors on regulation of amino acid catabolism under autophagy-induced plant cells, we examined the effect of nutrient starvation on the expression of low energy stress-related transcription factor homologs, GmbZIP53A and GmbZIP53B, and amino acid catabolism-related genes in soybean (Glycine max (L.) Merr.). Sucrose starvation treatment significantly enhanced the expressions of GmbZIP53A, but not GmbZIP53B asparagine synthase (GmASN1), proline dehydrogenase1(GmProDH), and branched chain amino acid transaminase 3 (GmBCAT3). GmbZIP53-related immunoreactive signals were upregulated under severe starvation with sucrose starvation and protease inhibitors, while 3% sucrose and sucrose starvation had no or marginal effects on the signal. Profiles of induction of GmASN1, GmProDH and GmBCAT3 under various nutrient conditions were consistent with the profiles of GmbZIP53 protein levels but not with those of GmbZIP mRNA levels. These results indicate that GmbZIP53 proteins levels are regulated by posttranslational mechanism in response to severe starvation stress and that the increased protein of GmbZIP53 under severe starvation accelerates transcriptional induction of GmASN1, GmProDH, and GmBCAT3. Furthermore, it is conceivable that decrease of branched chain amino acid level by the BCAT-mediated degradation eventually enhances autophagy under severe starvation.

INTRODUCTION

The perception and management of nutrient and energy levels in organisms are crucial for survival by adjusting metabolism to available resources. Recent studies revealed that sugar signals in higher plants activate various biological modules such as sugar sensor, transcription factors, sugar transporters, and metabolic enzymes of sugar and amino acids [1, 2]. In higher plants, sugar deprivation and/or low energy stress by decreased photosynthesis have appeared to induce protein degradation via autophagy and amino acid metabolism, leading to translocation of nutrients and senescence [3, 4]. Recently accumulated studies on bZIP type transcription factors of Arabidopsis concerning nutrient signal and amino acid metabolism have unveiled that a set of bZIP transcription factors, bZIP1 and bZIP53, classified to S-type subgroup among the bZIP superfamily, are master regulatory components in transcriptional induction of amino acid catabolism-related enzymes involved in low energy stress, sucrose starvation, and senescence-induced nutrient translocation [5–7]. Sucrose starvation and/or low energy stress activate bZIP1 and bZIP53a by transcriptional and translational regulation mechanisms. Resultantly, increased proteins of the bZIP1 and bZIP53 forming heterodimers with other bZIP members initiate the transcriptional activation of amino acid metabolism-related genes by binding to ACGT- or ACTCAT-like cis-elements within the promoters of the target genes [8].

Autophagy is a nonspecific and bulk protein degradation system of intracellular components and is induced under nutrient starvation. Molecular genetic studies using mutants of autophagy-related genes (ATG) unveiled that autophagy plays essential roles in growth, senescence, nutrient translocation, and stress responses of higher plants [9–11]. The expressions of ATG8- and ATG4-related genes are enhanced under sucrose starvation in Arabidopsis suspension cells, possibly generating intracellular amino acid pools and bioenergetic resources [12]. It is well known that status of nutrient conditions has pleiotropic effects on stress-related phytohormonal signals [13]. Our previous studies indicated that the expression of GmATG8cand GmATG8i in soybean leaf transiently increased at the senescence stage and that combination of sucrose starvation and protease inhibitors significantly enhanced the induction of GmATG8c,i and GmATG4 in soybean seedling [14, 15]. One of the most important roles of autophagy is regulation of intracellular amino acid pool which tends to be fluctuated by starvation and senescence via amino acid metabolism and translocation. Furthermore, intracellular amino acid pool, particularly branched chain amino acid (BCAA), which appeared to activate target of rapamycin (TOR) which negatively regulates autophagy [2]. Therefore, the regulation mechanism of amino acid catabolism plays an important role in autophagy induction under sucrose starvation.

By starvation-induced autophagy experiment and specific antibodies against GmbZIP53, BCAT, and ATG8i, we demonstrate that the nutrient starvation activates a sucrose-induced transcription factorGmbZIP53A at transcription and protein levels, which possibly play a key role in the regulatory network of amino acid catabolism and autophagy of soybean.

MATERIALS AND METHODS

Plant Materials

Seeds of soybean (Glycine max (L.) Merr. cv. Fukuyutaka) were sterilized and sown on 1% agarose in plastic box (10 × 10 × 15 cm) inside shading incubator set at 25°C. After 5 days of seeding, cotyledon was removed from soybean seedlings (about 8–10 cm length), before nutrient treatments.

Treatments of Sucrose Starvation and Sucrose-Rich Medium

Soybean seedlings were incubated at 25°C in a nutrient-rich, starvation, or starvation-protease inhibitor medium and harvested at indicated intervals as previously described [14]. The nutrient-rich medium contained 0.5 × Murashige Skoog (MS) medium with 3% sucrose. Sucrose starvation medium contained 0.5 × MS medium. Severe starvation medium was supplemented with protease inhibitors, 1 mM phenylmethylsulfonyl fluoride, 10 μg mL^{-1} leupeptin, 100 μM E64-d, and 10 mM quinacrine. The medium was adjusted to pH 7.0 by adding KOH. The resultant samples were frozen in liquid nitrogen and then were stored at −80°C.

RNA Preparation and RT-PCR

Soybean homologs of AtbZIP1, AtbZIP53, and amino acid catabolism-related genes were identified by searching at Phytozome V 9 with tBlastx program (http://www.phytozome.net/). Gene specific primers are shown in Table 1. GmActin was used as positive control. RNA preparation and RT-PCR were carried out by using Rever TraACE kit (Toyobo Co. Ltd. Osaka, Japan), GoTaq Kit (Promega Bioscience Inc., Madison, USA), as previously described [16]. PCR was performed with thermal cycler PC-816 (ASTEC Inc., Fukuoka, Japan) under a following thermal cycle condition: initial denaturing at 94°C for 2 min, followed by 3 steps of 24–29 cycles (as indicated) denaturing at 94°C for 10 sec, annealing at 58°C for 10 sec, extending at 72°C for 40 sec, and final extention at 72°C for 30 sec. PCR products were subjected to 1.5% agarose electrophoresis and ethidium bromide staining and then visualized by FluorChem (Cell Bioscience, Santa Clara, CA).

Table 1: Gene-specific oligo DNA primers used for RT-PCR and cloning

Gene	Accession number/ gene index	Primer sequence
GmbZIP53A	Glyma03g37790	F-AAAAGGATCCTCCTTCTCGGAAATGGCATCACC
		R-TAAAGTCGACAGGAATGATCAACGCAGAAACAT
GmbZIP53B	Glyma19g40390	F-CTCCGGATCCGAAATGGCGTCGTCAATCCAACA
		R-AAGAGTCGACTAACGCAGAAACATGTTTGCGGT
GmBCAT3	Glyma01g40420	F-AATGGATCCCATCGTGCAACTCCTGGTGGC
		R-AGTCCATGGATCTCAGCTCAACAGTCCAATTCA
GmProDH	Glyma13g07110	F-GAGGGGATCCTACATGTCTACAGAGAG
		R-GAACACTGCAGCTTTTAGTCTCCTTCC
GmASN1	Glyma18g06840	F-TCTAGGATCCAGAGTACCATTTTTGGA
		R-GCTCTATATGTCGACTCCTAGTGGAGA
GmASP3	Glyma4g08560	F-ACGTGTTGGATCCTTAAGCATTGTCTG
		R-CTGCGGTCGACAACATGTTTTAGGC
GmATG8i	AB453309/ Glyma02g01180	F-CCGGAATTCGTTGAGCTGCACAACAACCTA
		R-GATAGTCGACACTTCACAAAGTGTTGGATA
GmATG4	Glyma18g48380	F-GGCAAACCAGGTGCTTCAACATAC
		R-ATGCCAGGGATCAACCAAACTTGC
GmActin	V00450/ Glyma08g15480	F-GCGTGATCTCACTGATGCCCTTAT
		R-AGCCTTCGCAATCCACATCTGTTG

Raising Antibodies, Protein Preparation, and Immunoblot

Synthetic antigen epitopes (H_2N-VEIPEIPDPLLKPWQIPHP-COOH and H_2N-LANKRWVPPPGKGSLYLRP-COOH) were designed for soybean bZIP53 and plant BCAT shown in Figures 1(c) and 2(b), respectively. The antipeptide specific polyclonal antibodies were raised in a rabbit with those synthetic antigen oligopeptides (Sigma-Aldrich Co., St. Louis, USA) [16]. Protein extracts and immunoblot were carried out as previously described with minor modifications [14]. To detect GmATG8i, each 50 μg protein per lane was subjected to SDS-PAGE with a gel of 15% acrylamide containing 6 M urea to separate phosphatidylethanolamine-modified ATG8i. After electroblotting polypeptides onto PVDF membranes (Millipore, Billerica, USA) and blocking the PVDF membrane with TBS-Milk, the PVDF membrane was incubated in TBS-Milk-Tween with anti-GmAtg8i antibody, anti-bZIP53 antibody, or anti-BCAT antibody (dilution, 1/1,000 [v/v]) for 2 h at 4°C. The resultant PVDF membranes were incubated in TBS-Milk-Tween containing horse raddish peroxidase-conjugated anti-rabbit antibody (dilution, 1/5,000 [v/v], GE Healthcare Bio-Sciences LtD., Piscataway, NJ) for 1 h. Immunoreactive signals were visualized by ECL Plus Kit (GE Healthcare Bio-Sciences Ltd.) and FluorChem.

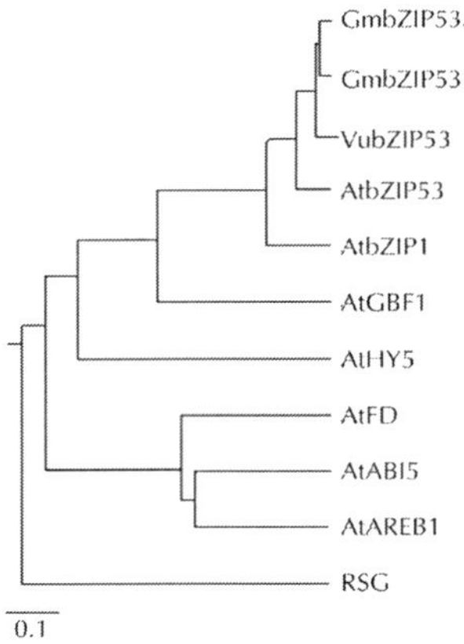

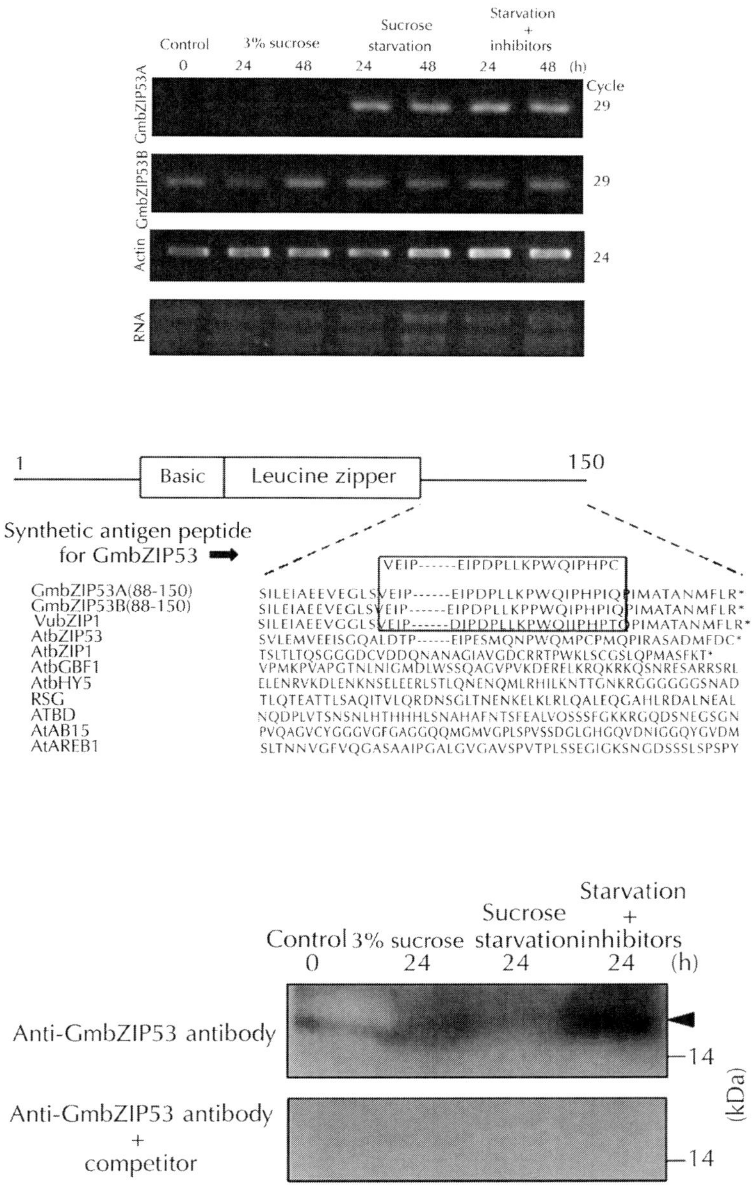

Figure 1: Effects of nutrient starvation on the expression and bZIP53 protein levels of soybean bZIP53 orthologs. (a) A phylogenic tree of bZIP type transcription factors from soybean, cowpea, Arabidopsis, and tobacco was con-

structed by ClustalW (UPGMA). The genes used were follows: GmbZIP53A and GmbZIP53B (Table 1),VubZIP53 (AB779667), AtbZIP53 (At3g62420), AtbZIP1 (At5g49450), AtGBF1(At4g36730), AtHY5 (At5g11260), AtFD (At4g35900), AtABI5 (At2g36270), AtAREB1 (At1g45249), and RSG (AB040471). Gene indics of soybean and Arabidopsis were referred to Phytozome V 9.0 (http://www.phytozome.net/) and TGI database (http://compbio.dfci.harvard.edu/tgi/plant.html), respectively. (b) The expression ofGmbZIP53A and GmbZIP53B was analyzed by RT-PCR. Actin was used as a standard. (c) Alignments of deduced amino acid sequences of bZIP53 orthologs and other bZIP type transcription factors. A highly conserved amino acid motif (H_2N-VEIPEIPDPLLKPWQIPHP-COOH) present in the carboxyl terminal region following bZIP domain was used as an epitope for raising anti-GmbZIP53-specific antibody. (d) Immunoblot of soybean seedlings treated with various nutrient conditions was carried out by anti-GmbZIP53 antibody. Each 50 μg protein per lane was subjected to SDS-PAGE (15% acrylamide gels). Nutrient treatments, semiquantitative RT-PCR, and immunoblot were carried out as described in Section 2.

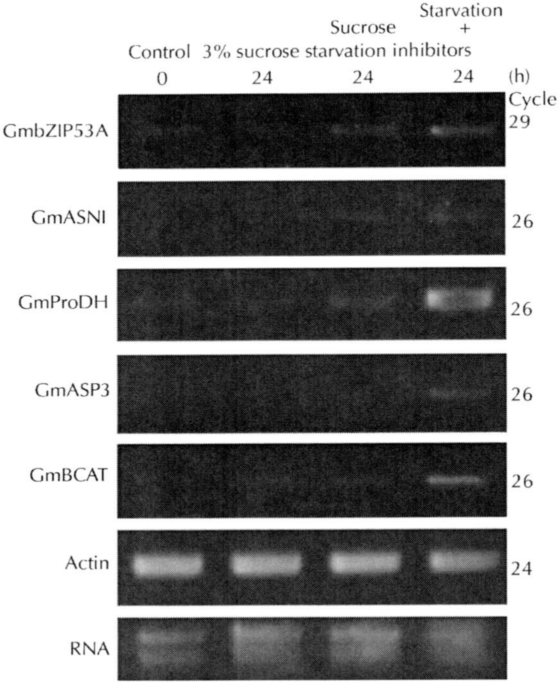

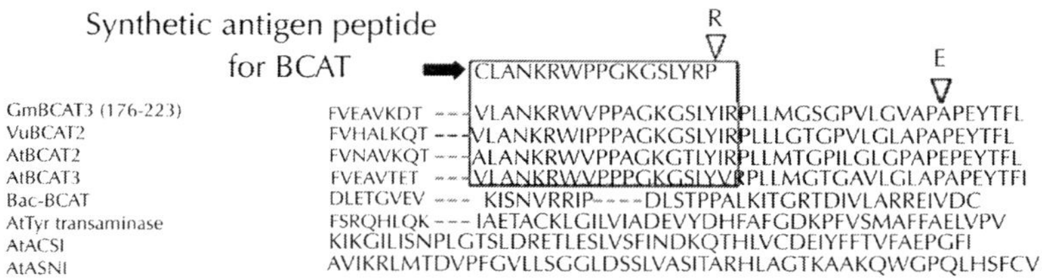

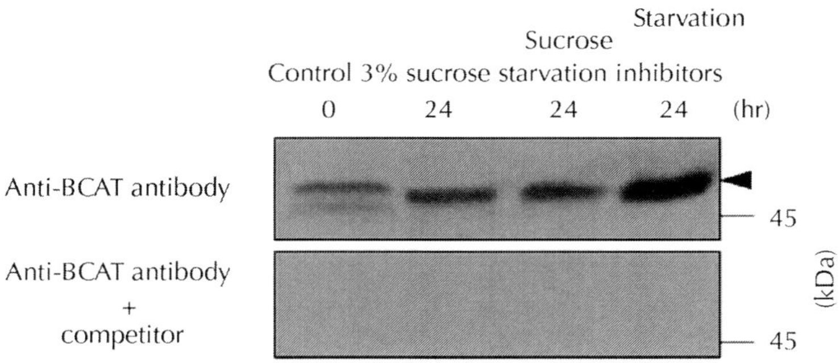

Figure 2: Effects of nutrient starvations on the expression of amino acid catabolism-related genes and BCAT protein levels. (a) The expression of GmbZIP53A, GmASN1,GmProDH, GmASP3, and GmBCAT2 was analyzed. (b) Alignments of deduced amino acid sequences of plant BCAT orthologs and other aminotransferase-related enzymes. A highly conserved amino Acid motif (H_2N-LANKRWVPPPGKGSLYLRP-COOH) present in the catalytic domain was used as an epitope for raising antiplant BCAT-specific antibody. (c) Immunoblot of soybean seedlings treated with various nutrient conditions was carried out by the antiplant BCAT-antibody. Each 50 µg protein per lane was subjected to SDS-PAGE (12.5% acrylamide gels). 14 Nutrient treatments, semiquantitative RT-PCR, and immunoblot were carried out as described in Section 2.

RESULTS

Nutrient Starvation Stress Induces a Bzip53 Ortholog in Soybean

Heterodimers of bZIP1 (At5g49450) and bZIP53 (At3g62420) have been identified to be crucial transcriptional regulators in Asn, Pro, and BCAA under sucrose starvation [7]. Among bZIP homolog genes in soybean genome, GmbZIP53A (Glyma03g37790) and GmbZIP53B (Glyma19g40390) were identified to have the highest similarities of 51% and 26% in amino acid sequences to bZIP53 (At3g62420), respectively. Difference of only 14 amino acids among total 150 amino acids between GmbZIP53A and GmbZIP53B was identified in their amino terminal regions. A phylogenic analysis of deduced amino acid sequences ofbZIP53 homologs of Arabidopsis thaliana, soybean (Glycine max (L.) Merr.), and cowpea (Vigna unguiculata L.) showed that GmbZIP53A, GmbZIP53B, and VubZIP53 (accession number AB77966) are classified into one clade containing bZIP1 and bZIP53 (Figure 1(a)) [17].

RT-PCR revealed that the expression profile of GmbZIP53A and GmbZIP53B is quite different under sucrose starvation treatment, while the two genes share high similarity. Sucrose starvation had significantly enhanced the expression of GmbZIP53A but not GmbZIP53B. The expression levels of GmbZIP53A in soybean seedling were significantly induced at 24 h after response to sucrose starvation treatment, while 3% sucrose had little or marginal effect on the expression of those genes (Figure 1(b)). Furthermore, sucrose starvation with inhibitors (leupeptin and E-64d) of proteases resulted in slightly higher induction ofGmbZIP53A at 24 and 48 h, compared to those in sucrose starvation treatment. The expression ofGmbZIP53B was relatively high level compared to GmbZIP53A in control and sucrose treatment. The expression levels of GmbZIP53B were constant in conditions of control, sucrose treatment, starvation, and starvation with inhibitors.

Synthetic antigen epitopes designed for bZIP53 are shown with alignments of bZIP type transcription factors (Figure 1(c)). A highly conserved motif of soybean bZIP53 was identified as 19 amino acids (VEIPEIPDPLLKPWQIPHP) at the carboxyl region following the conventional bZIP domain [5]. The immunoreactive signal at 24 h with molecular masses of about 16 kD is significantly higher in the presence of the inhibitors of proteases than in the absence, when soybean seedlings were subjected to sucrose starvation treatment. It suggests that a reduction of amino acid pool by suppressed proteolysis with the inhibitor is involved in induction mechanism of GmbZIP53-related protein. Treatments of 3% sucrose had no or marginal effects on the signals of bZIP53-related proteins, compared to those in control. In the presence of bZIP53 antigen competitor peptides, those immunoreactive signals disappeared, suggesting that the antibodies are highly specific to bZIP53-related polypeptides.

Induction of Amino Acid Catabolism-Related Genes

Recently, low energy stresses such as reduced photosynthesis and/or sucrose starvation appeared to induce dramatic change of amino acid catabolism for asparagine biosynthesis by inducing the expression ofasparagine synthase (AtASN1), aspartate aminotransferase (AtASP3), proline dehydrogenase 1 (AtProDH), and branched chain amino acid transaminase (AtBCAT2) [18–20]. Among soybean homolog genes amino acid catabolic-related enzymes, GmASN1 (Glyma18g06840), GmProDH (Glyma13g07110), and GmBCAT3(Glyma01g40420) were identified to have the highest similarites of 82%, 62%, and 62% in deduced amino acid sequences to AtASN1 (At3g47340), AtProDH (At3g30775), and AtBCAT2 (At1g10070), respectively.

Sucrose starvation induced GmbZIP53A, GmASN1, and GmProDH but not or marginally GmASP3 norGmBCAT. On the other hand, sucrose starvation with the protease inhibitors significantly upregulatedGmProDH, GmASP3, and GmBCAT, compared to those in the absence of inhibitors, while there is no difference of the expression of GmbZIP53A and GmASN1 between the presence or the absence of the inhibitor. Sucrose treatment had no or marginal effects on these genes (Figure 2(a)).

The synthetic antigen epitope designed for plant BCAT is shown with alignments of amino acid transaminase-related enzymes (Figure 2(c)). A highly conserved motif of plant BCATs was also identified as 19 amino acids (LANKRWVPPPGKGSLYLRP). Open arrowheads indicated with R (arginine) and E (glutamate) are conserved amino acid residues critical for interaction with pyridoxal phosphate and substrate, respectively [2]. Immunoreactive signals of BCAT-related proteins were detected at significant levels in severe starvation treatment with molecular masses of about 50 kDa (Figure 2(c)). On the other hand, treatments of 3% sucrose had little or marginal effects on the signals of BCAT-related proteins. Notably, the profile of the immunoreactive signals of the BCAT-related proteins in nutrient conditions was similar to that of the soybean bZIP53. Immunoblot with BCAT antigen competitor peptides indicates that the antibody cross-reacted to plant BCAT-related proteins specifically. In control, two immunoreactive signals at molecular masses close to each other were detected weakly in control. Accordingly, there are possibilities that the two immunoreactive signals in control indicate the presence of multiple polypeptides derived from other BCAT orthologs, or alternatively that the difference among two BCAT-related signals in control was derived by posttranslational modification.

Sucrose Treatment and Sucrose Starvation Oppositely Regulate Autophagy in Soybean

Autophagy status in soybean seedling under various nutrient conditions was examined by analyzing the expressions of GmATG8i and GmATG4 and protein levels of ATG8i-related proteins. The expression levels of GmATG8i and GmATG4 in soybean seedling were significantly induced at 24 h in response to sucrose starvation treatment, while 3% sucrose had little or marginal effect on the expression of those genes (Figure3(a)). Furthermore, sucrose starvation with protease inhibitors resulted in relatively higher induction ofGmATG8i and GmATG4 at 12 h, compared to those in sucrose starvation treatment. In contrast to starvation treatments, 3% sucrose treatment transiently suppressed the expression of GmATG8i andGmATG4i at 12 h. Previous studies with Arabidopsis and soybean revealed that the upregulation of ATG8sand ATG4 mRNA was accompanied with induction of autophagy in response to sucrose starvation stress [12]. The present data that sucrose starvation induced GmATG8i and GmATG4 is consistent with the previous observation on autophagy induction in Arabidopsis and soybean [14, 21].

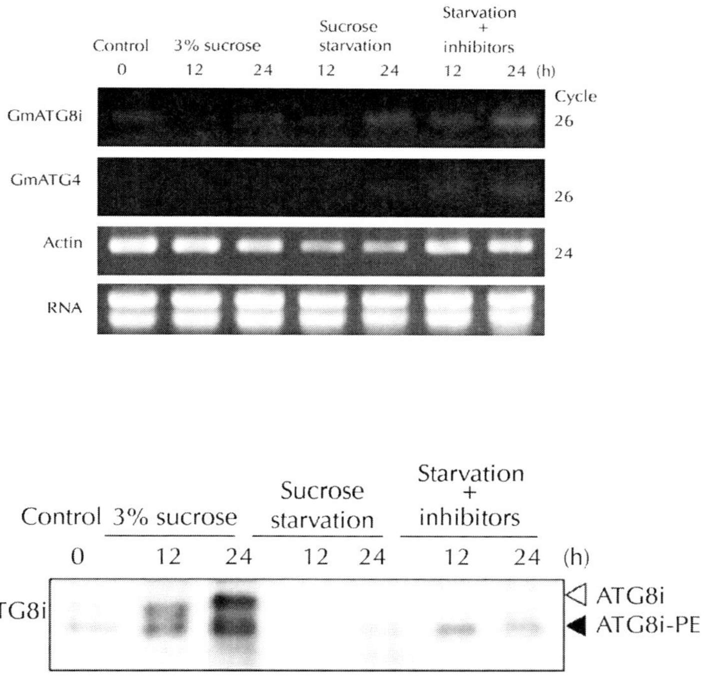

Figure 3: Induction of autophagy of soybean seedlings in response to nutrient starvation. (a) The expression profiles of GmATG8i and GmATG4 under various nutrient condition as indicated. (b) The immunoreactive signals of ATG8i-related polypeptides under various nutrient conditions as indicated. Mobility shift between GmATG8i (open triangle) and phosphatidyl ethanolamine- (PE-) conjugated GmATG8i (closed triangle) is indicated. Nutrient treatments, semiquantitative RT-PCR, and immunoblot were carried out as described in Section 2.

The profiles of ATG8i-related proteins in response to nutrient conditions were analyzed with an anti-ATG8i-specific antibody (Figure 3(b)). Combination of sucrose starvation and protease inhibitors, leupeptin and E-64d, significantly increased immunoreactive signals of ATG8i protein at 12 and 24 h, compared to sucrose starvation in the absence of leupeptin and E-64d (Figure 3(b)). The significant difference of ATG8i levels in sucrose-starved soybean seedlings between the presence and the absence of protease inhibitors indicates that sucrose starvation stimulated autophagy leading to translocation of ATG8-associated autophagosomes to vacuole. Sucrose treatment (3%) increased immunoreactive signals of ATG8i-related protein in soybean seedling at 12 and 24 h, while the treatment had little or marginal effects on the expression of GmATG8i (Figures 3(a) and 3(b)). In addition of significant signals of GmATG8i at 14 kDa, upper-shifted signals at 16 kDa of ATG8i-related proteins appeared in 3% sucrose treatment, suggesting that phosphatidylethanolamine-unconjugated GmATG8i was accumulated due to suppression of autophagy.

DISCUSSION

Recently, low energy stress on higher plants appeared to induce significant changes of amino acid metabolism via transcriptional and translational regulation mechanisms [7]. The present data indicates that nutrient starvation of soybean seedlings induces the expression of the bZIP53 ortholog and amino acid catabolism-related genes, accompanied with enhanced degradation of intracellular components via autophagy. Previous studies indicated that the intracellular-free amino acid pool generated by autophagy are utilized for amino acid recycle, respiration substrate and/or translocation of nitrogen to other tissues [22]. Among bZIP type transcription factors in Arabidopsis, heterodimer bZIP transcription factors, bZIP1 and bZIP53, appeared to be activated, transcriptionally and translationally, in response to low energy stress. Furthermore, it was shown that the bZIP1 and bZIP53 are master transcription factors regulating the expression of ASN1, ProDH, and BCAT2, which are involved in amino acid catabolism and nitrogen translocation [7]. The expression of GmZIP53A but not GmbZIP53B was significantly induced by sucrose starvation and severe starvation with inhibitors of proteases, while the GmbZIP53-related protein levels were upregulated by severe starvation (Figures 1(b) and 1(d)). Recent studies of plant transcription factors revealed that protein levels and transcriptional activities of Inducer of CBF (C-box Binding Factor) Expression 1 (ICE1) and drought responsive element binding factor 2A (DREB2A) are regulated by the protein stability via ubiquitin-proteasome system [23]. It can be assumed that the severe starvation with protease inhibitors on soybean seedlings suppresses amino acid recycle via autophagy, resulting in decreased amino acid pool. It is conceivable that the a starvation signal of decreased amino acid pool increases the protein stability of GmbZIP53A and/or GmbZIP53B via an independent pathway from sucrose starvation.

Our previous study indicates that leaf senescence accompanied with enhancing of ATG expression and nitrogen translocation at the pod filling stage is induced by increased uptake of photoassimilate from source organs [24]. In higher plants, ProDH mediates the degradation of the compatible osmolyte, proline, which is used to support the demand for nitrogen, carbon, and energy under carbohydrate limitation [7]. ASN1 and ASP3 are key enzymes of the biosynthesis of asparagine, which is utilized for intracellular nitrogen storage and nitrogen transport via vascular tissue especially when carbohydrate supply is limited [18]. It is conceivable that nutrient demand from the developing pod promotes a signal of decreased sugar level in leaf, leading to GmbZIP53A-mediated induction of amino acid catabolism-related genes which are necessary for nutrient translocation from senescing leaf to developing seed.

Starvation signals enhanced the expressions of ATGs and autophagy (Figure 3). Recent studies revealed that intracellular levels of sugar and BCAA play important roles in autophagy induction via AMP-activated protein kinase (AMPK)/Snf1-related protein kinase (SnRK) and target of rapamycin (TOR), respectively [2]. It has been shown that decrease of intracellular BCAA level accelerates autophagy by inactivation of TOR. BCATs appeared to function in the degradation of BCAA to provide an alternative carbon source under stress conditions in plants, animal, and bacteria [25]. Barley BCATs and Arabidopsis BCAT2 are induced by drought stress and sucrose starvation, respectively [7, 20]. It was reported that AtbZIP1 and AtbZIP53 bind to specific sequence motif elements, ACGT-box and ACTCAT-box, in the promoter regions ofProDH, ASN1, and BCAT [7]. According to soybean genome sequence data base, the 5′-promoter regions within 300 bp from starting ATG codon of GmBCAT and GmATG8i genes possesses 3 boxes similar to consensus motives of the ACGT-box and ACTCAT-box. It is possible that GmbZIP53A plays pivotal roles in the expression of both GmBCAT and GmATG8i via binding to ACGT-box and ACTCAT-box. A significant induction of GmBCAT in soybean seedling under severe starvation possibly resulted in an increased degradation of BCAA by GmBCAT. It can be assumed that decreased BCAA pool of soybean seedling under nutrient starvation induces autophagy via TOR signal. Increasing evidence suggests that the network of various nutrient signals, such as sugars and amino acids, regulate plant development, senescence, and primary metabolisms. Therefore, identification and molecular dissection of the signaling molecules involved in sugar signaling are important for research of autophagy and amino acid catabolism in higher plants.

ACKNOWLEDGMENTS

This work was supported in part by Grant-in-Aids from the Ministry of Education, Sports, Culture, Science and Technology of Japan, no. 25560038 (to T. Yuasa) and no. 23380013 (to M. Iwaya-Inoue), and by the Ito Foundation (to T. Yuasa).

REFERENCES

1. E. Baena-González and J. Sheen, "Convergent energy and stress signaling," Trends in Plant Science, vol. 13, no. 9, pp. 474–482, 2008.
2. C. Robaglia, M. Thomas, and C. Meyer, "Sensing nutrient and energy status by SnRK1 and TOR kinases," Current Opinion in Plant Biology, vol. 15, no. 3, pp. 301–307, 2012.
3. L. M. Weaver and R. M. Amasino, "Senescence is induced in individually darkened Arabidopsis leaves, but inhibited in whole darkened plants," Plant Physiology, vol. 127, no. 3, pp. 876–886, 2001.
4. H. Ishida, K. Yoshimoto, M. Izumi et al., "Mobilization of Rubisco and stroma-localized fluorescent proteins of chloroplasts to the vacuole by an ATG gene-dependent autophagic process," Plant Physiology, vol. 148, no. 1, pp. 142–155, 2008.
5. M. Jakoby, B. Weisshaar, W. Dröge-Laser et al., "bZIP transcription factors in Arabidopsis," Trends in Plant Science, vol. 7, no. 3, pp. 106–111, 2002.
6. S. G. Kang, J. Price, P.-C. Lin, J. C. Hong, and J.-C. Jang, "The Arabidopsis bZIP1 transcription factor is involved in sugar

signaling, protein networking, and DNA binding," Molecular Plant, vol. 3, no. 2, pp. 361–373, 2010

7. K. Dietrich, F. Weltmeier, A. Ehlert et al., "Heterodimers of the Arabidopsis transcription factors bZIP1 and bZIP53 reprogram amino acid metabolism during Low energy stress," Plant Cell, vol. 23, no. 1, pp. 381–395, 2011.
8. F. Weltmeier, F. Rahmani, A. Ehlert et al., "Expression patterns within the Arabidopsis C/S1 bZIP transcription factor network: availability of heterodimerization partners controls gene expression during stress response and development," Plant Molecular Biology, vol. 69, no. 1-2, pp. 107–119, 2009.
9. K. Yoshimoto, H. Hanaoka, S. Sato et al., "Processing of ATG8s, ubiquitin-like proteins, and their deconjugation by ATG4s are essential for plant autophagy," Plant Cell, vol. 16, no. 11, pp. 2967–2983, 2004
10. Y. Kabeya, N. Mizushima, T. Ueno et al., "LC3, a mammalian homologue of yeast Apg8p, is localized in autophagosome membranes after processing," The EMBO Journal, vol. 19, no. 21, pp. 5720–5728, 2000.
11. D. J. Klionsky, J. M. Cregg, W. A. Dunn Jr. et al., "A unified nomenclature for yeast autophagy-related genes," Developmental Cell, vol. 5, no. 4, pp. 539–545, 2003.
12. T. L. Rose, L. Bonneau, C. Der, D. Marty-Mazars, and F. Marty, "Starvation-induced expression of autophagy-related genes in Arabidopsis," Biology of the Cell, vol. 98, no. 1, pp. 53–67, 2006.
13. H. Rouached, A. B. Arpat, and Y. Poirier, "Regulation of phosphate starvation responses in plants: signaling players and cross-talks," Molecular Plant, vol. 3, no. 2, pp. 288–299, 2010.
14. N. M. P. S. Htwe, H. Tanigawa, Y. Ishibashi, S.-H. Zheng, T. Yuasa, and M. Iwaya-Inoue, "Nutrient starvation differentially regulates the autophagy-related gene GmATG8i in soybean seedlings," Plant Biotechnology, vol. 26, no. 3, pp. 317–326, 2009.
15. M. Okuda, M. P. S. H. Nang, K. Oshima et al., "The ethylene signal mediates induction of GmATG8i in soybean plants under starvationstress," Bioscience, Biotechnology, and Biochemistry, vol. 75, no. 7, pp. 1408–1412, 2011.
16. J. Nakamura, T. Yuasa, T. T. Huong et al., "Rice homologs of inducer

of CBF expression (OsICE) are involved in cold acclimation," Plant Biotechnology, vol. 28, no. 3, pp. 303–309, 2011.

17. A. Kaneko, E. Noguchi, Y. Ishibashi, T. Yuasa, and M. Iwaya-Inoue, "Sucrose starvation signal mediates induction of autophagy- and amino acid catabolism-related genes in cowpea seedling,"American Journal of Plant Sciences, vol. 4, no. 3, pp. 647–653, 2013.
18. L. H. M. Lam Hon Ming, S. S. Y. Peng, and G. M. Coruzzi, "Metabolic regulation of the gene encoding glutamine-dependent asparagine synthetase in Arabidopsis thaliana," Plant Physiology, vol. 106, no. 4, pp. 1347–1357, 1994.
19. K. Nakashima, R. Satoh, T. Kiyosue, K. Yamaguchi-Shinozaki, and K. Shinozaki, "A gene encoding proline dehydrogenase is not only induced by proline and hypoosmolarity, but is also developmentally regulated in the reproductive organs of arabidopsis," Plant Physiology, vol. 118, no. 4, pp. 1233–1241, 1998.
20. M. Malatrasi, M. Corradi, J. T. Svensson, T. J. Close, M. Gulli, and N. Marmiroli, "A branched-chain amino acid aminotransferase gene isolated from Hordeum vulgare is differentially regulated by drought stress," Theoretical and Applied Genetics, vol. 113, no. 6, pp. 965–976, 2006.
21. S. Slávikovả, G. Shy, Y. Yao et al., "The autophagy-associated Atg8 gene family operates both under favourable growth conditions and under starvation stresses in Arabidopsis plants," Journal of Experimental Botany, vol. 56, no. 421, pp. 2839–2849, 2005.
22. D. C. Bassham, M. Laporte, F. Marty et al., "Autophagy in development and stress responses of plants," Autophagy, vol. 2, no. 1, pp. 2–11, 2006.
23. K. Miura, B. J. Jing, J. Lee et al., "SIZ1-mediated sumoylation of ICE1 controls CBF3/DREB1A expression and freezing tolerance in Arabidopsis," Plant Cell, vol. 19, no. 4, pp. 1403–1414, 2007.
24. N. M. P. S. Htwe, T. Yuasa, Y. Ishibashi et al., "Leaf senescence of soybean at reproductive stage is associated with induction of autophagy-related genes, GmATG8c, GmATG8i and GmATG4," Plant Production Science, vol. 14, no. 2, pp. 141–147, 2011.
25. S. Binder, "Branched-chain amino acid metabolism in Arabidopsis thaliana," Arabidopsis Book, vol. 8, article e0137, 2010.

Mohammad Anwar Hossain[1,2] Pukclai Piyatida[3] Jaime A. Teixeira da Silva[4], nd Masayuki Fujita[1]

[1]Laboratory of Plant Stress Responses, Department of Applied Biological Science, Faculty of Agriculture, Kagawa University, Miki-cho, Kita-gun, Kagawa 761-0795, Japan

[2]Department of Genetics and Plant Breeding, Bangladesh Agricultural University, Mymensingh 2202, Bangladesh

[3]Department of Applied Biological Science, Faculty of Agriculture, Kagawa University, Miki-cho, Kita-gun, Kagawa 761-0795, Japan

[4]Laboratory of Ornamental Floriculture, Department of Bioproduction Science, Faculty of Agriculture, Kagawa University, Miki-cho, Kita-gun, Kagawa 761-0795, Japan

ABSTRACT

Heavy metal (HM) toxicity is one of the major abiotic stresses leading to hazardous effects in plants. A common consequence of HM toxicity is the excessive accumulation of reactive oxygen species (ROS) and methylglyoxal (MG), both of which can cause peroxidation of lipids, oxidation of protein, inactivation of enzymes, DNA damage and/or interact with other vital constituents of plant cells. Higher plants have evolved a sophisticated antioxidant defense system and a glyoxalase system to scavenge ROS and MG. In addition, HMs that enter the cell may be sequestered by amino acids, organic acids, glutathione (GSH), or by specific metal-binding ligands. Being a central molecule of both the antioxidant defense system and the glyoxalase system, GSH is involved in both direct and indirect control of ROS and MG and their reaction products in plant cells, thus protecting the plant from HM-induced oxidative damage. Recent plant molecular studies have shown that GSH by itself and its metabolizing enzymes—notably glutathione S-transferase, glutathione peroxidase, dehydroascorbate reductase, glutathione reductase, glyoxalase I and glyoxalase II—act additively and coordinately for efficient protection against ROS- and MG-induced damage in addition to detoxification, complexation, chelation and compartmentation of HMs. The aim of this review is to integrate a recent understanding of physiological and biochemical mechanisms of HM-induced plant stress response and tolerance based on the findings of current plant molecular biology research.

INTRODUCTION

The molecular and physiological basis of crop plant interactions with the environment has attracted considerable interest in recent years. Being sessile organisms, plants are constantly exposed during their life cycle to adverse environmental conditions that negatively affect growth, development, or productivity. The presence of toxic compounds, such as heavy metals (HMs), is one important factor that can cause damage to plants by altering major plant physiological and metabolic processes [1–5]. In a strict sense, the term HM includes only elements with specific gravity above five but frequently biologist's use this term for a vast range of metals and metalloids which are toxic to plants such as copper (Cu), iron (Fe), manganese (Mn), zinc (Zn), nickel (Ni), cobalt (Co), cadmium (Cd), and arsenic (As) etc. Importantly, few HMs and transition metals such as sodium (Na), potassium (K), calcium (Ca), magnesium (Mg), Fe, Cu, Zn, Co, or Ni are, at certain concentrations, essential micronutrients that are critically involved in the functional activities of large numbers of proteins involved in sustaining growth and development of living organisms. However, at excess concentrations, these metal ions can become detrimental to living organisms, including plants. Although HMs are natural constituents of soils and occur naturally in the environment, nowadays, contamination of soils by toxic metals and metalloids is of major concern worldwide [4, 5]. The problem of HM pollution is continuously worsening due to a series of human activities, leading to an intensification of research dealing with the phytotoxicity of these contaminants and with the mechanisms used by plants to counter their detrimental effects [4]. Transfer of toxic elements to the human food chain is a concrete danger that has to be faced in the near future. Therefore, a complete understanding of the molecular mechanisms and genetic basis of phytoremediation is an important aspect of developing plants as agents for phytoremediating contaminated sites.

Depending on their oxidation states, HMs can be highly reactive, resulting in toxicity of plant cells in many ways. At the cellular and molecular level, HM toxicity results in alterations of different plant physiological processes, including inactivation and denaturation of enzymes, proteins, blocking of functional groups of metabolically important molecules, displacement/substitution of essential metal ions from biomolecules and functional cellular units, conformational modifications and disruption of membrane integrity [5, 6], which is finally attributed to altered plant metabolism, inhibition of photosynthesis, respiration, and alerted activities of several key enzymes [2, 3, 7–13]. In addition, HMs are known to disturb redox homeostasis by stimulating the formation of free radicals and reactive oxygen species (ROS) such as singlet oxygen ($^1O_2$), superoxide radicals ($O_2^{\bullet -}$), hydrogen peroxide (H_2O_2), and hydroxyl radicals ($^{\bullet}OH$) [3, 8, 13, 14]. Recently, methylglyoxal (MG), a cytotoxic compound, was also found to increase in response various stresses including HMs [2, 15–17]. An increase in MG level in plant cells further intensifies the production of ROS by interfering with different plant physiological and metabolic processes such as inactivation of the antioxidant defense system [18, 19] and interfering with vital plant physiological processes such as photosynthesis [20]. This increase in ROS and MG exposes cells to oxidative stress leading to lipid peroxidation, biological macromolecule deterioration, membrane dismantling, ion leakage, and DNA-strand cleavage and finally death of plants [3, 4, 21–23].

Plant tolerance mechanisms require the coordination of complex physiological and biochemical processes, including changes in global gene expression [24]. Plants employ various strategies to cope with the toxic effects of metals or metalloids. Resistance to HM stress can be achieved by "avoidance" when plants are able to restrict metal uptake, or by "tolerance" when plants survive in the presence of high internal metal concentration. Avoidance involves reducing the concentration of metal entering the cell by extracellular precipitation, biosorption to cell walls, reduced uptake, or increased efflux. In a second type of situation, HMs are intracellularly chelated through the synthesis of amino acids, organic acids, GSH, or HM-binding ligands such as metallothioneins (MTs), phytochelatins (PCs), compartmentation within vacuoles, and upregulation of the antioxidant defense and glyoxalase systems to counter the deleterious effects caused by ROS and MG [2, 3, 17, 25–31]. A large number of recent studies in plants involving sensitive, tolerant, mutant, transgenic, and hyperaccumulator-adopting strategies in the fields of physiology, genomics, proteomics, and metabolomics suggest that GSH by itself and its related metabolizing enzymes, proteins, and peptides play a pivotal role in HM tolerance by controlling different plant physiological processes, including ROS and MG detoxification, HM uptake, translocation, chelation, and detoxification. The present paper represents a comprehensive understanding of physiological and biochemical mechanisms of HM-induced plant stress response and tolerance. However, special emphasis will be given to ROS and MG metabolism in HM-stressed plants in relation to GSH and its related enzymes, proteins, and genes and their consequence in HM-induced ROS and MG detoxification based on the findings of current plant molecular biology research.

MODE OF ACTION OF TOXIC HMS IN PLANT CELLS

The toxicity of HMs is manifested in many ways when plant cells accumulate them at high levels. HMs can be divided into two groups: redox active (Fe, Cu, Cr, Co) and redox inactive (Cd, Zn, Ni, Al, etc.). The redox active HMs are directly involved in the redox reaction in cells and result in the formation of $O_2^{\bullet -}$ and subsequently in H_2O_2 and $^{\bullet}OH$ production via the Haber-Weiss and Fenton reactions [32, 33]. Exposure of plants to redox inactive HMs also results in oxidative stress through indirect mechanisms such as interaction with the antioxidant defense system, disruption of the electron transport chain, or induction of lipid peroxidation. The latter can be due to an HM-induced increase in lipoxygenase (LOX) activity. Another important mechanism of HM toxicity is the ability of HMs to bind strongly to oxygen, nitrogen, and sulphur atoms [34]. This binding affinity is related to free enthalpy of the formation of the product of the HM and ligand with low solubility of these products. Because of these features, HMs can inactivate enzymes by binding to cysteine residues. For example, Cd binding to sulfhydryl groups of structural proteins and enzymes leads to misfolding and inhibition of activity and/or interference with redox-enzymatic regulation [1, 27].

Many enzymes need cofactors to work properly for both HM ions (such as Fe^{2+}, Mg^{2+}, Cu^{2+}, Ca^{2+}) and organic molecules (such as haem, biotin, FAD, NAD, or coenzyme A). The displacement of one HM ion by another leads to the inhibition or loss of enzyme activities. Divalent cations such as Co^{2+}, Ni^{2+}, and Zn^{2+} displace Mg^{2+} in ribulose-1,5- bisphosphate-carboxylase/oxygenase (RuBisCO) and result in a loss of activity [35, 36]. Displacement of Ca^{2+} by Cd^{2+} in calmodulin, an important protein in cellular signaling, led to the inhibition of calmodulin-dependent phosphodiesterase activity in radish [37]. Additionally, HMs cause membrane damage through various mechanisms, including the oxidation of and cross-linking with protein thiols, inhibition of key membrane protein such as H^+-ATPase, or causing changes in the composition and fluidity of membrane lipids [38]. Accumulation of MG, a cytotoxic compound, was found to increase in response to HM stress in plants due to impairment of the glyoxalase system that finally elicits oxidative stress by reducing the GSH content [2, 3, 17, 19, 39, and 40].

Based on the aforementioned, it can be concluded that HM toxicity is attributed to three mean reasons: (a) stimulation of ROS and MG production by auto-oxidation and the Fenton reaction or by modification of the antioxidant defense system and the glyoxalase system, (b) direct interaction with proteins due to their affinities for thioyl-, histidyl-, and carboxyl-groups, causing the HMs to target structural, catalytic, and transport sites of the cell, and (c) displacement of essential metal ions from specific binding sites, causing function to collapse [8, 33]. The possible sequential events of ROS-induced damage development in sensitive plants in response to HM stress are summarized in Figure 1.

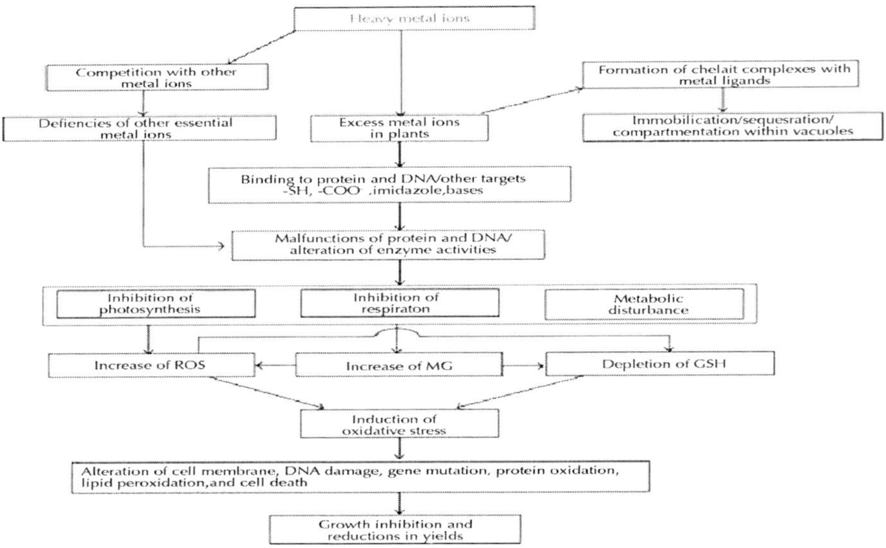

Figure 1: Possible biochemical and molecular mechanisms of heavy metal-mediated ROS induction and damage to the development of higher plants.

HEAVY METAL STRESS AND PLANT RESPONSES

The exposure of plants to toxic levels of HMs triggers a wide range of physiological and metabolic alterations [5, 13]. However, as different HMs have different sites of action within the plant, the overall visual toxic response differs between HMs. The most widespread visual evidence of HM toxicity is a reduction in plant growth [7] including leaf chlorosis, necrosis, turgor loss, a decrease in the rate of seed germination, and a crippled photosynthetic apparatus, often correlated with progressing senescence processes or with plant death [1, 24, 41–43]. All these effects are related to ultrastructural, biochemical, and molecular changes in plant tissues and cells brought about by the presence of HMs [44]. Moreover, HMs influence homeostatic events, including water uptake, transport, transpiration, and nutrient metabolism [45, 46] and interfering with the uptake of Ca, Mg, K, and P [47]. Elevated levels of HMs usually decrease photosynthesis due to their direct effect on the photosynthetic apparatus, including thylakoids. Several studies report that HMs can determine the release of proteins, lipids, and element components of thylakoid membranes, causing damage to light-harvesting complexes and Photosystem II (PS II) [48, 49]. Additionally, some HMs can replace Mg in the chlorophyll (Chl). Chl synthesis reduction, which is usually observed after HM stress, may be a consequence of enzyme inhibition involved in its synthetic pathway [50, 51]. In addition, HM decreases carbon assimilation by inhibiting enzymes involved in CO_2 fixation [52]. A time- and dose-dependent decrease in photosynthesis in maize (Zea mays L.) has been observed in response to HMs like Cu, Ni, Pb, and Zn [53]. Cd-induced inhibition of respiration has also been reported in rice (Oryza sativa L.) [54]. HMs also perturb carbohydrate metabolism and their partitioning in growing plants. The activity of α-amylase, β-amylase, and sucrose phosphate synthase decreased whereas starch phosphorylase, acid invertase, and sucrose synthase activity increased in rice (Oryza sativa L.) seedlings in response to as [55]. Additionally, nitrogen assimilation decreased in rice seedlings in response to as and Al due to inhibition in the activities of nitrate reductase, nitrite reductase, and glutamine synthetase [13, 55, and 56]. In rice seedlings, Ni stress suppresses the hydrolysis of RNA and proteins by inhibiting the activity of ribonuclease (RNase) and protease [13, 57].

HMs that bind to the cell nucleus because promutagenic damage including DNA base modification, inter- and intramolecular crosslinking of DNA and proteins, DNA strand breaks, rearrangement, and depurination [58]. HMs can also affect microtubule assembly-disassembly, thereby altering the cell cycle and cell division [59]. High mutation rate and malformed embryos have been observed inArabidopsis plants exposed to Cd [24, 60]. The exposure of plants to stressful conditions raises the ethylene level, a gaseous hormone, which affects several plant responses, including senescence and stress [61]. In higher plants, Cu-induced ethylene synthesis can increase senescence [62], inhibit cell growth and increase cell wall rigidity by means of lignification [63].

HM toxicity results in the accumulation of excess ROS inside the cell (this will be described in more detail later). ROS production is dependent upon the particular HM element Heavy metal toxicity results in the accumulation of excessive ROS inside the cell; Cu can directly generate ROS, whereas Cd is a redox-inactive HM and can only generate ROS indirectly by enzyme inactivation and by inducing the expression of LOX in plant tissues and therefore causing oxidation of polyunsaturated fatty acids [64, 65]. ROS action may result in cell disturbances which enhance senescence processes in cooperation with ethylene and jasmonic acid (JA), although the activation of the antioxidant machinery can help plants to overcome HM stress [44].

The plant's molecular response to HM stress is characterized by the synthesis of stress-related amino acids, protein, genes, and signaling molecules [1, 24]. Several recent studies also showed that HMs increase the synthesis of heat shock proteins (HSPs) [13, 66, 67]. A higher proline (Pro) level was found in the Cd hyperaccumulator plant Solanum nigrum more than in the nonaccumulator plant (Solanum melongena) indicating its role in HM detoxification [68]. Phytosiderophores, nicotianamine, and organic acids are a few examples of chelating compounds that are released by roots and might influence HM uptake [24]. Different kinetics of mitogen-activated protein kinase (MAPK) cascades in response to HM stress have also been reported [69]. Alfalfa (Medicago sativa) seedlings subjected to Cu or Cd ions result in the activation of four distinct MAPKs: SIMK, MMK2, MMK3, and SAMK. Importantly, Cu stress rapidly activated SIMK, MMK2, MMK3, and SAMK while Cd showed a similar but delayed MAPK activation [13, 69].

MOLECULAR MECHANISMS OF HEAVY METAL TOLERANCE IN PLANTS

Plant tolerance to a particular HM is governed by an inter-related network of physiological and molecular mechanisms and understanding of these mechanisms and their genetic basis is an important aspect to developing plants as agents of phytoremediation [1, 3, 24, and 70]. Different plant species may have evolved different mechanisms to tolerate excess HMs, and even within the one plant species more than one mechanism could be in operation. Plants have both constitutive and adaptive mechanisms to withstand excess HMs [71]. Physiological, biochemical, and molecular approaches continue to be employed to identify the underlying mechanisms of HM accumulation, tolerance, and adaptive mechanisms to cope with HM stress. Some adaptive mechanisms evolved by tolerant plants include immobilization, plasma membrane exclusion, restriction of uptake and transport, synthesis of specific HM transporters, chelation and sequestration of HMs by particular ligands (PCs and MTs), induction of mechanisms contrasting the effects of ROS and MG (such as upregulation of antioxidant and glyoxalase system), induction of stress proteins, the biosynthesis of Pro, polyamines, and signalling molecule such as salicylic acid and nitric oxide [1–3, 8, 9, 11, 17, 19, 24, 26, 28, 29, 40, 72, 73].

Restriction of Uptake and Transport of HMS

Exclusion of HMs from the Plants

Uptake of HMs by plants involves root interception of HM ions, entry of HM ions into roots, and their translocation to the shoot. Once an HM is bioavailable to the plant, the entry of HM ions inside the plant, either through the symplast (intracellular) or through the apoplast (extracellular), depends on the type of HM. Most HM ions enter plant cells by an energy-dependent process via specific or generic HM-ion carriers or channels [74]. One mechanism (avoidance strategy) of preventing or lessening the toxic effects of HMs is by preventing excess HMs entering the plant. There are two main ways in which a plant could do this, either by precipitating or by complexing HMs in the root environment. Plants can precipitate HMs by increasing the pH of the rhizosphere or by excreting anions such as phosphate. Root exudation of phosphate in maize have been detected in response to Al stress in the Al-tolerant cultivar (South American 3) with no toxicity symptoms whereas sensitive cultivars (Tuxpeño and South American 5) showed symptoms [75]. Additionally, root exudation of malate in sorghum (Sorghum bicolor L.) and exudation of citrate in maize have also been reported in response to Cd stress [76]. These findings support the idea that the HM-binding capabilities of root exudates may be an important mechanism for stabilizing HMs in the vicinity of the root thus making them unavailable to the plant and lessening the experienced toxicity. Zhu et al. [73] showed that the oxalate secreted from the root apex helps to exclude Cd from entering tomato (Lycopersicon esculentum L.) roots, thus contributing to Cd resistance in the Cd-resistant tomato cultivar (Micro-Tom). Their results suggest that the tolerant cultivar may have utilized exclusion of the HM via precipitation as an avoidance mechanism. Al-tolerant wheat (Triticum aestivum) genotypes have been found to produce more malic acid than sensitive genotypes upon exposure to Al [77, 78]. Additionally, iron plaque formation on the root surface (due to the release of O_2 and oxidants in the rhizosphere) has been suggested as a mechanism for excluding toxic HM in plants and particularly in Oryza sativa [79, 80]. Liu et al. [80] proposed that iron plaque adsorbs and sequesters Cd onto the root surface and can prevent Cd uptake by rice plants. While excluding excess HMs from the plant holds merit, some tolerant plants and hyperaccumulator plants actually have a higher uptake of HMs when compared to sensitive plants [4, 81, 82], so other mechanisms must also exist.

Cellular Exclusion of HMs

Cellular exclusion of HMs is an important adaptive strategy for HM tolerance in plants. A large fraction of HMs in plant roots are found in the apoplastic space. For example, at equal external Al concentrations, a sensitive wheat cultivar had more symplastic Al than the tolerant cultivar suggesting an exclusion mechanism [83]. HM transporter proteins are potentially involved in the exclusion of toxic HM ions from the symplastic to the apoplastic space.

HM Complexation at the Cell Wall-Plasma Membrane Interface

The cell wall-plasma membrane interface accumulates large portions of HMs and it is therefore believed that this could be the potential site of HM tolerance. In Italian ryegrass (Lolium multiflorum), 60% of Cu in the roots was bound by the cell wall and plasma membrane [84]. Plant cation exchange capacity (CEC) is largely determined by the exchange sites in cell walls [85]. Sensitive wheat cultivars have much lower cell wall CECs than tolerant cultivars [86] indicating that tolerant cultivars use a high CEC to complex HMs at the cell wall and prevent entry to the cell. However, the role of the cell wall in HM tolerance remains to be clarified.

Distribution of HM

Plants are able to minimize the adverse effects of excess HMs by regulating the distribution and translocation of HMs within their organs or cells. The most common forms of this are the much higher amounts of HMs found in plant roots than in shoots, except for hyperaccumulators [87–89], and reduced translocation of HM to the shoots is a widespread tolerance mechanism. Probably, this acts to exclude metals from sensitive metabolism in the shoots and, therefore, tolerant genotypes have a lower HM concentration in the shoot than sensitive genotypes. Additionally, plants also showed a resistant reaction through uniform distribution of HMs that entered a plant tissue or organ. Tolerant cowpea (Vigna unguiculata) genotypes have a relatively uniform distribution of Mn throughout the leaf [90]. In contrast, sensitive genotypes accumulate Mn in localized areas in the leaf which are seen as dark brown spots caused by manganese oxide precipitates [90].

Complexation and Compartmentation of HMs within the Plant Cell

Intracellular Sequestration or Compartmentation within Vacuoles

Once an HM has entered the cell, a plant uses various strategies to cope with its toxicity. One such strategy consists of transporting the HM out of the cell or sequestrating it into the vacuole, thereby removing it from the cytosol or other cellular compartments where sensitive metabolic activities takes place [1, 24, 91]. Therefore, the central vacuole seems to be a suitable storage reservoir for excessively accumulated HMs. In fact, two vacuolar proton pumps, a vacuolar proton-ATPase (V-ATPase) and vacuolar proton-pyrophosphatase (V-Ppase), energize vacuolar uptake of most solutes. Grasses can actively pump Zn into vacuoles with the more tolerant clones being able to continue the process at higher external Zn levels than sensitive clones [92]. Uptake can be catalyzed by either channels or transporters. The application of powerful genetic and molecular techniques has now identified a range of gene families that are likely to be involved in transition HM ion uptake into cells, HM vacuolar sequestration, HM remobilization from the vacuole, xylem loading, and unloading of HMs (Figure 2). Some well-characterized HM transporter proteins are zinc-regulated transporter (ZRT), iron-regulated transporter (IRT) like protein ZIP family, ATP-binding cassette (ABC) transporters, the P-type metal ATPases, the natural resistance-associated macrophage protein (NRAMP) family, multidrug resistance-associated proteins (MRP), ABC transporters of the mitochondria (ATM), cation diffusion facilitator (CDF) family of proteins, copper transporter (COPT) family proteins, pleiotropic drug resistance (PDR) transporters, yellow-stripe-like (YSL) transporter and Ca^{2+}: cation antiporter (CAX), and so forth [13, 93–95]. Plant cell walls are a continuous matrix, acting as a cation exchanger, holding variable quantities of metal, and providing for some HM exclusion [96]. In barley (Hordeum vulgare L.) leaves, the increase in cellular Zn with increasing exposure to external Zn was fully accounted for by an increase in vacuolar Zn with the cytoplasm exhibiting perfect homeostasis [97]. Vacuole isolation or compartmental flux analysis, although less frequently used in the study of HM accumulation, may also provide valuable information. In order to determine the localization of Cd and the potential Cd-binding peptides, protoplasts and vacuoles were isolated from leaves of Cd-exposed tobacco (Nicotiana rusticavar Pavonii) seedlings [98]. Purified vacuoles contained virtually all of the Cd-binding peptides and Cd found in protoplasts. Computer simulation studies with Zn [99] and Cd [100] showed that the vacuole is the probable site of HM sequestration and detoxification in Nicotiana tabacum. Evidence for plant vacuoles as the site of HM sequestration appears to be quite conclusive. However, many plant HM transporters remain to be identified at the molecular level and the transport function, specificity, and molecular location of most of these proteins in plants remain unknown.

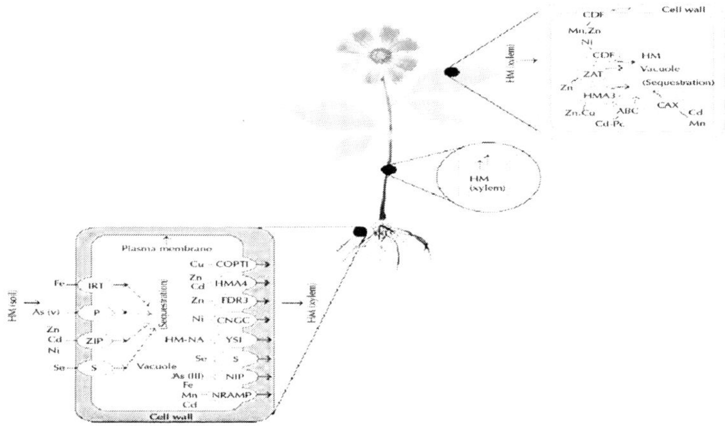

Figure 2: Diagrammatic representation of uptake and transport of heavy metals in plants through metal transporters (modified from) [4, 70].

Formation of Metal Complex by Phytochelatins

Chelation of HMs in the cytosol by high affinity ligands is potentially a very important mechanism of HM detoxification and tolerance in plants under HM stress. Plants make two types of peptide metal binding ligands: phytochelatins (PCs) and metallothioneins (MTs). Recent advances in the understanding of different aspects of biosynthesis and function of PCs are derived predominantly from molecular genetic approaches using model organisms. PCs are a family of Cys-rich polypeptides with the general structure $(\gamma\text{-Glu-Cys})_n$-X, in which X is Gly, γ-Ala, Ser, Gln, or Glu and n= 2–11 depending on the organism, although the most common forms have 2–4 peptides [28]. PCs are synthesized from GSH; the metal binds to the constitutively expressed enzyme γ-glutamylcysteinyl dipeptidyl transpeptidase (PC synthase), thereby activating it to catalyze the conversion of GSH to phytochelatin [101]. The biosynthesis of PCs is induced by many HMs, including Cd, Hg, Ag, Cu, Ni, Au, Pb, as, and Zn; however, Cd is by far the strongest inducer [102, 103]. PCs complex Cd ions through the thiolic group (–SH) of Cys and the PC-Cd complexes are accumulated in the vacuole through the activity of ABC transporters, thus limiting the circulation of free Cd^{2+} inside the cytosol [72]. Additionally, plants are not able to metabolize or eliminate Cd. Rather, they adopt the strategy of making Cd-GSH and Cd-PCs complexes to sequester Cd within vacuoles efficiently [1, 72] and also to transport Cd over a long distance through xylem and phloem vessels [104]. Further confirmation of the induction of PCs under HM stress was provided by the fact that PC production was found to be positively correlated with HM accumulation in both above-ground and below-ground tissues. In our study, with water hyacinth (Eichhornia crassipes) and Cd stress, we found that Cd^{2+} taken up by water hyacinth roots was present as Cd-binding complexes whereas, in the absence of Cd^{2+}, no such complexes were observed [105]. This indicates that these Cd^{2+}-binding complexes are formed in response to Cd^{2+} and play an important role in the accumulation of Cd in water hyacinth. Next, we characterized that the latter Cd-binding complex with a molecular weight of 4000 was composed of two cadystins A, (γ-Glu-Cys)$_3$-Gly, two cadystins B, (γ-Glu-Cys)$_2$-Gly, and inorganic sulfur, which is identical to the fission yeast Cd-BPI [106]. Iglesia-Turiño et al. [107] studied Hg accumulation in rape (Brassica napus) plants grown under an Hg concentration gradient (0–1,000 μM) and found that Hg accumulation was strongly correlated with PC2 concentration. PCs are also involved in the homeostasis of Zn^{2+} and Cu^{+}/Cu^{2+} by providing a transient storage form for the ions [108, 109]. The induction of PCs by the anion arsenate has been observed in a survey for peptide-inducing metal ions [102] and suggests a unique mode of PC synthase activation. Nouairi et al. [110] observed that Cd treatment registered a consistent increase in PC synthesis up to 50 μM (94.66 μmol g^{-1} FW) in Brassica juncea leaves with respect to the control (10.57 μmol g^{-1} FW). However, after a 15-day exposure, the PC content in B. napus leaves was not significantly different to the control at any concentration of Cd. In addition, phytochelatin synthase (PCS) gene expression studies in garlic (Allium sativum L.) plants exposed to HMs [111] and an in vitro analysis in the marine alga Dunaliela tertiolecta [112] also suggested a role of PCs in the detoxification of HMs and in the mitigation of oxidative stress. Both HM-resistant and HM-sensitive plants produce PCs; however, several reports have concluded that PCs are not primary responsible for the hyperaccumulation of Zn, Ni, or Pb [113–115]. The HM-detoxification process is not limited to the chelation of HM ions. After the activation of PC synthase by the HM ions and HM chelation by the PCs synthesized, the HM ion complex is transported to the vacuole and stabilized there by forming a complex with sulfides or organic acid [96]. It has also been shown that PCs have a role in HM transport [116], so that their detoxifying capabilities may actually be secondary or part of a more complex mechanism. Although it is clearly demonstrated that PCs can have an important role in HM detoxification and accumulation in higher plants, formation of HM complexes provides insufficient explanation for either the HM specificity or species specificity of hyperaccumulation [117]. Therefore, it remains to be determined what exact role PCs play in the HM-tolerance mechanism at the cellular level and this requires more thorough research.

Complexing by Metallothioneins

MTs are low molecular weight (4–8 kDa), Cys-rich, HM-binding, gene-encoded polypeptides that can bind HMs via the thiol groups of their Cys residues [118]. Although the precise physiological function of MTs has not yet been fully elucidated, proposed roles include (a) participation in maintaining the homeostasis of essential transition HMs, (b) sequestration of toxic HMs, and (c) protection against intracellular oxidative damage [119]. Plant MTs are extremely diverse and have been subdivided into three classes based on the arrangement of the Cys residue [120]. The Cys-Cys, Cys-X-Cys, and Cys-X-X-Cys motifs (in which X denotes any amino acid) are characteristic and invariant for MTs. The organization or distribution of cysteine residues confers different MT isoforms and their ability to bind and sequester different HM ions for detoxification and homeostasis. The biosynthesis of MTs is regulated at the transcriptional level and is induced by several factors, including hormones, cytotoxic agents, and HMs, such as Cd, Zn, Hg, Cu, Au, Ag, Co, Ni, and Bi [28, 121]. Gene expression studies were performed to quantify mRNA levels in different tissues, at different developmental stages and under stress conditions such as HM exposure. MT genes appear to be differentially regulated in a tissue-specific manner and in relation to the developmental stage and also in response to a number of stimuli, including HMs [122]. Ahn et al. [123] showed that three Brassica Rapa MT genes (BrMT1, BrMT2, andBrMT3) are differentially regulated under various HM stresses. In Fe-treated seedlings, BrMT1 andBrMT2 were not appreciably induced, while elevated expression of BrMT3 was apparent at 6 and 24 h. In Cu-treated seedlings, BrMT1 expression was increased at 3 h and it remained at the increased level thereafter, while BrMT2 expression was downregulated and BrMT3 remained unchanged. Similarly, in Zn-treated seedlings, BrMT1 increased slightly while BrMT2 was downregulated and BrMT3 remained unchanged. In Mn-treated seedlings, BrMT1 and BrMT3 genes showed increase expression until 12 h and then were downregulated. BrMT2 expression did not change until 12 h and then was gradually downregulated. Several data demonstrated the role of MTs in HM detoxification and homeostasis, but metal-inducibility of plant MTs has not always been demonstrated. Further information regarding the structures and properties of MTs could clarify their mechanism(s) of action and functions. The use of a model system and a model hyperaccumulator like Arabidopsis halleri and some Thlaspi species is likely to further elucidate the molecular mechanisms of HM transport, trafficking, tolerance, and homeostasis in plants [124].

Metal Chelation by Organic Acids, Amino Acids, and Phosphate Derivatives

Mechanism of HM tolerance and detoxification in plants can be divided into two categories: external exclusion and internal tolerance. In the external detoxification process, organic acids excreted from plant roots may form stable HM-ligand complexes with HM ions and change their mobility and bioavailability, thus preventing the HM ions from entering plants or avoiding their accumulation in the sensitive sites of roots. In internal HM detoxification, organic acids may chelate with HM in the cytosol, where the ions can be transformed into a nontoxic or less toxic form [27, 28, and 91]. Plants produce a range of ligands for Al, Cd, Cu,

Ni, Co, and Zn. Carboxylic and amino acids, such as citrate, malate, and oxalate, histidine (His) and nicotianamine (NA), and phosphate derivatives (phytate) are potential ligands for HMs and are found to play a role in tolerance and detoxification [8, 27, 28, 91, 96, 119, 125].

Citrate has a high capacity to chelate HM ions and has been well documented in the case of Fe [91]. However, other HMs such as Cd, Ni, Co, and Zn also have a strong affinity for citrate. Long before the discovery of PCs in plants, research on several Ni hyperaccumulators had shown that Ni is predominantly bound to citrate and that the amount of citrate produced is strongly correlated with the accumulated Ni [126, 127]. Since then, there have been a large number of studies in which carboxylic acids have been found to be associated with HMs. A correlation between citric acid exudation and Al tolerance was detected by Miyasaka et al. [128] in snapbean (Phaseolus vulgaris L.). Citric acid has been considered to be a major ligand at low Cd concentrations [129] and also found to contribute to Zn accumulation and tolerance [130]. Exposure of cells of the Co hyperaccumulator Crotalaria cobalticolaand non-accumulators Raufolia serpentina and Silene cucubalus to Co ions resulted in an increase of citrate, further denoting its involvement in the complexation of HM ions [131].

Root secretion of oxalate and detoxification of HMs like Al have also been reported [132]. Buckwheat (Fagopyrum esculentum Moench.) secretes oxalic acid from the roots in response to Al stress and accumulates nontoxic-Al-oxalate in the leaves [133]; thus detoxification occurs both externally and internally. Zheng et al. [134] found that buckwheat had higher resistance to Al compared with an Al-tolerant cultivar of wheat, Atlas 66. In wheat and maize, there is evidence that such secretion from the roots is mediated by Al-activated anion channels in the plasma membrane [132]. Similarly, malate was proposed to cytosolic Zn chelators in Zn tolerant plants [135]. Removal of Al from the roots results in a rapid decline in malate secretion to the non-Al level, indicating responsive Al and malate-secreting mechanisms. The differences in the degree of tolerance to Al depend on the transport of malate out of the apical root cells via an Al-activated malate-permeable channel [136]. Salazar et al. [137] observed that Al-tolerant genotypes secreted about 10-fold higher malate and about 3–5-fold higher succinate than Al-sensitive seedlings over 24 h exposure to 50 μM Al. Li et al. [138] found that alteration in the metabolism of organic acid was involved in the Al-induced secretion of organic acids in rye (Secale cereale L.) but only activation of an anion channel seems to be responsible for the rapid secretion of malate in wheat.

Histidine and NA also play roles in the chelation of HM ions both within plant cells and in the xylem sap [139]. Salt at al. [140] identified putative Zn-His complexes in the roots of the closely related Zn hyperaccumulator Thlaspi caerulescens. Additionally, free His in the xylem exudates of a Ni hyperaccumulator (Alyssum lesbiacum) was found to increase 36-fold in response to Ni stress [141]. Transgenic plants overexpressing the Salmonella typhimurium ATP phosphoribosyl transferase gene (StHisG) accumulated about 2-fold higher histidine levels than wild type plants and showed more than 10-fold increased biomass production in the presence of toxic Ni in the growth medium [142], convincingly demonstrating a mechanistic link between His and moderate Ni tolerance.

The plant HM chelator NA is a free nonproteinogenic amino acid. NA is mobile in the plant and has been detected in root and leaf cells as well as in phloem sap. It can bind HM ions like Fe, Zn, Cu, and Ni. NA is synthesized by a one-step condensation reaction of three molecules of S-adenosyl-Met (SAM) by the enzyme NA Synthase [143]. In Solanaceous plants, NA was shown to act in Fe homeostasis [144]. In addition to its role in long-distance HM transport, NA is proposed to be involved in the regulation of HM transfer within plant cells [145, 146]. Weber et al. [147] used Arabidopsis Gene Chips to identify those genes that are more active in roots of A. halerii than A. thaliana under controlled conditions. Two genes showing highest levels of expression in A. halerii roots code for an NA Synthase and a putative Zn^{2+} uptake system. In addition, roots of A. halerii also showed higher levels of both NA Synthase and NA. Taken together, these observations suggest active roles of NA in plant Zn homeostasis and NA Synthase in hyperaccumulation of Zn in A. halerii.

Phytate has been suggested as a possible complexing agent for HMs. Phytate tends to be associated with HM deposits in plants, including in Al-tolerant Zea mays vacuoles [148]. Zinc-phytate globules appeared to be quantitatively more frequent in tolerant ecotypes of Deschampsia caespitosa roots than in sensitive ecotypes [149]. However, it is unclear if HM phytate complexes cause a tolerance or stress response.

Based on the aforesaid, it could be stated that organic acids, amino acids, and phosphate are derivatives of HM detoxification. However, further research is needed to quantify the organic acids being produced as a toxicity versus tolerance response over a range of plant species by using various HMs.

Hyperaccumulating Mechanisms

HM-tolerant plants are often excluders, limiting the entry of root-to-shoot translocation, or retain the uptaken HM into the root cells or detoxifying them by their chelation in the cytoplasm or storing them in vacuoles. However, a class of rare plants called hyperaccumulators rapidly and efficiently translocate the HMs to the shoot via the xylem (Figure 2). Translocation in the xylem is probably transpiration driven [150]. They hyperaccumulate HMs even from low external HM concentrations, and most of the HM is translocated to the shoot [151]. At the root membrane level, HM uptake is unusually high in hyperaccumulators. This may be due to constitutive high expression of an HM transporter in the plasma membrane, as was found for the Zn and Cd hyperaccumulator Thlaspi caerulescens [152, 153].T. caerulescens shows reduced HM accumulation in root vacuoles, enhanced root-shoot translocation, enhanced uptake into leaf cells, and higher HM tolerance [154]. The high HM tolerance may in part be due to highly efficient intracellular compartmentalization and chelation [155]. Different chelators may be involved in translocation of HM cations through the xylem (Figure 2), such as organic acid chelators [150, 156], or nicotianamine [156]. Uptake of HM ions from the xylem apoplast into the shoot symplast is mediated by HM transporters in the shoot cell membrane.

Heat Shock Proteins (HSPs)

Heat shock proteins (HSPs) act as molecular chaperones in normal protein folding and assembly but may also function in the protection and repair of protein under stress conditions. Induction of HSPs by several transition HMs (Zn, Cu, Cd, Hg, Al, and Cr) has been reported [13, 157–161]. Increased accumulation of a large HSP (HSP70) was reported in response to Cd stress [157, 159]. Ireland et al. [161] also reported the increased accumulation of HSP70 protein in response to Cd stress in marine macro algae and fresh water plant species. In addition, expression of a small HSP (sHSP) was also found in the roots of Armeria maritima plants grown in Cu-rich soil [158]. Similarly, accumulation of HSP25 increased in response to Al stress in soybean (Glycine max (L.) Merr) [160]. A short burst of heat stress given prior to HM stress induces Cd tolerance. The molecular chaperones induced during HM stress could prevent irreversible protein denaturation resulting from the oxidative stress linked to HM exposure or help to channel their proteolytic degradation. However, the putative role of HSPs in HM tolerance is largely unknown.

Chemical Modification

Metal-modifying enzymes may be involved in the assimilation of HMs into organic molecules (e.g., selenate is metabolized to dimethylselenide [162]), or in changing the oxidation state of metals (e.g., toxic Cr (VI) is reduced to nontoxic Cr (III)). Lytle et al. [163] showed that Eichhornia crassipes (water hyacinth), supplied with Cr(VI) in nutrient culture, accumulated nontoxic Cr(III) in root and shoot tissues. This suggests that E. crassipes detoxified Cr (VI) upon root uptake and transported a portion of the detoxified Cr to leaf tissues. In dicots, Fe (and possibly also Cu) is reduced by a reductase at the root cell membrane before uptake [155, 164]. Reduction of HMs in situ by plants may be a useful detoxification mechanism for phytoremediation.

Modulation of Transcription Factor (TF)

Metal response element binding transcription factor 1, also called metal-responsive transcription factor 1 (MTF-1), plays an important role in the cellular response and tolerance to HM stress by triggering the activation of genes responsible for HM uptake, transport, and detoxification. TFs involved in HM stress response and tolerance have already been identified in different plant species [165, 166]. Importantly, Cd-responsive TFs share the same signal transduction pathway with other stress-related TFs [167]. TFs belonging to different families, such as WRKY, basic leucine zipper (bZIP), ethylene-responsive factor (ERF), and myeloblastosis protein (MYB), play a significant role in controlling the expression of specific stress-related genes in response to Cd stress [28].

Induction of Antioxidant Defense and Glyoxalase System

HM stress invariably induces oxidative stress and antioxidative defense systems, composed of free radical scavenging molecules such as AsA and GSH, and the enzymes involved in their biosynthesis and reduction [3, 8, 13, and 168]. Additionally, MG, a cytotoxic compound, was found to increase in response to various abiotic stresses, including HM stress [2, 17, 169], which is mainly detoxified by the glyoxalase system in plants and enhanced oxidative stress tolerance (discussed in more detail later in the next section).

Synthesis of Salicylic Acid and HM Tolerance

Salicylic acid (SA) is a natural signal molecule which plays an important role in regulating a number of physiological and biochemical process making plants resistant to biotic and abiotic stresses [170, 171]. Freeman et al. [172] showed that constitutively elevated SA signals GSH-mediated Ni tolerance inThlaspi, a Ni hyperaccumulator. SA pretreatment induces enhanced accumulation of GSH and protection against Ni toxicity, since GSH is known to protect against Ni-induced lipid peroxidation inArabidopsis and Thlaspi hyperaccumulators [172]. A wealth of research has shown that exogenous SA application significantly enhanced HM tolerance in plants [170, 171, and 173]. However, the protective function of SA mainly includes the regulation of ROS and antioxidants, induction of gene expression, and adsorption and distribution of elements [170, 171, 173, and 174].

Synthesis of Proline

Accumulation of Pro in response to HM stress has also been widely reported [8, 175–177]. Siripornadulsil et al. [178] demonstrated that increased Pro levels provide enhanced protection against Cd in microalgae. Moreover, many HM-tolerant plants have also been reported to possess substantially elevated constitutive Pro levels in the absence of excess HM ions when compared with their nontolerant relatives [8]. Importantly, Pro reduces Cd stress not by sequestering Cd but by reducing Cd-induced free radical damage and maintaining a stringent reducing environment (higher GSH levels) within the cell. In seeking additional supporting evidence for this, we recently showed in mung bean (Vigna radiata L.) [3] That exogenous Pro induces HM (Cd) tolerance by maintaining a higher GSH level and GSH metabolizing enzymes. However, further research towards integration of the growth inhibiting and protecting properties of Pro is needed. Huang et al. [179] studied the physiological and biochemical responses in the leaves of two mangrove plant seedlings (Kandelia candel and Bruguiera gymnorrhiza) exposed to multiple HMs (Cd^{2+}, Pb^{2+} and Hg^{2+}) and concluded that Pro, GSH and PCs-SH in K. candel may play a more important role in ameliorating the effect of HM toxicity than in B. gymnorrhiza.

Polyamines

Polyamines (PAs) are low molecular weight organic cations and are ubiquitous in all living organisms. The common PAs in plants are spermidine (Spd), spermine (Spm), and their diamine precursor, putrescine (Put). They influence a variety of growth, and development processes in plants and have been suggested to be a class of plant growth regulators and to act as second messengers (for a recent comprehensive reviews see Hussain et al. [180]). PAs are essential for normal growth and development through transcriptional and translational regulations [181], although their physiological significance derives from their involvement in various kinds of biotic and abiotic stress responses including HM stress in plants [182]. They have the capacity for scavenging free radicals and ROS [183], thus exerting a strong antioxidant function during various types of stress [184]. PAs and Pro are part of the "general adaptation syndrome" (GAS) response to environmental adversities, such as low temperature, HMs, or nutrient deficiency [184–188]. Several genes coding for the enzymes involved in PA metabolism have been characterized and cloned from different plant species. Engineered plants, overexpressing PAs biosynthetic genes, confer increased tolerance to multiple environmental stresses [189, 190], including HMs [191].

Synthesis of Nitric Oxide and HM Tolerance

Nitric oxide (NO), a ubiquitous bioactive signaling molecule, plays an important role in a broad spectrum of multiple physiological processes in plants by regulating the level and toxicity of ROS and hormones [192, 193] and by inducing transcriptional changes that permit the identification of genes involved in different functional processes such as signal transduction, defense and cell death, transport, basic metabolism, and ROS production and degradation [194]. NO protects plants from oxidation damage by regulating general mechanisms for cellular redox homeostasis and promoting the transformation of O2 • − to H_2O_2 and O_2 and also by enhancing H_2O_2-scavenging enzyme activities [192,195–197], although the NO molecule itself possesses antioxidant properties [198]. In addition to a direct ROS scavenging activity and to the modulation of lipid peroxidation by LOX inhibition, NO may also protect cells against oxidative processes by stimulating GSH synthesis. Solanum nigrum, a Cd hyperaccumulator, has received extensive attention in recent years. Xu et al. [199] reported that NO levels rapidly increased in S. nigrum roots under Cd stress and Cd-induced NO can both maintain intracellular antioxidative capacity and reduce oxidative damage. These results suggest that Cd-induced NO accumulation is advantageous for protecting against Cd stress in plants. Application of exogenous NO also modulated ROS-scavenging enzymes, induced the activity of H^+-ATPase and H^+-PPase in the plasma membrane or tonoplast, and also significantly alleviated the growth inhibition induced by $CuCl_2$in tomato plants. These results indicate that exogenous NO could effectively induce tomato seedlings to adjust physiological and biochemical mechanisms against Cu toxicity and maintain fundamentally metabolic capacity and normal growth under HM stress [200]. Exogenous NO also improves Cd tolerance by increasing the production of Pro and total GSH and reducing oxidative damage in the roots of Madicago truncatula seedlings [201]. Laspina et al. [202] reported that NO, when applied before Cd exposure, significantly attenuates Cd-induced oxidative damage in sunflower (Helianthus annuus L.) leaves. This result was mainly attributed to the prevention of growth inhibition and Chl degradation, the recovery of CAT activity and GSH levels, and the enhancement of AsA content and APX activity. These components of the antioxidant machinery allow the plant to cope better with HM stress. Based on the previous discussion it is therefore concluded that NO induces HM tolerance by regulating different physiological and biochemical mechanisms.

The previous discussions clearly demonstrate that plant HM tolerance is highly a complex and involves interrelated physiological, biochemical, and metabolic plant processes.

REACTIVE OXYGEN SPECIES PRODUCTION IN PLANT CELLS

Higher plants produce ROS during different metabolic processes in cellular organelles. However, during HM stress, their rate of production is dramatically elevated. Organelles with a highly oxidizing metabolic activity or with an intense rate of electron flow, such as Chl, mitochondria, and peroxisomes, are the predominant sources of ROS production in plant cells [19, 203–205]. The Chl is the prime source of ROS having the capacity to produce high amounts of superoxide ($O_2^{\bullet -}$) through the Mehler reaction and H_2O_2, essentially during the reduced rate of photosynthetic carbon fixation, a typical situation during abiotic stresses [206]. It is generally accepted that the water oxidizing system of PS II is affected by HM (Cd) by replacing the Ca^{2+} and Mn^{2+} ions in the PS II reaction centre; thereby inhibiting the reaction of PS II leads to the uncoupling of the electron transport in the Chl [207–211]. Cd was also found to inhibit the electron flow on the reducing side of PS I [212]. The negative effects of HM can also be observed in the carboxylating phase of photosynthesis. The main targets of the influence of HM are two key enzymes of CO_2 fixation, ribulose 1, 5-bisphosphate carboxylase (RuBisCO) and phosphoenol pyruvate carboxylase (PEPC). Cd^{2+} ions lower the activity of RuBPC and damage its structure by substituting for Mg^{2+} ions, which are important cofactors of carboxylation reactions, and may also shift RubisCO activity towards oxygenation reactions [211–213], and the glycolate that is produced moves from the Chl to peroxisomes, where it is oxidized by glycolate oxidase (GO) forming H_2O_2 [206]. The peroxisomal activities of xanthine oxidase (XO) generate $O_2^{\bullet -}$ and glycolate oxidase, flavin oxidase, and β-oxidation yield H_2O_2 [205, 214, 215]. HM stress enhanced peroxisomal mobility correlated with an increase in ROS [8, 31, and 216]. There is a rapid induction of H_2O_2

in leaf peroxisomes in response to Cd stress [217]. Plant mitochondria are suggested to be an important source of abiotic stress-induced ROS generation. The mitochondrial electron transport chain consists of several dehydrogenase complexes which reduce a common pool of ubiquinone (Q). ROS production is likely to occur mainly in complex I and the Q zone [218]. Susceptibly of mitochondrial respiratory chain pathways and HM-induced loss of functions have also been reported [12, 219]. The presence of free HM cations, redox active or inactive, in mitochondria may significantly contribute to the initiation and perpetuation of oxidative stress [217, 220]. Plasma membrane-bound NADPH oxidases as well as cell-wall associated peroxidases are the main sources of $O_2^{\bullet -}$ and H_2O_2 producing apoplastic enzymes [205]. NADPH oxidase-dependent ROS induction has been reported in wheat in response to Ni stress [221], in pea (Pisum sativum) in response to Cd stress [222], in Vicia faba in response to Pb stress [223], and in Arabidopsis in response to Cd and Cu [224]. In addition to these metabolic ROS resources, in the presence of redox-active HMs (Fe, Cu, Cr, V, and Co), $^{\bullet}OH$ can be formed from H_2O_2 through the Fenton reaction or from H_2O_2 and $O_2^{\bullet -}$ through the Haber-Weiss reaction [225]. The extremely reactive $^{\bullet}OH$ in the cell causes extensive oxidative damage of membranes and other macromolecules, including photosynthetic pigments, protein, DNA, and lipids [19, 226, 227]. ROS are normally scavenged immediately at their sites of production by locally present antioxidants. However, when this local antioxidant capacity cannot cope with ROS production, H_2O_2 can leak in the cytosol and diffuse to other compartments. Importantly, beyond their harmful effects on cells, ROS (especially H_2O_2) have been proposed to act as signals (Figure 3) in stress response and modulate the activation of stress responsive pathways, proteins, and genes [8, 11, 19, 204]. As H_2O_2 is immediately produced under HM stress, it is probably a key molecule triggering signal transduction and HM tolerance in plants [31].

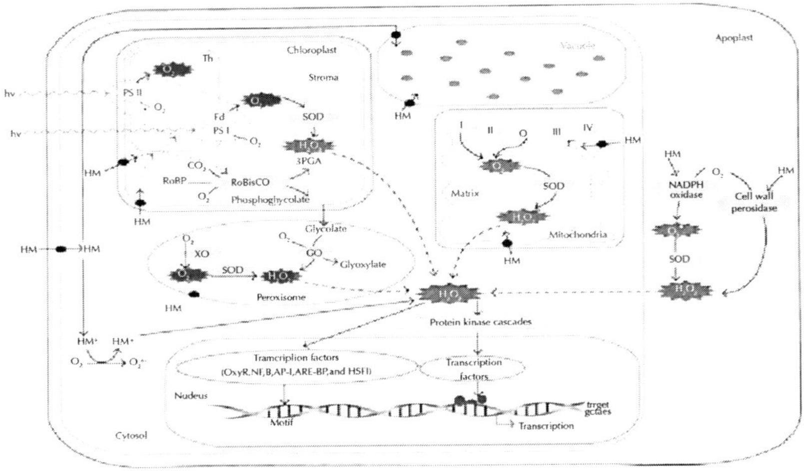

Figure 3: Heavy metal-induced ROS production in different organelles of plant cells and ROS-induced signaling in defense gene expression (modified from Hossain et al. [19]; Sharma and Dietz [8]). Yellow circles in the vacuole designate the deposit of HMs.

METHYLGLYOXAL PRODUCTION IN HIGHER PLANTS IN RESPONSE TO STRESS AND ITS NEGATIVE CONSEQUENCE ON CELLULAR SYSTEMS

MG is a highly reactive αβ-dicarbonyl aldehyde. Extensive research has been carried out on mammalian and animal systems, and different pathways have been proposed for endogenous MG formation from metabolic intermediates of carbohydrates, protein, and lipid metabolism. However, very little work has been done in plant systems regarding the endogenous production of MG. MG is formed spontaneously in plants by nonenzymatic mechanisms under physiological conditions from glycolysis and from photosynthesis intermediates, glyceraldehyde-3-

phosphate (G3P), and dihydroxyacetone phosphate (DHAP) [15, 228]. Under stress conditions, the rate of glycolysis increases, leading to an imbalance (in the initial and latter five reactions) in the pathway. Triosephosphates are very unstable metabolites, and removal of the phosphoryl group by β-elimination from 1, 2-enediolate of these trioses leads to the formation of MG [229, 230]. Therefore, spontaneous production of MG is an unavoidable consequence of the glycolysis pathway during stress. MG can also be formed enzymatically from G3P and DHAP. Triosephosphate isomerase hydrolyzes G3P and DHAP and removes phosphate to yield MG [231]. Many studies in plants have demonstrated the rapid induction of MG levels in response to various stresses in plants, including HM stress [2, 15–17, 232–234]. Singla-Pareek et al. [17] reported a significant increase (1.7-fold) of MG level in tobacco in response to HM stress (5 mM $ZnCl_2$, 24 h). In our report, we also found a significant increase (1.5-fold) in MG level in pumpkin (Cucurbita maxima Duch.) seedlings subjected to HM stress (1 mM $CdCl_2$, 24 h). Importantly, we also observed a rapid increase in MG levels in response to salinity, drought, high temperature, and low temperature stresses [2]. The rapid increase in the level of MG in plants due to different stresses (including HM stresses) clearly suggests that it is a general stress response. There is a possibility that MG could therefore act as a signal for plants to respond to stress [19].

MG is both a mutagen and a genotoxic agent. At high cellular concentrations, it inhibits cell proliferation [235] and results in a number of adverse effects such as increasing the degradation of proteins and inactivating the antioxidant defense system [18, 19, and 236]. Additionally, MG causes increased sister chromatic exchange and endoreduplication [237]. It also induces DNA strand breaks and increases point mutations [237]. MG can also cause oxidative stress indirectly through the formation of advanced glycation end products (AGEs), the irreversible chemical modifications, and cross-links in proteins [199, 238]. Very recently, Saito et al. [20] showed that unless MG is detoxified immediately after its production in the chloroplast, it will act as an intrinsic mediator that catalyses the photoreduction of O_2 at PS I, leading to the production of O2 • −. Therefore, efficient detoxification of MG overproduced during normal physiological processes or various abiotic stresses is one of the most important adaptive strategies of plant stress tolerance [15, 16, and 19].

ROS AND MG SCAVENGING AND DETOXIFICATION SYSTEMS IN PLANTS

Plants use an intrinsic mechanism known as the plant antioxidant system as a defense mechanism to regulate ROS levels according to the cellular needs at a particular time. These antioxidants include the enzymes superoxide dismutase (SOD; EC 1.15.1.1), ascorbate peroxidase (APX; EC 1.11.1.11), monodehydroascorbate reductase (MDHAR; EC 1.6.5.4), dehydroascorbate reductase (DHAR; EC 1.8.5.1), glutathione reductase (GR; EC 1.6.4.2), catalase (CAT; EC 1.11.1.6), glutathione peroxidase (GPX; EC 1.11.1.9), glutathione S-transferase (GST; EC 2.5.1.18), and water-soluble compounds such as AsA and GSH [11, 19, 40, 203]. Although AsA and GSH function as cofactors of enzymes of the antioxidant and glyoxalase pathways, both can also directly quench ROS and regulate gene expression associated with biotic and abiotic stress responses to optimize defense and survival. Importantly, sustaining the ROS concentration (especially H_2O_2) at an appropriate level can promote plant development and reinforce resistance to environmental stressors by modulating the expression of genes and redox signaling pathways [239]. Similarly, the glyoxalase system is an integral component and major pathway of cellular metabolism of MG in living systems present in the cytosol of cells and cellular organelles, particularly mitochondria. The function of the glyoxalase pathway in abiotic stress tolerance has widely been reported (for a recent review see Hossain et al. [19]). The ubiquitous glyoxalase pathway consists of two enzymes: glyoxalase I (Gly I; lactoylglutathione lyase; EC 4.4.1.5) and glyoxalase II (Gly II; hydroxyacylglutathione hydrolase; EC 3.1.2.6). These enzymes act coordinately to convert MG and other 2-oxoaldehydes to their 2-hydroxyacids using GSH as a cofactor in a two-step reaction [240]. The spontaneous reaction between GSH and MG forms hemithioacetal, which is then converted to S-D-lactoylglutathione (SLG) by Gly I. The second reaction is the hydrolysis of SLG to D-lactate catalyzed by Gly II and GSH is recycled back into the system (Figure 4). MG detoxification is therefore strongly dependent on the availability of cellular GSH. Deficiency of GSH limits the production of hemithioacetal, leading to the accumulation of MG.

The reactions catalyzed by the glyoxalase system are irreversible. The existence and widespread distribution of this shunt pathway document its fundamental importance in biological systems. Recent investigations in plants have brought new developments in the involvement of the glyoxalase system in stress tolerance and its involvement with oxidative defense systems (Figure 4). Further insights into the biological function of the glyoxalase system came from the molecular cloning of their respective genes that provides a potential framework for interpreting the physiological roles of the glyoxalase system in higher plants against various abiotic stresses, including HMs [2, 3, 11, 15, 16, 19, 40, 169, 232, and 241].

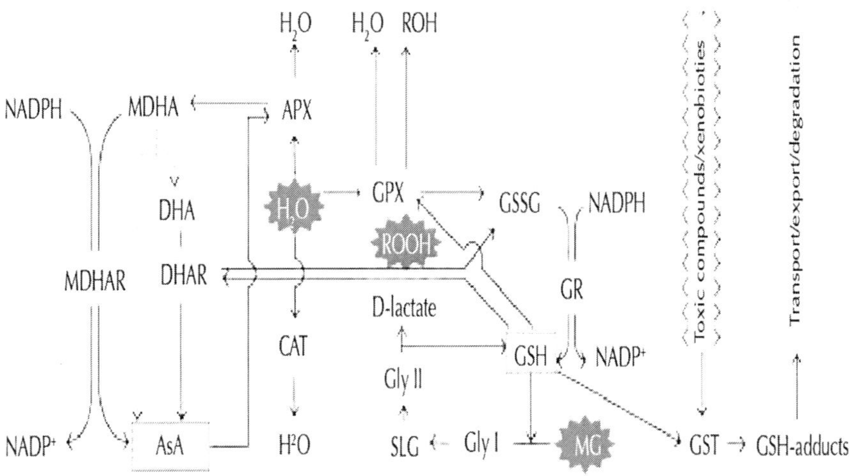

Figure 4: Schematic illustration of possible metabolic interaction of AsA- and GSH-based antioxidant system and GSH-based glyoxalase system in plant cells (modified from Hossain et al. [19]). Dotted lines indicate nonenzymatic reactions. For further discussion, see text.

PLANT METAL TOLERANCE AND INVOLVEMENT OF ANTIOXIDATIVE AND GLYOXALASE SYSTEMS

Plenty of research has been conducted by using HM accumulator, hyperaccumulator, HM-tolerant, and HM-sensitive genotypes and also involving transgenic approaches, to specify the constitutive and induced higher levels of key nonenzymatic and enzymatic antioxidants which could be correlated with HM sensitivity and HM tolerance [8, 13]. Encouraging observations have been made showing a strong relationship between HM tolerance and antioxidative and glyoxalase systems (Table 1), whereas plants with low antioxidant capacity show sensitivity of HM toxicity. A large number of HM-hyperaccumulating and HM-tolerant plants reflected that higher GSH biosynthesis and GSH utilizing and regenerating enzymes can have a significant impact on HM tolerance [172, 173, 242–248]. However, an increase of at least GSH biosynthesis was found to play a central role in Ni tolerance in Ni hyperaccumulators (Thlaspi spp.) [172]. additionally, the potential role of nonprotein thiols (NPTs) such as GSH and PCs in Cd uptake and complexation in the Zn/Cd hyperaccumulator Arabis paniculata has also been reported [247]. However, in these studies, no indication was made regarding HM-induced ROS metabolism. The biochemical mechanism of Cd tolerance in Phragmites australis (a rhizomatous plant of the Poaceae family possessing interesting characteristics useful in phytoremediation and phytostabilisation processes) was studied to decipher the role of ROS metabolism in HM tolerance. The roots of Cd-treated ($CdSO_4$, 50 μM, 21 d) plants had higher APX, CAT, GR, GST, and GPX activities and AsA and GSH content. Importantly, a sharp increase in the $NDAPH/NADP^+$ ratio was observed in roots, leaves, and stolons. Increased activity of antioxidant enzymes in Cd-treated plants suggests that HM tolerance in Phragmites plants might be associated with the efficient regulation of AsA and GSH and their metabolizing enzymes [242]. Additionally, a study with Sedum alfredii, a new Zn hyperaccumulator and Pb accumulator, showed an increase in GSH content in response to Zn and Pb treatments [249]. The authors suggest that GSH, rather than PCs, may be involved in Zn and Pb transport, hyperaccumulation/accumulation, and tolerance. The potential of such adaptive behavior of HM-tolerant and -accumulator plants in relation to GSH and ROS metabolism was also tested in a large number studies by using exogenous chemicals in which it was found that GSH metabolism and/or their metabolizing enzymes played a vital role in HM-induced oxidative stress tolerance (detailed later on).

Table 1: Heavy metal tolerant plant species and traits associated with GSH and their metabolizing enzymes in ROS and MG metabolism

Plants	**Phenotype/genotype**	**Components of defense system related to GSH**	**References**
Arabis paniculata	Zn/Cd hyperaccumulator	High GSH level	[247]
Phragmites australis	Cd accumulator	High GR, GST, GPX activities and AsA and GSH level	[242]
Sedum alfredii	Zn hyperaccumulator	High GSH level	[249]
Thlaspi caerulescens	Cd tolerant Cd accumulator	High GSH level	[243]
Pisum sativum	Cd tolerant	High GSH level	[173]
Arabidopsis thaliana	Ni tolerant	High GSH level	[172]
Thlaspi goesingense	Ni tolerant/Ni accumulator	High GSH level	[172]
Triticum aestivum	Al tolerant	High GST activity	[244]
Pteris vittata	As hyperaccumulator	High GSH level	[245]
Sedum alfredii	Cd/Zn hyperaccumulator	High GSH level and high DHAR and GR activities	[246]
Nicotiana tabacum	Zinc tolerant Gly I and Gly II inserted transgenic plants	High Gly I and Gly II activities and high GSH level	[17]
Sedum alfredii	Pb accumulator	High GSH level	[248]

Although antioxidant protection is determined by large families of nonenzymatic and enzymatic antioxidants, a detailed description of the function of each component in relation to HM tolerance is largely beyond the scope of this paper. Additionally, among the different components of antioxidant defense and MG detoxifying systems, a constitutively high level of GSH metabolizing enzymes has a substantial influence on HM tolerance in plants [17, 242, 246, and 247]. In this paper, we will describe a few important GSH-utilizing and regenerating enzymes such as GST, GPX, DHAR, GR, Gly I, and Gly II that largely participate in the efficient metabolism of major ROS and MG and their products, hence tightly controlling the major plant responses and H_2O_2-induced stress signaling to HMs and metalloids conferring metal-induced oxidative stress tolerance. The potential role of GSH in HM stress tolerance is described hereinafter.

Glutathione

Glutathione (GSH), a nonenzymatic antioxidant, is a low molecular weight thiol implicated in a wide range of metabolic processes and constitutes an important plant defense system against environmental stresses, including HMs [3, 10, 19, 40, and 250]. It is one of the major antioxidant and redox buffers in plants found abundantly in all cell compartments. GSH takes part in the control of H_2O_2 levels through the AsA-GSH cycle [226]. It can also function directly as a free radical scavenger by reacting with $^1O_2$, $O_2^{\bullet -}$, and $^{\bullet}OH$ [251]. Upregulation of the GSH level is of pivotal importance, because it induces the signal transduction and defense against ROS and MG which is achieved through different pathways with various control points (Figures 4 and 5) which include orchestrated activation of genes encoding enzymes related to GSH and AsA [10, 11, 19]. GSH protects proteins against denaturation caused by the oxidation of protein thiol groups under stress. It also plays an indirect role in protecting membranes by maintaining -tocopherol and zeaxanthin in the reduced state. GSH is a substrate for GPX and GST, which are also involved in the removal of ROS and endogenous toxic compounds [226]. It is also involved in the transfer and storage of sulfur [252] and in the detoxification of HMs where PC derived from GSH forms HM complexes [253]. It was also reported that GSH is involved in MG detoxification, a toxic metabolite produced in ample amounts under stressful conditions [3, 9, 10, 15,

16, 18, and 39]. The property of GSH is of great biological importance since it allows fine-tuning of the cellular redox environment under normal conditions and upon the onset of stress (Figure 5). Along with its oxidized form (GSSG), it acts as a redox couple important for maintaining cellular homeostasis, playing a key role in diverse signaling systems in plants [254]. GSH accumulates in response to increased ROS, or to compensate for decreases in the defense capability of other antioxidants and GSH levels are constitutively higher in plants adapted to stress conditions [255–257]. Recent experiments indicate that GSH may function as a cellular sensor to ensure maintenance of the NADPH pool [258] further illustrating the intertwining of the metabolism of different redox compounds. However, the concentration of cellular GSH has a major effect on its antioxidant function and it varies considerably under Cd stress. An increase in GSH concentration has been observed with increasing Cd concentration in Brassica juncea, Brassica campestris [259, 260], and Pisum sativum [173]. The adaptive behavior of hyperaccumulating plants reflects the pivotal importance of GSH and its metabolizing enzyme in HM tolerance (described previously, Table 1). In our recent study [2] with mung bean seedlings, we found a significant increase in GSH content in response to Cd (1 mM $CdCl_2$, 24 h). Schützendübel et al. [261] showed that a low level of short-term Cd stress significantly induced the level of GSH while its level decreased when both the duration and level of Cd concentration increased. However, Cd-induced depletion of GSH has been mainly attributed to PC synthesis [262]. The decline in the levels of GSH might also be attributed to an increased utilization from regeneration of AsA from DHA or from the direct interaction with Cd [263]. Zhou et al. [264] reported a sharp increase in GSH accumulation inMedicago sativa roots treated with different concentrations of $HgCl_2$. A marked increase in GSH content in the leaves, stems, and roots under multiple HM (Cd^+, Pb^+, and Hg^+) stress was reported by Huang et al. [265]. As induces a significant increase in GSH levels in as tolerant Holcus lanatus [266]. Compared to As-sensitive plant species, stimulated GSH synthesis may play a major role in maintaining a high GSH/GSSG ratio in As-tolerant plant species [267, 268]. However, the GSH level in response to various HMs depends on the plant species, the age of the plant, and duration of the treatment. GSH redox couple is an information-rich redox buffer that interacts with numerous cellular components [226]. Recent studies in plants have demonstrated that NO influenced GSH synthesis, as demonstrated in Medicago truncatula roots in which the levels of GSH, γ-EC, and glutathione synthetase (GS) gene expression were increased by NO [199, 269]. During the interaction of GSH with NO, S-nitrosoglutathione (GSNO) is formed in a reaction that may interconnect the ROS- and reactive nitrogen-based signaling pathways [239]. The participation of GSNO in the stress response was shown in Cd-treated plants [270]. The general regulatory role of NO in HM-stressed plants has been demonstrated in several studies. Xiong et al. [271] provide excellent coverage on this aspect in their review.

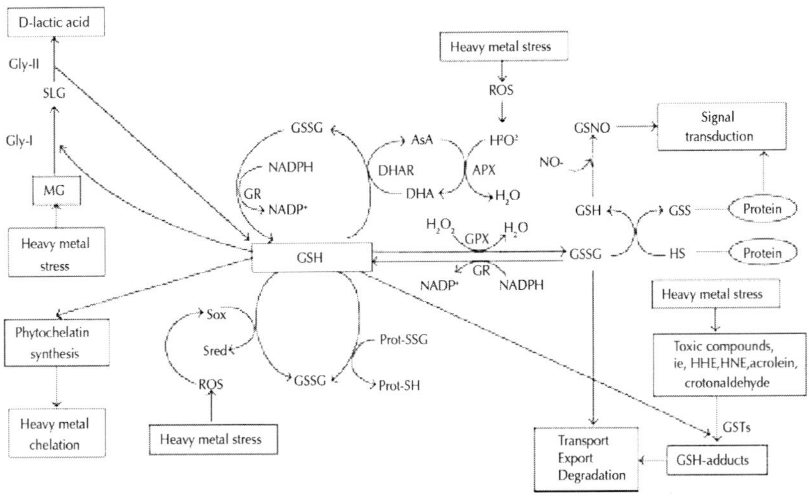

Figure 5: Multiple functions and regulation of GSH and its metabolizing enzyme in heavy metal-induced ROS and MG metabolism in plant cells (adapted from Hossain and Fujita [10]). For further discussion see text.

Glutathione S-Transferases

GSTs are a superfamily of multifunctional enzymes best known for their role in enzymatic detoxification of xenobiotics. GST acts by catalyzing the conjugation of GSH with electrophilic, often hydrophobic toxic compounds to form derivatives that can be secreted from the cell, sequestered in the vacuole, or catabolized [272]. In addition, plant GSTs also provide protection against oxidative stress induced by abiotic stresses and oxidants [272–277]. Functioning as GPX and DHAR, plant GSTs can catalyze the reduction of hydroperoxides to less harmful alcohols and safeguard protein function from oxidative damage and maintain redox homeostasis by regenerating AsA from DHA [272]. Among the various aldehydes, αβ-unsaturated aldehydes contribute more to various reactions than do other aldehydes and are more likely to form stable adducts. In plants, alkenals arise from lipid hydroperoxides after cleavage by hydroperoxide lyase [275, 278]. Previously, we reported that alcohol (ethanol), aldehydes (acetaldehyde), and saturated chain aldehydes (formaldehyde, acetaldehyde, and propionaldehyde, butyraldehyde, pentanal, 1-hexanal) induce GST activity whereas αβ-unsaturated aldehydes are the most effective inducer [275]. In sorghum, for example, phi GSTs (SSP-7108) actively detoxify 4-hydroxynonenal, a degradation product of oxidative membrane damage [279]. Environmental chemicals and their metabolites detoxified by GST include acrolein, atrazine, and others [280]. An increase in GST activity was observed in pumpkin (Curbita maxima Duch.) seedlings subjected to Cd, Cr, Mn, and As stress [274, 276] and in rice (Oryza sativa L. cv. N07-63) seedlings in response to Cd stress (50 μM Cd, 7 d) [281]. Halusková et al. [282] showed that barley (Hordeum vulgare cv. Jubilant) subjected to several HMs (Cd, Pb, Cu, Hg, Co, and Zn) showed a significant increase in GST activity. Similarly, our results showed a significant increase in GST activity in mung bean seedlings subjected of Cd stress (1 mM $CdCl_2$, 24 h) which was accompanied by higher oxidative damage [3]. In another study, there was a significant increase in GST activity in response to Cd stress (1 mM $CdCl_2$) in onion callus [11, 283]. Szollosi et al. [284] found a time- and dose-dependent increase in GST activity in Brassica juncea. Therefore, it is possible that in the case of strong stress, where basal antioxidant mechanisms are exhausted, more effective responses are activated like GST [3,19, 40]. Zhang and Ge [285] found a close relationship between Cd level and GSH content as well as GST activity, suggesting that these two parameters of antioxidant defense system may be used as biomarkers of Cd-induced stress. A sharp increase in GST activity in both root and shoot tissues was also observed in response to Ni stress (100 and 200 μM) in wheat [286]. Pb-induced GST activity was also observed in Macrotyloma uniflorum and Cicer arietinum [287]. Skórzy ska-Polit et al. [278] found a significant increase in GST activity in response to Cu stress (100 μM, 7 d) in Arabidopsis thaliana.

The activity of GST was also greater in the Al-tolerant lines, suggesting that this enzyme may be important not only for the detoxification of certain HMs [288] but also for that of Al phytotoxicity, perhaps via the elimination of the lipid peroxidation induced by Al stress, as found by Cakman and Horst [289]. In plants exposed to HMs, the role of GST may be related not only to the removal of toxic products of lipid peroxidation but also to its possible involvement in the transport of phytochelatin-metal complexes to the vacuole [290]. Dixit et al. [291] noted that tobacco plants overexpressing the GST gene (TvGST) showed better Cd tolerance, as indicated by lower Cd accumulation and lipid peroxidation, than WT plants. Most importantly, the transgenic plants showed significantly higher SOD, GST, GPX, APX, and CAT activities under Cd stress than WT plants, further proving that GSH metabolizing enzymes act coordinately in inducing stress tolerance, including to HM stress [11, 14, 19].

Glutathione Peroxidase (GPX)

Glutathione peroxidases (GPXs) are key enzymes of the antioxidant network in plants present in different subcellular organelles. Their principal activity is thought to catalyze the reduction of H_2O_2, organic hydroperoxides (ROOHs), and lipid hydroperoxides to H_2O and alcohol using GSH and/or other reducing equivalents [292]. The most identified plant GPX genes were shown to have high homology to the mammalian phospholipid hydroperoxide glutathione peroxidases (PHGPXs), which have a higher affinity to lipid hydroperoxides than to H_2O_2. However, at least two plant PHGPXs probably represent novel isoforms of TRX peroxidase, which are generally more active against H_2O_2than lipid peroxides [293]. Differential regulation of GPX activity was observed in response various HMs and metalloids stress treatments (see Anjum et al. [14]). Cuypers et al. [294] suggested that peroxidase activity can be used as a potential biomarker for HM toxicity. Besides disproportionation of H_2O_2, GPX also catalyses the quenching of organic peroxides. Our recent study showed that a significant increase in GPX activity was in response to Cd stress (1 mM $CdCl_2$, 24 h) in mung bean seedlings [3]. Similarly, a significant increase in GPX activity was also observed in Allium cepa exposed to a variety of HMs (Hg, Pb, Cr, Cu, Zn, or Cd) suggesting that the elevated activity of GPX was a result of HM-induced free radical generation [295]. Arabidopsis thaliana subjected to Cd stress (1 and 10 μM, 7 d) showed a significant increase in GPX activity. Higher GPX activity might be an indication of protection against lipid peroxidation caused by Cd exposure [296]. Differential response of GPX activity was observed in Hordeum vulgare cv. jubilant root tips subjected to Cd, Hg, Cu, Zn, Co, and Pb stresses. Cd stress increased GPX activity in a dose-dependent manner. Additionally, Ni, Hg, and Cu also showed a significant increase in GPX activity whereas Pb, Co, and Zn treatment had no significant effect on GPX activity [282]. GPX activity also increased in As-tolerant Brassica juncea var. TPM-1 in response to both As (V) and As (III) stress [297]. The enhancement of GPX activity upon exposure to both As (V) and As (III) in TPM-1 suggested that this enzyme served as an intrinsic defense tool to resist As-induced oxidative stress. An induction of GPX activity has also been previously reported under as stress in other plants [298, 299]. On the other hand, TM-4 demonstrated a decline in GPX activity after 15 days in response to both As (V) and As (III), suggesting that downstream processing of H_2O_2 was not being efficiently taken care of. There are reports of a significant increase in GPX activity in leaf tissues of different plant species exposed to Se and Cu, under both laboratory and field conditions [300, 301]. Aravind and Prasad [302] studied the influence of exogenous Zn on Cd-induced antioxidant defense inCeratophyllum demersum. Cd stress (10 μM, 7 d) showed a 40% decrease in GPX activity whereas Zn-supplementation under Cd stress efficiently restored as well as increased the activity of GPX accounting for increased ROS scavenging induced by Zn.

Therefore, GPX activity in response to HM exposure is diverse and depends on the HM in question, the species, and pivotal role it plays in HM-induced ROS detoxification.

Dehydroascorbate Reductase

The univalent oxidation of AsA leads to the formation of monodehydroascorbate (MDHA) which is reduced to AsA by MDHAR. In the chloroplast, the MDHA radical can also be reduced to AsA by thylakoid-associated, reduced ferredoxin that is more effective than reduction by MDHAR [303]. If MDHA is not rapidly reduced again to AsA by MDHAR, it will spontaneously disproportionate into AsA and DHA. DHA is then reduced to AsA by Dehydroascorbate reductase (DHAR) in a reaction requiring GSH [304]. In the absence of sufficient DHAR, however, DHA undergoes irreversible hydrolysis to 2, 3-diketogulonic acid. Rapid regeneration is necessary in order to maintain the antioxidative capacity of AsA. DHAR allows the plant to recycle DHA, thereby recapturing AsA before it is lost. Because AsA is the major reductant in plants, DHAR serves to regulate its redox state. Thus, DHAR is a physiologically important reducing enzyme in the AsA-GSH recycling reaction for higher plants [304]. The following section will update readers on the modulation of DHAR activity by HM stress in different plants species based on reports over the past decade.

Schützendübel et al. [261] showed that DHAR activity decreased in Populus x canescens roots subjected to 50 μM Cd stress within 15 h of treatment but later the activity increased significantly with the progression of the stress period. A consistent increase in DHAR activity was observed in two barely genotypes (Weisuobuzhi, relatively-Cd tolerant and Dong 17, relatively-Cd sensitive) exposed to 5 μM Cd for 1, 5, 10, 15, and 25 d [305]. Similarly, DHAR activity increased in wheat roots and leaves subjected to different concentrations of Cd (10, 20, 40 μM) but the increase in AsA level was not enough to prevent protein or lipid oxidation [306]. In contrast, a decrease in DHAR activity was also observed in our study with mung bean seedlings subjected to Cd-stress (1 mM $CdCl_2$, 24 h) which was accompanied by lower AsA content in leaf tissues [3]. Sharma and Dubey et al. [7] reported that DHAR activity increased significantly in Al^{3+}-treated rice seedlings compared to controls, and the increase was more pronounced in the 80 μM Al^{3+}-treated seedlings than in the 160 μM Al^{3+}-treated seedlings. The enhanced activity of DHAR observed in Al^{3+}-stressed rice seedlings appears to be due to the need of maintaining a favorable redox status, by maintaining a sufficient amount of AsA and GSH and to overcome the possible problems of oxidation. As tolerant cultivar (TPM-1) showed significant increase in DHAR activity in response to As (V) and As (III); in contrast, the sensitive cultivar (TM4) showed a significant increase in response to As (V) but the activity declined upon exposure to As (III) [297]. Additionally, DHAR activity increased in rice shoots and roots subjected to Ni stress, suggesting that Ni activates the AsA regenerating system to maintain an elevated level of AsA [307]. Two tomato cultivars ("Kosaco" and "Josefina") subjected to B stress (0.50 and 2.00 mM H_3BO_3) showed a significant increase in DHAR activity in addition to a total and reduced AsA content although the level of $H_2O_2$and MDA increased, indicating that the increase enzymatic activity was high enough to reduce the oxidative stresses [308]. Transgenic tobacco plants overexpressing the DHAR gene showed better root growth than WT plants after exposure to toxic levels of Al accompanied by lower H_2O_2 content, less lipid peroxidation, and a lower level of oxidative DNA damage. Furthermore, DHAR-overexpressing plants maintained a higher AsA level both with and without Al exposure when compared to WT plants. Based on their findings, the authors suggest that the overexpression of DHAR that maintained a higher AsA level was essential in acquiring Al tolerance [309]. Similarly, transgenic Arabidopsis plants overexpressing CAT and GST genes subjected to Cd $(NO_3)_2$ (8 mM) stress showed a significant increase (151%) in DHAR activity, whereas the activity declined in a nontransgenic line. The higher activity of DHAR in transgenic plants was consistent with the AsA redox state under Cd stress suggesting an activated redox cycling of AsA [310].

Glutathione Reductase

Glutathione reductase (GR) belongs to a family of flavin-containing pyridine nucleotide-disulphide oxido-reductases which catalyzes the reduction of GSSG to GSH with the accompanying oxidation of NADPH [311]. GR plays an essential role in a cell's defense against reactive oxygen metabolites by sustaining the reduced status of glutathione and ascorbate pools which in term maintain the cellular redox state under various abiotic stresses including HM [11, 19, 39, 40, 312–314]. Biochemical and molecular studies have shown that GR is an important enzyme for scavenging ROS, which are continuously generated in different compartments during various abiotic stresses. Accumulated evidence suggests that GR plays an important role in the detoxification of Cd-induced ROS, possibly via the AsA-GSH cycle. Nouairi et al. [110] reported that GR activity in leaves of Brassica napus was strongly affected by Cd treatment. GR activity increased significantly up to 10 μM Cd and then decreased at higher concentrations after 15 d of treatment. In contrast, GR activity levels in the leaves ofBrassica juncea plants were unaffected by changing the Cd supply. Aravind and Prasad [302] showed a sharp decrease in GR activity in Ceratophyllum demersum in response to Cd stress (10 μM). Importantly, Zn supplementation in Cd-treated plants improved the impairment of GR activity.Arabidopsis thaliana subjected to Cd stress

(1 and 10 μM, 7 d) showed a significant increase in GR activity [296]. In our study [3], in mung bean seedlings, there was a significant decrease in GR activity in response to Cd Stress (1 mM $CdCl_2$, 24 h) whereas there was an increase in GR activity in both Cd-tolerant (Pusa 9531) and -sensitive (PS 16) cultivars of mung bean [315], indicating that the response of GR varies greatly due to differences in genotype. Similar to Cd stress, a significant increase in GR activity was also observed in response to As (V) and As (III) in both tolerant (TPM-1) and sensitive cultivar (TM4) after 15 days of treatment [297]. In Hydrilla verticillata, GR activity increased in response to a low concentration of Cu whereas at higher concentrations (5 and 25 μM) the activity declined, manifested by higher lipid peroxidation [316]. Israr and Sahi [317] showed a significant increase (62.8%) in GR activity in response to Hg stress in Sesbania drummondii whereas the activity of GR drastically declined with 50 μM Hg. Similarly, Medicago sativa roots exhibited a strong inhibition of GR activity when exposed to Hg (10 μM) for 24 h [264]. Russo et al. [318] reported that GR activity inBrassica napus exposed to Cu (100 μM) decreased in roots, whereas in leaves, GR activity was induced during the whole stress period. In earlier reports, a reduction in cellular GSH levels as well as GR activity was observed in plants exposed to toxic levels of Cu [319]. Verma and Dubey [320] showed a significant increase in GR activity in shoot tissues of rice after 15 d of Pb treatment (1000 μM). Increased GR activity in Pb-treated rice seedlings suggested the possible involvement of GR in regenerating GSH from GSSG under Pb toxicity conditions to increase the GSH/GSSG ratio and the total glutathione pool. These authors concluded that GR appears to play a pivotal role in combating oxidative stress in plants. Finally, it can be concluded that increased GR activity under HM stress contributes to the maintenance of a higher GSH/GSSG ratio and increased GSH level which is used during phytochelatin synthesis and by DHAR and other GSH-dependent enzymes involved in ROS and MG metabolism [3, 11, 14, 19, and 40].

Glyoxalase I

Glyoxalase I (Gly I) is the first enzyme of the GSH-based glyoxalase pathway involved in detoxification of MG. Induction of Gly I activity in response to various stresses has been reported over a range of plant species under both biotic and abiotic stresses, including HM [2, 3,

10, 11, 15–17, 19, 232, 233]. In our study with pumpkin seedlings, we found a slight increase in Gly I activity in response to Cd stress (1 mM $CdCl_2$, 24 h). Similar to our results, Singla-Pareek et al. [17] also observed an increase in Gly I activity in tobacco seedlings subjected to Zn stress (5 mM $ZnCl_2$, 24 h). Later on, we studied the regulation of Gly I enzyme activities in response to various stresses in the callus of onion which had very high Gly I activity than other plant species [321]. We observed a significant increase in Gly I activity in response to various stresses where low-temperature stress showed the highest induction (2.57-fold) followed by salinity (2.52-fold) and drought stress (1.43-fold). Importantly, we observed a decrease in Gly I activity in response to HM stress (0.5 mM $CdCl_2$, 7 d) in onion callus [9]. Therefore, it can be concluded that Gly I activity varies due to differences in genetic background, stress conditions, type of HM used, or differential regulation of different Gly I isoforms [322]. It is conceivable that an elevated level of Gly I activity is required to remove excessive MG produced in ample amounts under normal and various stressful conditions [2, 3, 9, 11, 15, 16, 19, 40]. El-Shabrawi et al. [323] found that salt-tolerant rice cultivar (Pokkali) maintained higher Gly I activity under salt stress than a sensitive cultivar (IR64). Importantly, the other antioxidant enzymes (SOD, APX, CAT, GST, GPX, GR and POX) studied also showed a significant increase relative to Gly I, which indicates a balance among the glyoxalase pathway and antioxidant pathway enzymes to increase abiotic stress tolerance.

Glyoxalase II

Glyoxalase II (Gly II) catalyzes the second step of MG detoxification and forms D-lactate while GSH is regenerated. Two Gly II isoforms were identified from Arabidopsis and they are differentially expressed, suggesting their different roles in plant cells [324]. The cytoplasmic form of Arabidopsis Gly II contains a Fe-Zn binuclear metal centre that is essential for its activity. Both metals participate in substrate binding, transition site stabilization, and the hydrolysis reaction. Subtle alterations in the geometry and/or electrostatics of the binuclear center have profound effects on the activity of the enzyme [325,326]. In our study [9] with onion callus we found a decrease in Gly II activity in response to Cd stress (0.5 mM $CdCl_2$, 7 d). Similarly, we also observed a decrease in Gly II activity in mung bean seedlings subjected to Cd stress (1 mM

$CdCl_2$, 24 h) [3]. Based on our results, we conclude that Cd stress might interact with the Fe-Zn binuclear centre and decreased the catalytic activity of the enzymes. Additionally, Gly II activity increased slightly in WT tobacco seedlings and even more increased in the transgenic plants overexpressing both Gly I and II under Zn stress (5 mM $ZnCl_2$, 24 h) [17] and maintained a higher GSH level. As GSH is recycled by the glyoxalase system, it is assumed that an increased level of MG is efficiently detoxified in transgenic plants overexpressing glyoxalase pathway enzymes, thus creating the probability of upregulation of GSH levels at least during stress, which would help to form PCs under HM stress, thus proving HM stress tolerance. Transgenic tobacco plants overexpressing both Gly I and Gly II genes also showed higher salinity tolerance and by additional increase of GSH metabolizing enzyme (GST, GPX, and GR) activities further denoting the close interaction between the antioxidant system and the glyoxalase system in imparting stress tolerance [11,16, 19, 39, 40, 327, 328].

EXOGENOUS CHEMICAL TREATMENT AND HEAVY METAL TOLERANCE IN PLANTS IN RELATION TO GSH AND THEIR METABOLIZING ENZYMES

Some of the adaptive biochemical responses of HM stress in plants are characterized by the synthesis of GSH, amino acids like Pro and PAs, and singling molecules like SA and NO (described previously). A wealth of studies have demonstrated that treatment of plants with such compounds influences plant HM uptake and HM-induced oxidative stress tolerance by efficient regulation of ROS and MG detoxification pathways.

We investigated the regulatory effects of exogenous Pro and glycinebetaine (betaine) in inducing HM tolerance of mung bean (cv. Binamoog-1) seedlings with special reference to ROS and MG metabolism [3]. Seven-day-old seedlings were subjected to Cd stress (1 mM $CdCl_2$, 48 h) with or without Pro and betaine. Cd stress caused a profound increase in GSH and GSSG content, while the AsA content decreased with a sharp increase in H_2O_2 and lipid peroxidation (MDA). APX, GST, GPX, and Gly I activities increased in response to Cd stress while the activity of CAT, MDHAR, DHAR, GR, and Gly II decreased. Exogenous application of Pro and betaine (5 mM) showed an increase in GSH and AsA content, maintained a high GSH/GSSG ratio, and increased the activity of APX, DHAR, MDHAR, GR, GST, GPX, CAT, Gly I, and Gly II involved in ROS and MG detoxification systems more than the control and most Cd-stressed plants, with a concomitant decrease in GSSG content, H_2O_2, and MDA levels. These findings suggest that both betaine and Pro provide protection against Cd-induced oxidative stress by reducing H_2O_2 and MDA levels and by increasing the antioxidant defense and MG detoxification systems. Additionally, Xu et al. [201] showed that improvement of Cd-tolerance caused by Pro pretreatment was correlated with increase antioxidant enzyme activity and higher intracellular GSH content.

The influence of exogenous GSH in the antioxidant defense system against Cd-induced oxidative stress was studied in two barley genotypes (sensitive, Dong 17, and tolerant, Weisuobuzhi) by Chen et al. [305]. Seedlings at the two-leaf stage (10 d) were subjected to 5 μM Cd with basal nutrient solution for up to 25 d. Cd stress significantly inhibited plant height, root length, and biomass whereas accumulation of O2 • −, H_2O_2, and MDA increased sharply in both tolerant and sensitive genotypes. The addition of GSH (20 mg/L) in 5 μM Cd culture medium significantly alleviated Cd-induced growth inhibition, especially for the sensitive genotype Dong 17, and dramatically depressed O2 • −, H_2O_2, and MDA accumulation. The GSH content decreased significantly in response to Cd stress, whereas exogenous GSH application increased the endogenous GSH content more than in seedlings subjected to Cd stress alone. External GSH counteracted Cd-induced alterations of certain antioxidant enzymes. The activities of DHAR, MDHAR, and GPX activities were downregulated and reached the control level in seedlings treated with GSH whereas depressed APX and CAT activity was elevated specially in susceptible genotype (Dong 17). Therefore, exogenous GSH not only influences the GSH metabolizing enzymes but also substantially influences other ROS-metabolizing enzymes while playing an imvportant role in Cd-induced alleviation of oxidative stress.

The role of GSH in heat shock- (HS-) and H_2O_2-induced protection of rice (cv. Taichung 1) seedlings against Cd stress (5 μM Cd, 7 d) was investigated by Chao et al. [329]. HS (45°C, 3 h) and H_2O_2 (0.5 mM, 3 h) pretreatment resulted in an increase in GSH content in the leaves of rice seedlings. Pretreatment of seedlings with exogenous GSH (1 mM, 3 h) resulted in an increase in GSH in leaves, enhancing subsequent Cd tolerance. Pretreatment with buthionine sulfoximine (BSO), a specific inhibitor of GSH synthesis, effectively inhibited GSH content induced by HS and H_2O_2 and reduced subsequent Cd tolerance. Furthermore, the effect of BSO on HS- and H_2O_2-induced GSH accumulation and toxicity by subsequent Cd stress can be reversed by the addition of GSH. Time-course analyses of HS in rice seedlings demonstrated that the accumulation of H_2O_2 preceded the increase in GSH. Finally, they concluded that the early accumulation of H_2O_2 during HS signals an increase in GSH content, which in turn protects rice seedlings from ROS-induced damage under Cd stress.

To understand the mechanism by which PAs relieve the damage of plants by HMs and improve the phytoremediation efficiency, Tang et al. [330] sprayed Spd (0.25 to 0.50 mM) on Typha latifolia L. leaves and then subjected them to Cd stress (0.1 and 0.2 mM Cd^{2+}). Cd^{2+} stress induced oxidative injury, as evidenced by an increase in the generation of O2 • −, as well as the H_2O_2 and MDA contents in both leaves and caudices. An increase in the activities of CAT, GPX, and GR was observed in both leaves and caudices, SOD activity increased in caudices, and APX activity increased in leaves following Cd^{2+} treatment. The GSH content in both leaves and caudices and the reductive ascorbate content in leaves increased obviously, prompted by the application of exogenous Spd. Spraying with Spd increased the activity of GR and APX in both leaves and caudices, whereas the activity of SOD, CAT, and GPX increased only in caudices following a spray with Spd. The generation of O2 • − and the H_2O_2 and MDA content in both leaves and caudices decreased after spraying with Spd. Finally, it was assumed that exogenous Spd elevated the tolerance of T. latifolia under Cd^{2+} stress primarily by increasing GR activity and the GSH level.

Application of exogenous silicon (Si) was also found to increase Cd tolerance in two Brassica chinensisL. Cultivars (Shanghaiqing, a Cd-sensitive cultivar; and Hangyoudong, a Cd-tolerant cultivar). Plants were grown in 0.5 and 5 mg/L Cd with or without 1.5 mM Si. The activity of antioxidant enzymes such as SOD, CAT, and APX decreased at both concentrations of Cd, although MDA and H_2O_2 concentrations increased at the higher Cd level, which was counteracted by the addition of Si. The content of AsA, GSH, and nonprotein thiols increased at the higher Cd level, further intensified by the addition of Si. However, the protective influence of Si was more effective in enhancing Cd tolerance in the Cd-tolerant cultivar than in the Cd-sensitive cultivar. It can be concluded that Si-enhanced Cd tolerance in B. chinensis was attributed mainly to Si-suppressed Cd uptake and root-to-shoot Cd transport and Si-enhanced antioxidant defense activity [331].

H_2O_2 is a small, diffusible molecule and is widely considered to be a signal molecule in the regulation of the defense system [239]. H_2O_2 plays a dual role in plants: at low concentration, it acts as a signal that induces the expression of numerous defense genes and activates multiple defense responses to abiotic stresses, while excessive accumulation leads to cellular oxidative damage and even programmed death [19, 331, and 332]. The influence of exogenous H_2O_2 pretreatment on Cd tolerance and translocation was studied by Hu et al. [281] in rice (cv. N07-63) seedlings. Cd-stress (50 µM, 7 d) significantly decreased plant growth as indicated by decreased length and biomass of shoots and roots. Additionally, the level of H_2O_2 and MDA increased significantly in both shoot and root tissues of Cd-stressed seedlings due to impairment of antioxidant metabolism. Importantly, H_2O_2 pretreatment (100 µM, 24 h) mitigated Cd stress by increasing the activities of SOD, CAT, GPX, APX, and GST as well as nonenzymatic antioxidant such as GSH and AsA resulting in lower H_2O_2 and MDA content and less growth inhibition. H_2O_2 pretreatment also decreased Cd concentration in shoots indicating that H_2O_2 affected Cd distribution in rice seedlings. Therefore, improved Cd tolerance was partly due to an enhanced antioxidative system and higher Cd sequestration in roots that ultimately reduced Cd translocation.

Xu et al. [333] showed the H_2O_2-induced upregulation of AsA and GSH metabolism showing Al-induced oxidative stress tolerance in wheat seedlings. Al stress increased the O2 • − and H_2O_2 levels leading to more predominant lipid peroxidation, programmed cell death, and inhibition of root elongation in both Al-tolerant and -sensitive genotypes. Al stress increased the activities of SOD, POD, CAT, MDHAR, DHAR, GR, GPX and AsA and GSH content and their redox state. However, H_2O_2-pretreated Al-stressed seedlings showed higher SOD, POD, CAT,

MDHAR, DHAR, GR, GPX, and AsA and GSH content and their redox state than nontreated Al-stressed seedlings. Importantly, enhancement of antioxidant capacity was greater in the Al-sensitive genotype than in the tolerant one. Therefore, optimum H_2O_2 pretreatment makes the plant more tolerant to Al-induced oxidative stress in a synergistic fashion.

The protective effects of NO against the toxicity caused by excess Cu ($CuSO_4$, 50 μM) in Panax ginsengwere also studied by Tewari et al. [334]. The addition of an NO donor, SNP, was effective in reducing Cu-induced toxicity in P. ginseng adventitious roots. The protective effect of SNP, as indicated by the extent of lipid peroxidation, was reversed by incorporation of a NO scavenger, in the medium suggesting that the protective effect of SNP is attributable to NO released. Reduction of excess Cu-induced toxicity by SNP is most likely mediated through modulation in the activities of antioxidant enzymes involved in H_2O_2 detoxification (APX, CAT and POD) and in GSH regeneration (GR) and contents of molecular antioxidants (particularly nonprotein thiol, ascorbate and its redox status).

The role of SA in regulating Hg-induced oxidative stress in the roots of alfalfa (Medicago sativa) was investigated by Zhou et al. [171]. Plants were pretreated with 0.2 mM SA for 12 h and subsequently exposed to 10 μM Hg^{2+} for 24 h. The SA-promoted root growth was correlated with decreased lipid peroxidation in root cells. The ameliorating effect of SA was confirmed by histochemical staining for detecting the loss of membrane integrity in Hg-treated roots. Additionally, treatment with 0.2 mM SA increased the activity of NADH oxidase, APX, and POD in the roots exposed to Hg. However, slightly decreased SOD activity was observed in SA + Hg-treated roots than in roots treated with Hg alone. Importantly, SA-pretreated Hg-stressed seedlings accumulated more AsA, GSH, and Pro in the roots than seedlings subjected to Hg-stress without pre-treatment. Finally, the authors suggested that exogenous SA improved Hg tolerance by modulating antioxidant enzymes and also by increasing the AsA and GSH contents.

Therefore, the chemicals or compounds synthesized in plant cells in response to HM stress further prove their protective roles in HM tolerance in relation to ROS and MG metabolism.

PROTEOMIC CHANGES IN RESPONSE TO HM STRESS IN RELATION TO GSH METABOLISM

Proteomic offers a new platform for studying complex biological functions involving large numbers and networks of proteins and can serve as a key tool for revealing the molecular mechanisms that are involved in the interactions between toxic HMs and plant species. Quantitative analysis of proteins and metabolites is able to provide accurate information about molecular changes allowing the establishment of a range of more or less specific mechanisms, leading to the identification of major players in the considered pathways [5]. However, the detail description of plant proteomic changes under HM stress fall beyond the scope of this paper. For a few excellent recent reviews in plant proteomic changes in response to HM stress, see [335, 336]. In this paper, we will focus only on GSH and its related proteins which have a significant impact of HM-induced stress response and tolerance in plants.

A comparative proteomic analysis of the short-term responses of rice roots and leaves to Cd stress was made by Lee et al. [337]. A total of 36 proteins were up- or downregulated following Cd stress (100 μM $CdCl_2$, 24 h). Importantly, the upregulated proteins in rice roots belonged to the proteins involved in the oxidative stress response and GSH metabolism (GST, GS, GR, APX1, and POD) indicating that prompt antioxidative responses might be necessary for the reduction of Cd-induced oxidative stress. Previously, we reported a robust increase in CmGSTU1 and CmGSTU2 protein in response to As stress (0.1 mM, 24 h) and CmGSTU1, CmGSTU2, and CmGSTF1 in response to Cr (1 mM, 24 h), Mn (1 mM, 24 h), and Cd stress (100 μM Cd, 24 h) [274, 276]. In our more recent study with onion callus, we also observed significant accumulation of two GST protein (GSTe and AcGSTF1) under Cd stress (1 mM $CdCl_2$, 96 h) condition [283] but no increase in Gly I protein was observed in response to Cd stress (0.5 mM $CdCl_2$, 7 d) [9].

Zhao et al. [67] studied the Cd-induced changes of the leaf proteome by two-dimevv nsional gel electrophoresis (2-DE) in a Cd hyperaccumulator plant, Phytolacca Americana L. Among the 32 differentially expressed protein spots, 14 were enhanced/induced while 11 were reduced under Cd stress (400 μM, 48 h). However, major changes were found in the proteins involved in the photosynthetic pathway as well as the sulfur- and GSH-related metabolisms. Additionally, Wang et al. [338] made a comparative proteomic analysis of wheat (Triticum aestivum L. cv. Yangmai 15) roots to gain a comprehensive understanding of the plant's response to Cd stress exposed to a range of Cd concentrations (10, 100 and 200 μM, 1 week). Protein profiles analyzed by 2-DE revealed that 46 protein spots showed a 1.5-fold change in protein abundance following Cd exposure; 31 protein spots were upregulated and 15 protein spots were downregulated; 25 of them were identified by matrix-assisted laser desorption/ionization time-of-flight mass spectrometry (MALDI-TOF MS). Importantly, most of the upregulated proteins were related to sulfur and GSH metabolism (such as ECS, S-adenosyl-methionine synthase (SAMS), GST, DHAR, and Malate dehydrogenase) involved in HM detoxification and antioxidant processes indicating the role of GSH metabolism in Cd stress tolerance.

To elucidate the precise mechanisms underlying the toxicity of as (50 and 100 μM $Na_2HAsO_4 \cdot 7H_2O$) in rice (Oryza sativa L. cv. Dongjin), the metabolism and defense reactions were studied by Ahsan et al. [339] using a combination of comparative proteomic analysis in combination with physiological and biochemical analyses. Proteomic analyses identified 23 differentially expressed proteins in rice roots, including predicted ones such as S-adenosylmethionine synthetase (SAMS), CS, and novel proteins including tyrosine-specific protein phosphatase protein and GST containing an omega domain. These differentially expressed proteins are functionally involved in cell signaling, stress and detoxification, defense and development, and protein biosynthesis. Based on physiological and proteomic investigations, it is proposed that as stress in plants generates ROS, triggering signaling molecules such as JA and S-adenosyl-L-methionine (SAM), and activating the detoxification process, which mainly involves GSH/PC biosynthesis. Importantly, these pathways are coordinately regulated, where GSH plays a central role in the complex cellular network [339].

Ingle et al. [340] analyzed the root and shoot proteomes of a Ni hyperaccumulator plant (Alyssum lesbiacum) by using 2-DE coupled with LC-MS/MS to gain more information about Ni metabolism. The accumulation of several proteins involved with sulfur metabolism in response to Ni stress suggests that Cys or a downstream metabolite, such as GSH, may play a role in Ni tolerance. Similarly, a proteomic study identifies the metabolic pathways involved in Cr stress during germination of maize (Zea mays L.) seedlings exposed to high concentrations of potassium dichromate, and the proteome pattern and MT content were analyzed by Labra et al. [341]. Increases in the abundance of several antioxidant, detoxification, and glycolytic pathway-related enzymes suggested that under Cr stress, activation of oxidative stress mechanisms occurred together with an effect on sugar metabolism. Importantly, increases in the abundance of Gly I and MTs after exposure to Cr suggest that plants might use glyoxalase and MTs pathways to detoxify Cr or Cr-induced toxic by-products in plant cells. Involvement of Gly I during higher ATP synthesis under stress has also been reported [228].

Tuomainen et al. [342] compared the effects of Zn deficiency under increased concentrations of Zn and Cd on three genotypes of Thlaspi caerulescens, a tiny plant from the Brassicaceae family closely related to A. thaliana and a known hyperaccumulator of several HMs. These researchers used close homology of most protein sequences found in T. caerulescens to A. thaliana and were able to identify the vast majority of differentially expressed proteins. Under increased Zn or Cd concentrations, most upregulated proteins in shoots belonged to enzymes participating in ROS scavenging such as Cu/Zn-SOD, APX, DHAR, and peroxiredoxin. Later on, proteomic changes in response to excess Cu-stress were studied by Ahsan et al. [343] in rice (Oryza sativa L. cv. Hwayeong) seedlings by using $CuSO_4 \cdot 5H_2O$ (0.2 to 1.5 mM). Protein profiles analyzed by 2-DE revealed that 25 protein spots were differentially expressed in Cu-treated samples. Among them, 18 proteins were upregulated while 7 were downregulated. These differentially displayed proteins were identified by MALDI-TOF MS. The upregulation of some antioxidant, and stress-related proteins such as Gly I, peroxiredoxin, aldose reductase, and some regulatory proteins such as DnaK-type molecular chaperone, UlpI protease, and receptor-like kinase clearly indicated that excess Cu generates oxidative stress that might be disruptive to other important metabolic processes.

Zhen et al. [344] studied the molecular mechanism underlying Al stress (50 μM $AlCl_3$) responses by comparative proteome analysis of Al-stress-responsive proteins in an Al-resistant soybean (Glycine max (L.) Merr) cultivar (Baxi 10). A total of 39 differentially expressed spots in abundance were identified by MS, with 21 being upregulated, 13 newly induced, and 5 downregulated. Importantly, proteins related to sulfur metabolism (GST and OASTL) were upregulated and GST expression was found to be an early response to Al toxicity. They concluded that the expression of GSH-related proteins might detoxify the ROS elicited by Al toxicity. Similarly, Yang et al. [345] studied the Al stress- (100 and 250 μM $AlCl_3$) induced proteomic changes in an Al-resistant rice cultivar (Oryza sativa L. cv Xiangnuo 1). A total of 17 Al-responsive proteins were identified, with 12 of those being upregulated and 5 downregulated. Among the upregulated proteins were Cu/Zn SOD, GST, and SAMS, which are the consistently known Al-induced enzymes. Most of these proteins identified are functionally associated with signal transduction, antioxidation, and detoxification. Cysteine synthase (CS), consistently detected in both Al stress systems, was further validated by Western blot and CS activity assays. Moreover, the metabolic products of CS catalysis, that is, both the total glutathione pool and GSH, were also significantly increased in response to Al stress. Finally, they suggested that antioxidation and detoxification were ultimately related to sulfur metabolism, particularly to CS, and may play a functional role in rice's adaptation to Al.

From the foregoing examples, it is clear that proteins involved in GSH biosynthesis or the GSH-cycle, such as CS, -ECS, GR, GSTs, DHAR, and Gly I, are differentially regulated under HM-stress. However, such proteomic studies lead to identification of potential protein markers whose changes in abundance can be associated with quantitative changes in some physiological parameters used for the description of genotypes and level of stress tolerance. Importantly, among the different key identified proteins, those related to GSH synthesis or GSH utilizing and regenerating enzymes have a potential role in HM-induced stress response and tolerance.

GLUTATHIONE (GSH), PHYTOCHELATINS (PCS), METALLOTHIONEINS (MTS), REACTIVE OXYGEN SPECIES (ROS), AND GLYOXALASE SYSTEM-RELATED GENE EXPRESSION UNDER HEAVY METAL STRESS

Plants adaptation to HM stress is controlled by a well-balanced, genetically determined signaling system. The identification and characterization of plant genes which control responses to HM stresses is an essential step to elucidate the complex regulatory network, which determines the tolerance of plants to HMs. Although molecules related to HM transport, chelation, and sequestration such as transporters and HM chelators are good candidates for the function of tolerance and accumulation, knowledge about global gene expression remains limited. The molecular genetics of model organisms can lend much to our understanding of the basic components of essential HM metabolism in plants. The constitutive expression of genes may contribute, at least in part, to the molecular adoption of plants that might be useful for plant genetic transformation and phytoremediation.

The expression of GSH metabolic genes was studied by Ortega-Villasante et al. [346] in Cd- and Hg-treated alfalfa seedlings, including the level of GSH, homoglutathione (hGSH), and AsA. The expression of genes involved in their metabolism, that is, ECS, GS, hGS, GR1, GR2, GPX, and PCS, was analyzed by semiquantitative RT-PCR. Different exposure periods to 30 μM Cd did not alter the expression of the genes analyzed. However, 30 μM Hg clearly affected the accumulation of some of the transcripts. With the exception of PCS, GPX, and GS genes, which were expressed at similar levels at all periods analyzed, the genes were induced in a transient manner. Hg produced a clear induction of GR1 (cytosolic) andGR2 (plastidic) genes, with maximum expression being reached at 3 h after metal supply and decreasing thereafter. Similarly, ECS and hGS genes were expressed transiently, also reaching maximum transcript accumulation after 3 h. notably, this pattern can be linked to GSH/hGSH depletion. Similarly, induction of the GSH metabolic gene in response to Cd stress was studied in Camellia sinensis (L.) O. Kuntze by Mohanpuria et al. [347]. The transcript levels of GSH biosynthetic and regenerating genes, namely, -ECS, GS, and GR increased in response to Cd stress. However, the transcript levels of GST did not show any change following Cd treatment. This study in which Cd exposure induced oxidative stress in C. sinensis and upregulated the transcript levels of GSH metabolic genes, except for GST, suggests the role of these enzymes in the protection of plants from a high level of Cd exposure. Consequently, Arabidopsis thaliana (ecotype Columbia) plants treated with Cd or Cu responded by increasing the transcription of genes for GSH synthesis, -ECS and GS, as well as GR. The response was specific for those HMs whose toxicity is thought to be mitigated through PCs and other toxic and nontoxic HMs, which did not alter mRNA levels. JA also activated the same suite of genes, which suggests that it might be involved in the signal transduction pathway for Cu and Cd [348].

Lee et al. [337] showed that rice roots subjected to Cd stress (100 μM, 24 h) increased GST and APX gene expression. Three GST genes (GSTs: OsGSTU3, OsGSTU4, and OsGSTU12) that belong to the plant-specific tau class were markedly overexpressed in roots following Cd exposure. Ogawa et al. [349] also found strong induction of the OsGST gene (OsGSTU5) in response to Cd stress. GSTs have been reported to be involved in the direct quenching of Cd ions, forming GSH-Cd complexes [350]. Panda and Matsumoto [351] studied antioxidant gene expression in pea (Pisum sativum L.) root and shoot tissues under Al stress (10 and 50 μM). GST and CAT gene expression was higher in roots whereas cAPX increased in shoots. The Cu/Zn SOD gene showed higher expression after 24 h whereas MnSOD and FeSOD showed higher expression after 24 and 48 h.

Perennial ryegrass (Lolium perenne L.), widely used in temperate climates, is a turfgrass that enriches Cd. Candidate gene expression in perennial ryegrass to Cd stress was studied by Luo et al. [352]. Grasses were subjected to three levels of Cd (0, 0.2, and 0.5 mM, 7 d). The Cd stress-induced upregulated expression of FeSOD, MnSOD, Chl Cu/ZnSOD, Cyt Cu/ZnSOD, APX, GPX, GR, and POD was observed at 4–24 h after treatment. They concluded that genes involved in antioxidant enzymes could play important roles in Cd tolerance in perennial ryegrass.

Smeets et al. [353] studied the early antioxidative defense mechanisms by applying a range of realistic Cd concentrations (5, 10, 20 μM $CdSO_4$, 24 h) to the root and leaf tissues of Arabidopsis thaliana (Columbia ecotype). Highly coordinated oxidative stress-related defense gene (FeSOD, MnSOD, APX, DHAR, GR, CAT, and GPX) expression was observed at 5 and 10 μM $CdSO_4$; however, 20 μM $CdSO_4$stress gene expression of some antioxidant enzyme downward to control level was oberved. Importantly, GPX gene expression in leaf tissues showed a continual increase as Cd concentration increased. Transcript abundance of several genes involved in antioxidative defense in root and leaf tissues was observed. The generation of $O_2^{\bullet -}$ seems to be the main cause of oxidative stress in the roots, whereas in the leaves H_2O_2 appears to be an important player. Furthermore, an increased transcript level of LOX, a potential inducer of oxidative stress, suggests a central role of this gene in causing the Cd-related redox imbalance. Their findings suggest that the antioxidative defense system is moderated by the activation of different genes in different organs and cellular compartments.

The identification of genes involved in HM tolerance was studied by Goupil et al. [354] in cultivated tomato (Lycopersicon esculentum cv. VFN-8) plants subjected to As (V) or Cr (VI) stress by real-time PCR transcript analysis of genes using RNA isolated from tomato roots or shoots treated for 24 h with As (V) or Cr (VI) at concentrations ranging from 80 to 640 μM. An increase in Hsp90-1 transcript accumulation was observed due to both of the metallic stress treatments. MT2- and GR1-like transcripts accumulated in tomato roots treated with As (V) but were only slightly affected by Cr (VI) treatment. Tomatoes showed phenotypic symptoms of HM toxicity when plants were exposed to Cr (VI) but not to As (V), indicating that tomatoes are more tolerant to As than Cr.

PCs are synthesized from GSH by the action of the enzyme γ-glutamylcysteine dipeptidyl transpeptidase, trivially named phytochelatin synthase (PCS). The enzyme is constitutively expressed but may be regulated at transcriptional and translational levels by HMs and metalloids [355]. The relative contribution of PCs and MTs in HM

(Cd or Cu) tolerance was examined in black mangrove, Avicennia germinans. Seedlings were subjected to Cd and Cu stress hydroponic conditions and gene expression was evaluated. Importantly, a low concentration of Cd and Cu led to a significant increase inAvPCS expression within a short period of metal exposure (4 h) whereas a sharp increase in AvMt2expression was observed after 16 h of metal exposure. This strongly suggested that rapid induction ofAvMt2 and AvPCS expression might contribute to Cd and Cu detoxification either by sequestering Cd and/or by directly scavenging deleterious oxygen radicals [356]. Similar results were reported by Akashi et al. [357] and Zhang et al. [358] where Mt2 and PCS genes were shown to act as potent scavengers of hydroxyl radicals in Citrullus lanatus and A. sativum exposed to HMs, respectively.

To discover the genetic factor underlying the mechanism of Al tolerance, Duressa et al. [359] conducted a comprehensive transcriptomic study by using Al-tolerant soybean line (PI 416937) and found that GST and GR transcript increased in response to Al stress (10 μM Al) indicating that among the various genes involved in transcription activation, stress response, cell metabolism, and signaling, the ROS metabolic genes had a potential role in Al tolerance in plants. The induction of proteins related to GSH metabolism and its potential implication in as tolerance in rice seedlings was reported by Ahsan et al. [339]. They also studied the expression of CS, GST-omega, and GR genes by Northern blot analysis. Importantly, genes exhibited very similar expression patterns at the mRNA and protein levels. Thus, their results provided additional evidence for correlative expression/accumulation of the identified proteins and gene expression in as-stressed rice roots. Similarly, in our study [276] with pumpkin (Cucurbita maxima L.) seedlings, we found a significant increase in the induction of CmGSTU1, CmGSTU2, and CmGSTU3 transcripts in response to as stress (0.1 mM, 6 h). Additionally, in our previous study with pumpkin seedlings, we found an increase in Gly I transcript in response to Cd stress (1 mM Cd, 24 h). Lin et al. [360] found a significant increase in Gly I transcript in response to Zn stress (10 and 20 mM $ZnCl_2$, 24 h). Based on the previous evidence we, therefore, concluded that genes that utilize and regenerate GSH as well as PC- and MT-related genes play a predominant role in HM stress response and tolerance in plants.

PLANT MICROBE INTERACTION AND HEAVY METAL TOLERANCE IN RELATION TO GLUTATHIONE

The extraordinary genetic and physiological diversity of fungi makes them key components of almost all ecosystems, with a great potential for bioremediation purposes [3, 361, 362]. Arbuscular mycorrhizal (AM) fungi, which are ubiquitous root symbionts, are particularly attractive in this regard. It is becoming increasingly evident that the mutualistic or symbiotic association between plants and fungi significantly contributes to plant growth by improving mineral nutrition and also by protecting plants against a variety of biotic and abiotic stresses, including HM stress [25, 362, and 363]. Although biophysical methods do not allow discrimination of HM distribution between fungal and plant cell structures, the fungi obviously direct HM allocation within the roots. Several mechanisms for this allocation can be envisaged [25, 33, 364, 365]: (a) HMs may be bound to the cell wall and may be deposited in the vacuoles of the fungi, (b) HMs may be attached to siderophores and sequestered either into the root apoplasm or into the soil, (c) HMs may be bound to MTs or PCs inside the fungal or plant cells, or (d) HM transporters at the plasmalemma or tonoplast of both partners may catalyse the export of HMs from the cytoplasm. AMF (Glomus intraradices) that colonized Medicago truncatula plants subjected to Zn stress ($ZnSO_4$, 100 μM) showed a sharp transcriptional increase of a GSH-dependent GST gene [366]. Upregulation of ROS metabolic genes indicates that effective fungal HM-tolerance mechanism(s) could provide a pivotal contribution to the increased HM tolerance of mycorrhizal plants. The association of plants with mycorrhizal fungi was found to increase plant fitness under HM stress by utilizing GSH-dependent detoxification mechanisms. However, further investigations may reveal the more subtle regulatory role of AMF fungi on GSH-dependent HM tolerance of plants or other abiotic stresses. A complete understanding of the molecular bases of HM tolerance in this symbiotic system will, thus, also aid the selection of the most effective AM isolates [367] and plant-fungus combinations for bioremediation and soil protection [368]. Therefore, detailed mechanisms of AMF-induced HM tolerance in relation to GSH metabolism in plants are still a matter of intensive research.

TRANSGENIC PLANTS AND HEAVY METAL TOLERANCE WITH REFERENCE TO GLUTATHIONE

Molecular biology and genetic engineering are being increasingly considered as effective tools for better understanding and improving the phytoremediation capability of plants because several key steps have been identified at the molecular level, through molecular-genetic technologies [28, 70, 155]. Classic genetic studies demonstrated that many genes are involved in HM uptake, translocation, sequestration, chemical modification, and tolerance. Unless regulatory genes are identified that simultaneously induce many HM-related genes, it is feasible that more than one gene will need to be upregulated in order to substantially enhance HM phytoremediation capacity. Transfer and/or overexpression of genes responsible for HM uptake, translocation, and sequestration may allow for the production of plants which, depending on the strategy, can be successfully exploited in phytoremediation [31, 70, 369–371]. The overexpression of any (combination) of these genes is a possible strategy for genetic engineering. Depending on which phytoremediation application is to be used, the genetic engineering strategy may strive to create plants that accumulate more HMs in harvestable plant parts (phytoextraction) or adsorb more HMs at their root surface (rhizofiltration, phytostabilization).

Cysteine is the precursor molecule for the synthesis of GSH, the predominant nonprotein thiol, which plays an important role in plant stress responses. GSH has been implicated in plant responses to toxic levels of HMs, as it is the precursor for the synthesis of PCs, the thiolate peptides involved in the detoxification of Cd and other HMs. Stimulation of the synthesis of building blocks for GSH may have been found to increase Cd tolerance in transgenic plants. Arabidopsis thaliana overexpressing cytosolic O-acetylserine (thiol) lyase gene (OASTL) showed Cd tolerance by increasing bothcysteine and GSH availability [372]. Few research findings showed that transgenic plants overexpressing the GSH synthesis genes, γ-ECS or GS, showed enhance HM tolerance. Two E. coli genes involved in GSH biosynthesis, GS or γ-ECS, were constitutively expressed in Brassica juncea [373, 374]. The expression of either GS or γ-ECS in plant species resulted in higher accumulation of GSH and PCs as well as enhanced tolerance to Cd-induced oxidative stress. Additionally, transgenic poplar hybrid overexpressing a cytosol targeted γ-ECS gene showed a 3- to 4-fold increase in GSH levels [375]. However, transgenic tobacco overexpressing chloroplast-targeted γ-ECS contained elevated GSH levels but exhibited light-dependent chlorosis and necrosis [376]. This was attributed to the increased levels of GSSG and a failure of the redox sensing mechanism in chloroplasts of the transgenic plants. Transgenic poplar plants that overexpress γ-ECS in the cytosol accumulated 2.5–3 times higher Cd than WT plants in young leaves when grown on soil in laboratory conditions [250]. Reisinger et al. [377] showed that B. juncea overexpressing GS or γ-ECS showed enhanced tolerance to a variety of HMs (particularly As, Cd, Zn, and Pb) due to higher capacity of GSH synthesis as well as PC synthesis. Ivanova et al. [378] found that overexpression of the GSH1 gene in the cytosol led to a 2-fold increase of foliar GSH content and influenced the photosynthetic apparatus at different levels of organisation, that is, leaves, photosynthetic cells, and chloroplasts. Biomass accumulation of WT poplar hybrid (Populus tremula × Populus Alba) decreased in HM-contaminated soil by more than 30-fold, whereas transformants showed a 2-fold decrease. Thus, poplars overexpressing γ-ECS in the cytosol were more tolerant to HM stress.

However, Li et al. [379] showed that overexpression of either γ-ECS or GS in A. thaliana was not sufficient to increase accumulation of as or Hg in above-ground tissues, even though the plants were more resistant to these contaminants. Coexpression of two genes, γ-ECS and GS, showed an increase in the accumulation of Hg governed by enhanced concentrations of PCs as compared to single gene transformants. Based on their finding, it can be concluded that for the accumulation of some HMs, both GSH and PCs are a prerequisite for higher tolerance. Later on, Pomponi et al. [380] also showed that an increase in Cd tolerance and accumulation in tobacco plants over-expressing AtPCS1 was directly related to GSH availability. Increased tolerance to HM stress by overexpressing PCs synthetic genes has also been widely reported [381–384]. Poplar plants overexpressing the TaPCS1 gene produced more biomass and accumulated more Pb and Zn than WT plants when growing on mined soils [382]. Transgenic tobacco plants overexpressing the Thlaspi caerulescens phytochelatin synthetase gene (TcPCS) enhanced PC production in tobacco. Cd accumulation in the roots and shoots of TcPCS1 transgenic seedlings increased compared to the WT, while Cd translocation from roots to shoots was not affected under Cd treatment. The roots of the TcPCS1 transgenic tobacco seedlings were significantly longer than those of WT under Cd stress, indicating that TcPCS1 expression might increase Cd accumulation and tolerance in transgenic tobacco. Importantly, the MDA content in TcPCS1 plants was below that of WT.

Wojas et al. [383] further showed the fundamental role of GSH in transgenic tobacco plants overexpressing the AtPCS1 or CePCS gene. Transgenic plants under moderate stress (50 μM As) showed increased tolerance and accumulation; however, under a high level (200 μM), a reduction of γ-Glutamylcysteine and GSH was observed and PC level decreased in both leaf and root tissues indicating that HM tolerance depends on both the GSH content and the concentration of HMs. Importantly, transgenic Arabidopsis plants overexpressing the AtPCS1 gene showed high resistance to As but hypersensitivity to Cd [385], and critical analysis of different transgenic lines with various expression levels of PCS transgene expression has revealed that moderate expression of the transgene is desirable. Coexpression of the AsPCS1 and GSH1 genes in Arabidopsis thaliana (Columbia ecotype) showed better Cd and as tolerance as indicated by longer roots. Single-gene transgenic lines had higher tolerance to and accumulated more Cd and As than WT. Importantly, dual-gene transformants exhibited significantly higher tolerance and accumulated more Cd than single-gene transgenic lines. Simultaneous overexpression of AsPCS1 and GSH1 led to elevated total PC production in transgenicArabidopsis. These results indicate that coordinated induction of both GSH and PCs synthesis genes represents a highly promising new tool for phytoremediation [386]. Additionally, transgenic plants overexpressing the human MT-II gene in tobacco and oilseed rape (Brassica napus) enhanced Cd tolerance and Cd accumulation or modified Cd distribution. Growth of these transgenic seedlings was unaffected up to 100 μmol/L Cd [387] whereas control seedlings showed severe inhibition of root and shoot growth and leaf chlorosis. Sekhar et al. [388] showed that transgenic Arabidopsis plants overexpressing the pigeon pea type I MT gene (CcMT1) imparted increased Cu and Cd tolerance as evidence by higher plant biomass and Chl content besides profuse root growth compared to control plants. Transgenic plants expressing CcMT1 accumulated more Cu and Cd ions in roots than in shoots. However, no differences were observed in the expression levels of Cu (AtZIP2) and Cd (AtNRAMP) transporter genes of CcMT1-transgenics and control plants under both stressed and unstressed conditions. The overall results amply indicate that CcMT1 contributes to the detoxification of HM ions, thereby providing marked tolerance against metal stresses.

The involvement of the GHS-dependent glyoxalase pathway in HM tolerance was also reported by Singla-Pareek et al. [17]. Transgenic tobacco plants overexpressing both the Gly I and Gly II genes showed restricted MG accumulation and less lipid peroxidation at a high Zn concentration (5 mM $ZnCl_2$) and were able to survive, sustain growth, and flower and set normal viable seeds without any reduction in yield. The most striking observation was that double transgenic plants performed better than either the single-gene transformants as initially observed under salt stress. The endogenous ion content reflected that roots are the major sink for excess Zn accumulation, with a negligible amount in the seeds of transgenic plants. The use of the GSH biosynthetic inhibitor (BSO) revealed that higher PCs synthesizing capability and maintenance of the GSH/GSSG redox ratio in transgenic plants provided tolerance to Zn stress [17, 30]. They concluded that maintenance of glutathione homeostasis in transgenic plants was the possible mechanism behind tolerance to Zn toxicity.

The aforementioned results clearly indicate that use of transgenic plants overexpressing cysteine, GSH, PC, MT, or glyoxalase pathway-related genes showed substantial tolerance against HM stress. Therefore, transgenic approaches can be effectively used for the cleanup of HM pollution and as a promising approach for phytoremediation.

CONCLUSIONS AND FUTURE PERSPECTIVE

The numerous deleterious health effect upon exposure to toxic HMs in the environment is a matter of serious concern and a global issue. Therefore, molecular and cellular adaptation of plant cells in response to HM stress appears to be necessary to improve plant HM tolerance that ultimately reduces the chance of entering HM into the food chain. Plant HM tolerance is a multigenic traits and controlled at multiple levels. A wealth of recent studies demonstrated that plants protect themselves from HM toxicity, besides other mechanisms, through an elevated level of nonenzymatic and enzymatic components of antioxidant and glyoxalase defense systems [2, 3, 8, 11, 13, 17, 19, and 31]. Significant progress has been achieved in regards to HM toxicity and tolerance and different key components ensuring HM tolerance in plants have been identified; however, many key questions remain unanswered. Additionally, different HMs appear to have different mechanisms to elicit toxicity symptoms and plants employ different mechanisms as resistant reactions to particular HMs. Likewise, the reaction response and tolerance mechanisms are also different when a plant is grown under excess HMs in hydroponic or actual field conditions. Therefore, it is very difficult to hypothesize a common resistance mechanism against all HMs and metalloids. Importantly, among the various nonenzymatic antioxidant molecules that plants usually utilize against HM stress, GSH is the most vital one. Besides detoxification of ROS directly or detoxification of ROS and MG through antioxidative and glyoxalase systems, GSH also plays additional roles, including HM uptake, transport, and sequestration and formation of specific metal-binding ligands such as PCs. The reviewed literature confirms the central role of GSH metabolism in HM tolerance in plants as evidenced by proteomic, genomic, and transgenic studies over a range of plant species.

Although significant progress has been made in different aspects of GSH and its related enzymes (GST, GPX, DHAR, GR, Gly I, and Gly II) in HM-induced ROS and MG metabolism and HM chelation in a large number of plant species, most of the studies were conducted under laboratory conditions. Therefore, to get more realistic information we need to move from the laboratory to actual field conditions by using commercial plant species at various phases of plant growth. Likewise, the subcellular distribution of GSH and its related enzymes in model plant species may give deeper insight into compartment-specific HMs and metalloid-accrued reactions and resistance mechanisms and ROS and MG metabolism and other fundamental plant physiological process [13, 14]. Mutants, hyperaccumulators and naturally tolerant plant species give us important cues about a plant's response to HM stress, HM tolerance and metal-induced ROS and MG metabolism. Recently, a few genes responsible for acquired tolerance, to HM stress have already been identified. Ectopic expression of these genes related to GSH and its metabolizing enzymes in different species plant showed promising results. However, very few transgenic plants showed resistant reactions over a range of HMs [389, 390]. The increasing identification and study of the remarkable natural variation in the capacity of plants to accumulate and tolerate HMs is and will continue to provide a wealth of information. Therefore, concerted efforts by various research domains will further increase our understanding of the fundamental mechanisms involved in hyperaccumulation processes that naturally occur in metal hyperaccumulating plants. This should allow us to develop plants that are more ideally suited for phytoremediation of HM- contaminated soils [391].

ACKNOWLEDGMENTS

The authers apologize to their colleagues whose primary work could not be cited due to space constraints. Financial grant from Japan government (Monbukagakusho) is gratefully acknowledged.

REFERENCES

1. G. DalCorso, S. Farinati, S. Maistri, and A. Furini, "How plants cope with cadmium: staking all on metabolism and gene expression," Journal of Integrative Plant Biology, vol. 50, no. 10, pp. 1268–1280, 2008.
2. M. A. Hossain, M. Z. Hossain, and M. Fujita, "Stress-induced changes of methylglyoxal level and glyoxalase I activity in pumpkin seedlings and cDNA cloning of glyoxalase I gene," Australian Journal of Crop Science, vol. 3, no. 2, pp. 53–64, 2009.
3. M. A. Hossain, M. Hasanuzzaman, and M. Fujita, "Up-regulation of antioxidant and glyoxalase systems by exogenous glycinebetaine and proline in mung bean confer tolerance to cadmium stress," Physiology and Molecular Biology of Plants, vol. 16, no. 3, pp. 259–272, 2010.
4. N. Rascio and F. Navari-Izzo, "Heavy metal hyperaccumulating plants: how and why do they do it? And what makes them so interesting?" Plant Science, vol. 180, no. 2, pp. 169–181, 2011.

5. F. Villiers, C. Ducruix, V. Hugouvieux et al., "Investigating the plant response to cadmium exposure by proteomic and metabolomic approaches," Proteomics, vol. 11, no. 9, pp. 1650–1663, 2011.
6. G. Ramesh, Cloning and characterization of metallothionein genes of ectomycorrhizal fungus Hebeloma cylindrosporum, Ph.D. thesis, Thapar University, Punjab, India, 2008.
7. P. Sharma and R. S. Dubey, "Involvement of oxidative stress and role of antioxidative defense system in growing rice seedlings exposed to toxic concentrations of aluminum," Plant Cell Reports, vol. 26, no. 11, pp. 2027–2038, 2007.
8. S. S. Sharma and K. J. Dietz, "The relationship between metal toxicity and cellular redox imbalance," Trends in Plant Science, vol. 14, no. 1, pp. 43–50, 2009.
9. M. A. Hossain and M. Fujita, "Purification of glyoxalase I from onion bulbs and molecular cloning of its cDNA," Bioscience, Biotechnology and Biochemistry, vol. 73, no. 9, pp. 2007–2013, 2009.
10. M. A. Hossain and M. Fujita, "Regulatory role of components of ascorbate-glutathione (AsA-GSH) pathway in plant tolerance to oxidative stress," in Oxidative Stress in Plants: Causes, Consequences and Tolerance, N. A. Anjum, S. Umar, and A. Ahmed, Eds., IK International Publishing House Pvt. Ltd., New Delhi, India, 2011.
11. M. A. Hossain, M. D. Hossain, M. M. Rohman, J. A. T. da Silva, and M. Fujita, "Onion major compounds (flavonoids, organosulfurs) and highly expressed glutathione-related enzymes: possible physiological interaction, gene cloning and abiotic stress response," in Onion Consumption and Health, C. B. Aguirre and L. M. Jaramillo, Eds., Nova Science Publishers, New York, NY, USA, 2012.
12. Y. F. Tan, N. O'Toole, N. L. Taylor, and A. Harvey Millar, "Divalent metal ions in plant mitochondria and their role in interactions with proteins and oxidative stress-induced damage to respiratory function," Plant Physiology, vol. 152, no. 2, pp. 747–761, 2010.
13. R. S. Dubey, "Metal toxicity, oxidative stress and antioxidative defense system in plants," inReactive Oxygen Species and Antioxidants in Higher Plants, S. D. Gupta, Ed., pp. 177–203, CRC Press, Boca Raton, Fla, USA, 2011.

14. N. A. Anjum, I. Ahmad, I. Mohmood et al., "Modulation of glutathione and its related enzymes in plants' responses to toxic metals and metalloids-A review," Environmental and Experimental Botany, vol. 75, pp. 307–324, 2012.
15. S. K. Yadav, S. L. Singla-Pareek, M. Ray, M. K. Reddy, and S. K. Sopory, "Methylglyoxal levels in plants under salinity stress are dependent on glyoxalase I and glutathione," Biochemical and Biophysical Research Communications, vol. 337, no. 1, pp. 61–67, 2005.
16. S. K. Yadav, S. L. Singla-Pareek, M. K. Reddy, and S. K. Sopory, "Transgenic tobacco plants overexpressing glyoxalase enzymes resist an increase in methylglyoxal and maintain higher reduced glutathione levels under salinity stress," FEBS Letters, vol. 579, no. 27, pp. 6265–6271, 2005.
17. S. L. Singla-Pareek, S. K. Yadav, A. Pareek, M. K. Reddy, and S. K. Sopory, "Transgenic tobacco overexpressing glyoxalase pathway enzymes grow and set viable seeds in zinc-spiked soils," Plant Physiology, vol. 140, no. 2, pp. 613–623, 2006.
18. M. A. Hoque, M. Uraji, M. N. Akhter Banu, I. C. Mori, Y. Nakamura, and Y. Murata, "The effects of methylglyoxal on glutathione S-transferase from Nicotiana tabacum," Bioscience, Biotechnology and Biochemistry, vol. 74, no. 10, pp. 2124–2126, 2010.
19. M. A. Hossain, J. A. T. da Silva, and M. Fujita, "Glyoxalase system and reactive oxygen species detoxification system in plant abiotic stress response and tolerance: an intimate relationship," inAbiotic Stress in Plants-Mechanisms and Adaptations, A. K. Shanker and B. Venkateswarlu, Eds., pp. 235–266, INTECH-Open Access Publisher, Rijeka, Croatia, 2011.
20. R. Saito, H. Yamamoto, A. Makino, T. Sugimoto, and C. Miyake, "Methylglyoxal functions as Hill oxidant and stimulates the photoreductin of O_2 at photosystem I: a symptom of plant diabetes,"Plant, Cell and Environment, vol. 34, no. 9, pp. 1454–1464, 2011.
21. F. Navari-Izzo, "Thylakoid-bound and stromal antioxidative enzymes in wheat treated with excess copper," Physiologia Plantarum, vol. 104, no. 4, pp. 630–638, 1998.

22. M. C. Romero-Puertas, J. M. Palma, M. Gómez, L. A. Del Río, and L. M. Sandalio, "Cadmium causes the oxidative modification of proteins in pea plants," Plant, Cell and Environment, vol. 25, no. 5, pp. 677–686, 2002
23. D. Barconi, G. Bernardini, and A. Santucci, "Linking protein oxidation to environmental pollutants: redox proteome approaches," Journal of Proteomics, vol. 74, no. 11, pp. 2324–2337, 2011.
24. G. Dalcorso, S. Farinati, and A. Furini, "Regulatory networks of cadmium stress in plants," Plant Signaling and Behavior, vol. 5, no. 6, pp. 1–5, 2010.
25. C. Leyval, K. Turnau, and K. Haselwandter, "Effect of heavy metal pollution on mycorrhizal colonization and function: physiological, ecological and applied aspects," Mycorrhiza, vol. 7, no. 3, pp. 139–153, 1997.
26. C. S. Cobbett, "Phytochelatins and their roles in heavy metal detoxification," Plant Physiology, vol. 123, no. 3, pp. 825–832, 2000.
27. J. L. Hall, "Cellular mechanisms for heavy metal detoxification and tolerance," Journal of Experimental Botany, vol. 53, no. 366, pp. 1–11, 2002.
28. X. E. Yang, X. F. Jin, Y. Feng, and E. Islam, "Molecular mechanisms and genetic basis of heavy metal tolerance/hyperaccumulation in plants," Journal of Integrative Plant Biology, vol. 47, no. 9, pp. 1025–1035, 2005.
29. S. Clemens, "Toxic metal accumulation, responses toexposure andmechanisms oftolerance inplants," Biochimie, vol. 88, no. 11, pp. 1707–1719, 2006.
30. S. K. Yadav, "Heavy metals toxicity in plants: an overview on the role of glutathione and phytochelatins in heavy metal stress tolerance of plants," South African Journal of Botany, vol. 76, no. 2, pp. 167–179, 2010.
31. C. S. Seth, T. Remans, E. Keunen et al., "Phytoextraction of toxic metals: a central role for glutathione," Plant, Cell and Environment, vol. 35, no. 2, pp. 334–346, 2012.
32. K. J. Dietz, M. Bair, and U. Krämer, "Free radical and reactive oxygen species as mediators of heavy metal toxicity in plants," in

Heavy Metal Stress in Plants from Molecules to Ecosystems, M. N. V. Prasad and J. Hagemeyer, Eds., pp. 73–79, Spinger-Verlag, Berlin, Germany, 1999.

33. A. Schützendübel and A. Polle, "Plant responses to abiotic stresses: heavy metal-induced oxidative stress and protection by mycorrhization," Journal of Experimental Botany, vol. 53, no. 372, pp. 1351–1365, 2002.

34. E. Nieboer and D. H. S. Richardson, "The replacement of the nondescript term %heavy metals› by a biologically and chemically significant classification of metal ions," Environmental Pollution Series B: Chemical and Physical, vol. 1, no. 1, pp. 3–26, 1980.

35. G. F. Wildner and J. Henkel, "The effect of divalent metal ions on the activity of Mg^{2+} depleted ribulose-1, 5-bisphosphate oxygenase," Planta, vol. 146, no. 2, pp. 223–228, 1979.

36. F. Van Assche and H. Clijsters, "Inhibition of photosynthesis in Phaseolus vulgaris by treatment with toxic concentration of zinc: effect on ribulose-1,5-bisphosphate carboxylase/ oxygenase,"Journal of Plant Physiology, vol. 125, no. 3-4, pp. 355–360, 1986.

37. A. Rivetta, N. Negrini, and M. Cocucci, "Involvement of Ca^{2+}-calmodulin in Cd^{2+} toxicity during the early phases of radish (Raphanus sativus L.) Seed germination," Plant, Cell and Environment, vol. 20, no. 5, pp. 600–608, 1997.

38. A. A. Meharg, "The role of plasmalemma in metal tolerance in angiosperm," Physiologia Plantarum, vol. 88, no. 1, pp. 191–198, 1993.

39. M. A. Hossain and M. Fujita, "Evidence for a role of exogenous glycinebetaine and proline in antioxidant defense and methylglyoxal detoxification systems in mung bean seedlings under salt stress," Physiology and Molecular Biology of Plants, vol. 16, no. 1, pp. 19–29, 2010.

40. M. A. Hossain, M. Hasanuzzaman, and M. Fujita, "Coordinate induction of antioxidant defense and glyoxalase system by exogenous proline and glycinebetaine is correlated with salt tolerance in mung bean," Frontiers of Agriculture in China, vol. 5, no. 1, pp. 1–14, 2011.

41. C. D. Foy, R. L. Chaney, and M. C. White, "Physiology of metal toxicity in plants," Annual Review of Plant Physiology and Plant Molecular Biology, vol. 29, pp. 511–566, 1978.
42. L. M. Sandalio, H. C. Dalurzo, M. Gómez, M. C. Romero-Puertas, and L. A. Del Río, "Cadmium-induced changes in the growth and oxidative metabolism of pea plants," Journal of Experimental Botany, vol. 52, no. 364, pp. 2115–2126, 2001.
43. P. Carrier, A. Baryla, and M. Havaux, "Cadmium distribution and microlocalization in oilseed rape (Brassica napus) after long-term growth on cadmium-contaminated soil," Planta, vol. 216, no. 6, pp. 939–950, 2003.
44. E. Gamalero, G. Lingua, G. Berta, and B. R. Glick, "Beneficial role of plant growth promoting bacteria and arbuscular mycorrhizal fungi on plant responses to heavy metal stress," Canadian Journal of Microbiology, vol. 55, no. 5, pp. 501–514, 2009.
45. F. Fodor, "Physiological responses of vascular plants to heavy metals," in Physiology and Biochemistry of Metal Toxicity and Tolerance in Plants, M. N. V. Prasad and K. Strzalka, Eds., pp. 149–177, Kluwer Academic, Dortrech, UK, 2002.
46. C. Poschenrieder and J. Barceló, "Water relations in heavy metal stressed plants," in Heavy Metal Stress in Plants, M. N. V. Prasad, Ed., pp. 249–270, Springer, Berlin, Germany, 3rd edition, 2004.
47. M. P. Benavides, S. M. Gallego, and M. L. Tomaro, "Cadmium toxicity in plants," Brazilian Journal of Plant Physiology, vol. 17, no. 1, pp. 21–34, 2005.
48. Y. T. Hsu and C. H. Kao, "Cadmium toxicity is reduced by nitric oxide in rice leaves," Plant Growth Regulation, vol. 42, no. 3, pp. 227–238, 2004
49. M. Backor, P. Vaczi, M. Bartak, J. Budova, and A. Dzubaj, "Uptake, photosynthetic characteristics and membrane lipid peroxidation levels in the lichen photobiont Trebouxia erici exposed to copper and cadmium," Bryologist, vol. 110, no. 1, pp. 100–107, 2007.
50. B. Boddi, A. R. Oravecz, and E. Lehoczki, "Effect of cadmium on organization and photoreduction of protochlorophyllide in dark-grown leaves and etioplast inner membrane preparations of wheat," Photosynthetica, vol. 31, no. 3, pp. 411–420, 1995.

51. K. Shakya, M. K. Chettri, and T. Sawidis, "Impact of heavy metals (copper, zinc, and lead) on the chlorophyll content of some mosses," Archives of Environmental Contamination and Toxicology, vol. 54, no. 3, pp. 412–421, 2008. · ·
52. L. Perfus-Barbeoch, N. Leonhardt, A. Vavasseur, and C. Forestier, "Heavy metal toxicity: cadmium permeates through calcium channels and disturbs the plant water status," Plant Journal, vol. 32, no. 4, pp. 539–548, 2002. · ·
53. F. Vinit-Dunand, D. Epron, B. Alaoui-Sossé, and P. M. Badot, "Effects of copper on growth and on photosynthesis of mature and expanding leaves in cucumber plants," Plant Science, vol. 163, no. 1, pp. 53–58, 2002. · ·
54. A. Llamas, C. I. Ullrich, and A. Sanz, "Cd^{2+} effects on transmembrane electrical potential difference, respiration and membrane permeability of rice (Oryza sativa L) roots," Plant and Soil, vol. 219, no. 1-2, pp. 21–28, 2000.
55. A. Bhushan Jha and R. Shanker Dubey, "Arsenic exposure alters activity behaviour of key nitrogen assimilatory enzymes in growing rice plants," Plant Growth Regulation, vol. 43, no. 3, pp. 259–268, 2004. · ·
56. P. Sharma and R. S. Dubey, "Modulation of nitrate reductase activity in rice seedlings under aluminium toxicity and water stress: role of osmolytes as enzyme protectant," Journal of Plant Physiology, vol. 162, no. 8, pp. 854–864, 2005. · ·
57. R. Maheshwari and R. S. Dubey, "Nickel toxicity inhibits ribonuclease and protease activities in rice seedlings: protective effects of proline," Plant Growth Regulation, vol. 51, no. 3, pp. 231–243, 2007. · ·
58. K. S. Kasprzak, "Possible role of oxidative damage in metal-induced carcinogenesis," Cancer Investigation, vol. 13, no. 4, pp. 411–430, 1995.
59. A. Fusconi, O. Repetto, E. Bona et al., "Effects of cadmium on meristem activity and nucleus ploidy in roots of Pisum sativum L. cv. Frisson seedlings," Environmental and Experimental Botany, vol. 58, no. 1–3, pp. 253–260, 2006. · ·
60. W. H. O. Ernst, G. J. Krauss, J. A. C. Verkleij, and D. Wesenberg, "Interaction of heavy metals with the sulphur metabolism in

angiosperms from an ecological point of view," Plant, Cell and Environment, vol. 31, no. 1, pp. 123–143, 2008. ·

61. J. Deikman, "Molecular mechanisms of ethylene regulation of gene transcription," Physiologia Plantarum, vol. 100, no. 3, pp. 561–566, 1997. · ·

62. W. Maksymiec, "Effect of copper on cellular processes in higher plants," Photosynthetica, vol. 34, no. 3, pp. 321–342, 1997. · ·

63. A. J. Enyedi, N. Yalpani, P. Silverman, and I. Raskin, "Signal molecules in systemic plant resistance to pathogens and pests," Cell, vol. 70, no. 6, pp. 879–886, 1992. · ·

64. U. H. Cho and N. H. Seo, "Oxidative stress in Arabidopsis thaliana exposed to cadmium is due to hydrogen peroxide accumulation," Plant Science, vol. 168, no. 1, pp. 113–120, 2005. · ·

65. E. Skórzy ska-Polit, B. Pawlikowska-Pawl ga, E. Szczuka, M. Dr kiewicz, and Z. Krupa, "The activity and localization of lipoxygenases in Arabidopsis thaliana under cadmium and copper stresses," Plant Growth Regulation, vol. 48, no. 1, pp. 29–39, 2006. · ·

66. Y. Zhen, J. L. Qi, S. S. Wang et al., "Comparative proteome analysis of differentially expressed proteins induced by Al toxicity in soybean," Physiologia Plantarum, vol. 131, no. 4, pp. 542–554, 2007. · ·

67. L. Zhao, Y.-L. Sun, and S.-X. Cui, "Cd-induced changes in leaf proteome of the hyperaccumulator plant Phytolacca americana," Chemosphere, vol. 85, no. 1, pp. 56–66, 2011. ·

68. R. L. Sun, Q. X. Zhou, F. H. Sun, and C. X. Jin, "Antioxidative defense and proline/phytochelatin accumulation in a newly discovered Cd-hyperaccumulator, Solanum nigrum L," Environmental and Experimental Botany, vol. 60, no. 3, pp. 468–476, 2007. · ·

69. C. Jonak, H. Nakagami, and H. Hirt, "Heavy metal stress. Activation of distinct mitogen-activated protein kinase pathways by copper and cadmium," Plant Physiology, vol. 136, no. 2, pp. 3276–3283, 2004.

70. S. Clemens, M. G. Palmgren, and U. Krämer, "A long way ahead: understanding and engineering plant metal accumulation," Trends in Plant Science, vol. 7, no. 7, pp. 309–315, 2002. · ·

71. A. A. Meharg, "Integrated tolerance mechanisms: constitutive and adaptive plant responses to elevated metal concentrations in the environment," Plant, Cell and Environment, vol. 17, no. 9, pp. 989–993, 1994.
72. L. Sanità Di Toppi and R. Gabbrielli, "Response to cadmium in higher plants," Environmental and Experimental Botany, vol. 41, no. 2, pp. 105–130, 1999. · ·
73. X. F. Zhu, C. Zheng, Y. T. Hu et al., "Cadmium-induced oxalate secretion from root apex is associated with cadmium exclusion and resistance in Lycopersicon esulentum," Plant, Cell and Environment, vol. 34, no. 7, pp. 1055–1064, 2011. ·
74. J. M. Bubb and J. N. Lester, "The impact of heavy metals on lowland rivers and the implications for man and the environment," Science of the Total Environment, vol. 100, pp. 207–233, 1991.
75. D. M. Pellet, D. L. Grunes, and L. V. Kochian, "Organic acid exudation as an aluminum-tolerance mechanism in maize (Zea mays L.)," Planta, vol. 196, no. 4, pp. 788–795, 1995.
76. A. P. Pinto, I. Simões, and A. M. Mota, "Cadmium impact on root exudates of sorghum and maize plants: a speciation study," Journal of Plant Nutrition, vol. 31, no. 10, pp. 1746–1755, 2008. · ·
77. E. Delhaize, P. R. Ryan, and R. J. Randall, "Aluminum tolerance in wheat (Triticum aestivum L.). II.Aluminum-stimulated excretion of malic acid from root apices," Plant Physiology, vol. 103, no. 3, pp. 695–702, 1993.
78. J. W. Huang, D. M. Pellet, L. A. Papernik, and L. V. Kochian, "Aluminum interactions with voltage-dependent calcium transport in plasma membrane vesicles isolated from roots of aluminum-sensitive and -resistant wheat cultivars," Plant Physiology, vol. 110, no. 2, pp. 561–569, 1996.
79. C. C. Chen, J. B. Dixon, and F. T. Turner, "Iron coatings on rice roots: morphology and models of development," Soil Science Society of America Journal, vol. 44, pp. 1113–1119, 1980.
80. H. Liu, J. Zhang, P. Christie, and F. Zhang, "Influence of iron plaque on uptake and accumulation of Cd by rice (Oryza sativa L.) seedlings grown in soil," Science of the Total Environment, vol. 394, no. 2-3, pp. 361–368, 2008. · ·

81. N. Smirnoff and G. R. Stewart, "Nitrogen assimilation and zinc toxicity to zinc-tolerant and non-tolerant clones of Deschampsia cespitosa (L.) Beauv," New Phytologist, vol. 107, no. 4, pp. 671–680, 1987.

82. H. Harmens, N. G. C. P. B. Gusmão, P. R. D. Hartog, J. A. C. Verkeij, and W. H. O. Ernst, "Uptake and transport of zinc in zinc-sensitive and zinc-tolerant Silene vulgaris," Journal of Plant Physiology, vol. 141, no. 3, pp. 309–315, 1993.

83. K. R. Tice, D. R. Parker, and D. A. DeMason, "Operationally defined apoplastic and symplastic aluminum fractions in root tips of aluminum-intoxicated wheat," Plant Physiology, vol. 100, no. 1, pp. 309–318, 1992.

84. K. Iwasaki, K. Sakurai, and E. Takahashi, "Copper binding by the root cell walls of Italian ryegrass and red clover," Soil Science and Plant Nutrition, vol. 36, no. 3, pp. 431–439, 1990.

85. W. J. Horst and H. Marschner, "Effect of excessive manganese supply on uptake and translocation of calcium in bean plants (Phaseolus vulgaris L.)," Zeitschrift fur Planzenphysiologie, vol. 87, no. 2-3, pp. 137–148, 1978.

86. A. Masion and P. M. Bertsch, "Aluminium speciation in the presence of wheat root cell walls: a wet chemical study," Plant, Cell and Environment, vol. 20, no. 4, pp. 504–512, 1997.

87. J. A. Qureshi, D. A. Thurman, K. Hardwick, and H. A. Collin, "Uptake and accumulation of zinc, lead and copper in zinc and lead tolerant Anthoxanthum odoratum L," New Phytologist, vol. 100, no. 3, pp. 429–434, 1985.

88. D. M. Wheeler and I. L. Power, "Comparison of plant uptake and plant toxicity of various ions in wheat," Plant and Soil, vol. 172, no. 2, pp. 167–173, 1995.

89. S. J. Zheng, J. F. Ma, and H. Matsumoto, "High aluminum resistance in buckwheat: I. Al-induced specific secretion of oxalic acid from root tips," Plant Physiology, vol. 117, no. 3, pp. 745–751, 1998.

90. W. J. Horst, "Factors responsible for genotypic manganese tolerance in cowpea (Vigna unguiculata)," Plant and Soil, vol. 72, no. 2-3, pp. 213–218, 1983.

91. S. Clemens, "Molecular mechanisms of plant metal tolerance and homeostasis," Planta, vol. 212, no. 4, pp. 475–486, 2001. · ·
92. A. Brookes, J. C. Collins, and D. A. Thurman, "The mechanism of zinc tolerance in grasses,"Journal of Plant Nutrition, vol. 3, no. 1–4, pp. 695–705, 1981. ·
93. M. Lee, K. Lee, J. Lee, E. W. Noh, andY. Lee, "AtPDR12 contributes to lead resistance inArabidopsis," Plant Physiology, vol. 138, no. 2, pp. 827–836, 2005. · ·
94. H. C. Chiang, J. C. Lo, and K. C. Yeh, "Genes associated with heavy metal tolerance and accumulation in Zn/Cd hyperaccumulator Arabidopsis halleri: a genomic survey with cDNA microarray," Environmental Science and Technology, vol. 40, no. 21, pp. 6792–6798, 2006. · ·
95. U. Krämer, I. Talke, and M. Hanikenne, "Transition metal transport," Federation of European Biochemical Societies Letters, vol. 581, no. 12, pp. 2263–2272, 2007.
96. W. E. Rauser, "Structure and function of metal chelators produced by plants: the case for organic acids, amino acids, phytin, and metallothioneins," Cell Biochemistry and Biophysics, vol. 31, no. 1, pp. 19–48, 1999.
97. A. Brune, W. Urbach, and K. J. Dietz, "Compartmentation and transport of zinc in barley primary leaves as basic mechanisms involved in zinc tolerance," Plant, Cell and Environment, vol. 17, no. 2, pp. 153–162, 1994.
98. R. Vögeli-Lange and G. J. Wagner, "Subcellular localization of cadmium and cadmium-binding peptides in tobacco leaves: implication of a transport function for cadmium-binding peptides,"Plant Physiology, vol. 92, no. 4, pp. 1086–1093, 1990.
99. J. Wang, B. P. Evangelou, M. T. Nielsen, and G. J. Wagner, "Computer, simulated evaluation of possible mechanisms for sequestering metal ion activity in plant vacuoles: II. Zinc," Plant Physiology, vol. 99, no. 2, pp. 621–626, 1992.
100. J. Wang, B. P. Evangelou, M. T. Nielsen, and G. J. Wagner, "Computer-simulated evaluation of possible mechanisms for quenching heavy metal ion activity in plant vacuoles: I. Cadmium,"Plant Physiology, vol. 97, no. 3, pp. 1154–1160, 1991.

101. M. H. Zenk, "Heavy metal detoxification in higher plants - A review," Gene, vol. 179, no. 1, pp. 21–30, 1996. · ·

102. E. Grill, E. L. Winnacker, and M. H. Zenk, "Phytochelatins, a class of heavy-metal-binding peptides from plants, are functionally analogous to metallothioneins," Proceedings of the National Academy of Sciences of the United States of America, vol. 84, no. 2, pp. 439–443, 1987.

103. E. Grill, S. Loffler, E. L. Winnacke, and M. H. Zenk, "Phytochelatins, the heavy-metal-binding peptides of plants, are synthesized from glutathione by a specific γ-glutamylcysteine dipeptidyl transpeptidase (phytochelatin synthase)," Proceeding of the National Academy of Science of the United States of America, vol. 86, no. 18, pp. 6838–6842, 1989.

104. D. G. Mendoza-Cózatl, E. Butko, F. Springer et al., "Identification of high levels of phytochelatins, glutathione and cadmium in the phloem sap of Brassica napus. A role for thiol-peptides in the long-distance transport of cadmium and the effect of cadmium on iron translocation," Plant Journal, vol. 54, no. 2, pp. 249–259, 2008. · ·

105. M. Fujita, "The presence of two Cd-binding components in the roots of water hyacinth cultivated in a Cd^{2+}-containing medium," Plant Cell Physiology, vol. 26, no. 2, pp. 295–300, 1985.

106. M. Fujita and T. Kawanishi, "Purificatin and characterization of a Cd-binding complexes from the root tissues of various higher plants cultivated in Cd^{2+}-containing medium," Plant and Cell Physiology, vol. 27, no. 7, pp. 1317–1325, 1986.

107. S. Iglesia-Turiño, A. Febrero, O. Jauregui, C. Caldelas, J. L. Araus, and J. Bort, "Detection and quantification of unbound phytochelatin 2 in plant extracts of Brassica napus grown with different levels of mercury," Plant Physiology, vol. 142, no. 2, pp. 742–749, 2006. · ·

108. E. Grill, J. Thumann, E. L. Winnacker, and M. H. Zenk, "Induction of heavy-metal binding phytochelatins by inoculation of cell cultures in standard media," Plant Cell Reports, vol. 7, no. 6, pp. 375–378, 1988.

109. J. Thumann, E. Grill, E. L. Winnacker, and M. H. Zenk, "Reactivation of metal-requiring apoenzymes by phytochelatin-metal complexes," FEBS Letters, vol. 284, no. 1, pp. 66–69,

1991. ·

110. I. Nouairi, W. Ben Ammar, N. Ben Youssef, D. D. Ben Miled, M. H. Ghorbal, and M. Zarrouk, "Antioxidant defense system in leaves of Indian mustard (Brassica juncea) and rape (Brassica napus) under cadmium stress," Acta Physiologiae Plantarum, vol. 31, no. 2, pp. 237–247, 2009. · ·
111. H. Zhang, W. Xu, J. Guo, Z. He, and M. Ma, "Coordinated responses of phytochelatins and metallothioneins to heavy metals in garlic seedlings," Plant Science, vol. 169, no. 6, pp. 1059–1065, 2005. · ·
112. N. Tsuji, N. Hirayanagi, M. Okada et al., "Enhancement of tolerance to heavy metals and oxidative stress in Dunaliella tertiolecta by Zn-induced phytochelatin synthesis," Biochemical and Biophysical Research Communications, vol. 293, no. 1, pp. 653–659, 2002. · ·
113. U. Krämer, J. D. Cotter-Howells, J. M. Charnock, A. J. M. Baker, and J. A. C. Smith, "Free histidine as a metal chelator in plants that accumulate nickel," Nature, vol. 379, no. 6566, pp. 635–638, 1996. · ·
114. Z. G. Shen, F. J. Zhao, and S. P. McGrath, "Uptake and transport of zinc in the hyperaccumulatorThlaspi caerulescens and the non-hyperaccumulator Thlaspi ochroleucum," Plant, Cell and Environment, vol. 20, no. 7, pp. 898–906, 1997.
115. I. Leopold, D. Günther, J. Schmidt, and D. Neumann, "Phytochelatins and heavy metal tolerance," Phytochemistry, vol. 50, no. 8, pp. 1323–1328, 1999. · ·
116. D. E. Salt and W. E. Rauser, "MgATP-dependent transport of phytochelatins across the tonoplast of oat roots," Plant Physiology, vol. 107, no. 4, pp. 1293–1301, 1995.
117. A. J. M. Baker, S. P. McGrath, R. D. Reeves, and J. A. C. Smith, "Metal hyperaccumulator plants: a review of the ecology and physiology of a biological resource for phytoremedation of metal polluted soils," in Phytoremediation of Contaminated Soil and Water, N. Terry and G. Banuelos, Eds., pp. 85–107, Michael Lewis, Boca Raton, Fla, USA, 2000.
118. D. H. Hamer, "Metallothionein," Annual Review of Biochemistry, vol. 55, pp. 913–951, 1986.

119. K. Gasic and S. Korban, "Heavy metal stress," in Physiology and Molecular Biology of Stress Tolerance in Plants, K. V. M. Rao, A. S. Raghavendra, and K. J. Reddy, Eds., pp. 219–254, Springer, Dordrecht, the Netherlands, 2006.
120. G. Zhou, Y. Xu, J. Li, L. Yang, and J.-Y. Liu, "Molecular analyses of the metallothionein gene family in rice (Oryza sativa L.)," Journal of Biochemistry and Molecular Biology, vol. 39, no. 5, pp. 595–606, 2006.
121. J. H. R. Kagi, "Overview of metallothionein," Methods in Enzymology, vol. 205, pp. 613–626, 1991.
122. S. Castiglione, C. Franchin, T. Fossati, G. Lingua, P. Torrigiani, and S. Biondi, "High zinc concentrations reduce rooting capacity and alter metallothionein gene expression in white poplar (Populus alba L. cv. Villafranca)," Chemosphere, vol. 67, no. 6, pp. 1117–1126, 2007.
123. Y. O. Ahn, S. H. Kim, J. Lee, H. R. Kim, H.-S. Lee, and S.-S. Kwak, "Three Brassica rapametallothionein genes are differentially regulated under various stress conditions," Molecular Biology Reports, vol. 39, no. 3, pp. 2059–2067, 2012.
124. R. Jabeen, A. Ahmad, and M. Iqbal, "Phytoremediation of heavy metals: physiological and molecular mechanisms," Botanical Review, vol. 75, no. 4, pp. 339–364, 2009.
125. D. Singh and S. K. Chauhan, "Organic acids of crop plants in alumium detoxification," Current Science, vol. 100, no. 10, pp. 1509–1515, 2011.
126. J. Lee, R. D. Reeves, R. R. Brooks, and T. Jaffré, "Isolation and identification of a citrato-complex of nickel from nickel-accumulating plants," Phytochemistry, vol. 16, no. 10, pp. 1503–1505, 1977.
127. J. Lee, R. D. Reeves, R. R. Brooks, and T. Jaffré, "The relation between nickel and citric acid in some nickel-accumulating plants," Phytochemistry, vol. 17, no. 6, pp. 1033–1035, 1978.
128. S. C. Miyasaka, J. George Buta, R. K. Howell, and C. D. Foy, "Mechanism of aluminum tolerance in snapbeans: root exudation of citric acid," Plant Physiology, vol. 96, no. 3, pp. 737–743, 1991.

129. G. J. Wagner, "Accumulation of Cadmium in Crop Plants And Its Consequences to Human Health," Advances in Agronomy, vol. 51, no. C, pp. 173–212, 1993. · ·

130. D. L. Godbold, W. J. Horst, J. C. Collins, D. A. Thurman, and H. Marschner, "Accumulation of zinc and organic acids in roots of zinc tolerant and non-tolerant ecotypes of Deschampsia caespitosa," International Journal of Plant Physiology, vol. 116, no. 1, pp. 59–69, 1984.

131. M. Oven, E. Grill, A. Golan-Goldhirsh, T. M. Kutchan, and M. H. Zenk, "Increase of free cysteine and citric acid in plant cells exposed to cobalt ions," Phytochemistry, vol. 60, no. 5, pp. 467–474, 2002. · ·

132. J. F. Ma, P. R. Ryan, and E. Delhaize, "Aluminium tolerance in plants and the complexing role of organic acids," Trends in Plant Science, vol. 6, no. 6, pp. 273–278, 2001. · ·

133. Jian Feng Ma, Shao Jian Zheng, H. Matsumoto, and S. Hiradate, "Detoxifying aluminium with buckwheat," Nature, vol. 390, no. 6660, pp. 569–570, 1997.

134. S. Zheng, J. F. Ma, and H. Matsumoto, "High alumium resistance in buckwheat: I. Al induced specific secretion of oxalic acid from root tip tips," Plant Physiology, vol. 117, no. 3, pp. 745–751, 1997.

135. W. Mathys, "The role of malate, oxalate, and mustard oil glucosides in the evolution of zinc-resistance in herbage plants," Physiologia Plantarum, vol. 40, no. 2, pp. 130–136, 1977. · ·

136. W. H. Zhang, P. R. Ryan, and S. D. Tyerman, "Malate-permeable channels and cation channels activated by aluminum in the apical cells of wheat roots," Plant Physiology, vol. 125, no. 3, pp. 1459–1472, 2001. · ·

137. F. S. Salazar, S. Pandey, L. Narro et al., "Diallel analysis of acid-soil tolerant and intolerant tropical maize populations," Crop Science, vol. 37, no. 5, pp. 1457–1462, 1997.

138. X. F. Li, J. F. Ma, and H. Matsumoto, "Pattern of aluminum-induced secretion of organic acids differs between rye and wheat," Plant Physiology, vol. 123, no. 4, pp. 1537–1543, 2000.

139. V. K. Rai, "Role of amino acids in plant responses to stresses," Biologia Plantarum, vol. 45, no. 4, pp. 481–487, 2002. · ·

140. D. E. Salt, R. C. Prince, A. J. M. Baker, I. Raskin, and I. J. Pickering, "Zinc legands in the metal hyperaccumulatorThalspi caerulescens as determined using X-absorption spectroscopy,"Environmental Science and Technology, vol. 33, no. 5, pp. 713–717, 1999.
141. U. Krämer, J. D. Cotter-Howells, J. M. Charnock, A. J. M. Baker, and J. A. C. Smith, "Free histidine as a metal chelator in plants that accumulate nickel," Nature, vol. 379, no. 6566, pp. 635–638, 1996. · ·
142. K. Wycisk, E. J. Kim, J. I. Schroeder, and U. Krämer, "Enhancing the first enzymatic step in the histidine biosynthesis pathway increases the free histidine pool and nickel tolerance inArabidopsis thaliana," FEBS Letters, vol. 578, no. 1-2, pp. 128–134, 2004. ·
143. K. Higuchi, K. Suzuki, H. Nakanishi, H. Yamaguchi, N. K. Nishizawa, and S. Mori, "Cloning of nicotianamine synthase genes, novel genes involved in the biosynthesis of phytosiderophores,"Plant Physiology, vol. 119, no. 2, pp. 471–479, 1999.
144. H. Q. Ling, G. Koch, H. Bäumlein, and M. W. Ganal, "Map-based cloning of chloronerva, a gene involved in iron uptake of higher plants encoding nicotianamine synthase," Proceedings of the National Academy of Sciences of the United States of America, vol. 96, no. 12, pp. 7098–7103, 1999. · ·
145. A. Pich, R. Manteuffel, S. Hillmer, G. Scholz, and W. Schmidt, "Fe homeostasis in plant cells: does nicotianamine play multiple roles in the regulation of cytoplasmic Fe concentration?"Planta, vol. 213, no. 6, pp. 967–976, 2001.
146. M. Takahashi, Y. Terada, I. Nakai et al., "Role of nicotianamine in the intracellular delivery of metals and plant reproductive development," Plant Cell, vol. 15, no. 6, pp. 1263–1280, 2003.
147. M. Weber, E. Harada, C. Vess, E. V. Roepenack-Lahaye, and S. Clemens, "Comparative microarray analysis of Arabidopsis thaliana and Arabidopsis halleri roots identifies nicotianamine synthase, a ZIP transporter and other genes as potential metal hyperaccumulation factors," Plant Journal, vol. 37, no. 2, pp. 269–281, 2004.
148. M. D. Vazquez, C. Poschenrieder, I. Corrales, and J. Barcelo, "Change in apoplastic aluminum during the initial growth

response to aluminum by roots of a tolerant maize variety," Plant Physiology, vol. 119, no. 2, pp. 435–444, 1999.

149. R. F. M. van Steveninck, M. E. van Steveninck, D. R. Fernando, W. J. Horst, and H. Marschner, "Deposition of zinc phytate in globular bodies in roots of Deschampsia caespitosa; a detoxification mechanism?" Journal of Plant Physiology, vol. 131, no. 3-4, pp. 247–257, 1987.

150. D. E. Salt, M. Blaylock, N. P. B. A. Kumar et al., "Phytoremediation: a novel strategy for the removal of toxic metals from the environment using plants," Biotechnology, vol. 13, no. 5, pp. 468–474, 1995.

151. D. E. Salt and U. Krämer, "Mechanisms of metal hyperaccumulation in plants," inPhytoremediation of Toxic Metals-Using Plants to Clean Up the Environment, I. Raskin and B. D. Ensley, Eds., pp. 231–246, Wiley, New York, NY, USA, 2000.

152. N. S. Pence, P. B. Larsen, S. D. Ebbs et al., "The molecular physiology of heavy metal transport in the Zn/Cd hyperaccumulator Thlaspi caerulescens," Proceedings of the National Academy of Sciences of the United States of America, vol. 97, no. 9, pp. 4956–4960, 2000.

153. E. Lombi, F. J. Zhao, S. J. Dunham, and S. P. McGrath, "Phytoremediation of heavy metal-contaminated soils: natural hyperaccumulation versus chemically enhanced phytoextraction,"Journal of Environmental Quality, vol. 30, no. 6, pp. 1919–1926, 2001.

154. M. M. Lasat, N. S. Pence, D. F. Garvin, S. D. Ebbs, and L. V. Kochian, "Molecular physiology of zinc transport in the Zn hyperaccumulator Thlaspi caerulescens," Journal of Experimental Botany, vol. 51, no. 342, pp. 71–79, 2000.

155. E. Pilon-Smits and M. Pilon, "Phytoremediation of metals using transgenic plants," Critical Reviews in Plant Sciences, vol. 21, no. 5, pp. 439–456, 2002.

156. N. von Wiren, S. Klair, S. Bansal, et al., "Nicotinamide chelates both Fe III and Fe II. Implication of a transport function for cadmium-binding pepties," Plant Physiology, vol. 119, no. 3, pp. 1107–1114, 1999.

157. D. Neumann, O. Lichtenberger, D. Gunther, K. Tschiersch, and L. Nover, "Heat-shock proteins induce heavy-metal tolerance in higher plants," Planta, vol. 194, no. 3, pp. 360–367, 1994.

158. D. Neumann, U. Zur Nieden, O. Lichtenberger, and I. Leopold, "How does Armeria maritima tolerate high heavy metal concentrations?" Journal of Plant Physiology, vol. 146, no. 5-6, pp. 704–717, 1995.

159. J. E. Sarry, L. Kuhn, C. Ducruix et al., "The early responses of Arabidopsis thaliana cells to cadmium exposure explored by protein and metabolite profiling analyses," Proteomics, vol. 6, no. 7, pp. 2180–2198, 2006. · ·

160. Y. Zhen, J. L. Qi, S. S. Wang et al., "Comparative proteome analysis of differentially expressed proteins induced by Al toxicity in soybean," Physiologia Plantarum, vol. 131, no. 4, pp. 542–554, 2007. · ·

161. H. E. Ireland, S. J. Harding, G. A. Bonwick, M. Jones, C. J. Smith, and J. H. H. Williams, "Evaluation of heat shock protein 70 as a biomarker of environmental stress in Fucus serratus andLemna minor," Biomarkers, vol. 9, no. 2, pp. 139–155, 2004. ·

162. M. P. de Souza, E. A. H. Pilon-Smits, and N. Terry, "The physiology and biochemistry of selenium volatilization by plants," in Phytoremediation of Toxic Metal-Using Plants to Clean up the Environment, I. Raskin and B. D. Ensley, Eds., pp. 171–190, Wiley, New York, NY, USA, 2000.

163. C. M. Lytle, P. W. Lytle, N. Yang et al., "Reduction of Cr(VI) to Cr(III) by wetland plants: potential for in situ heavy metal detoxification," Environmental Science and Technology, vol. 32, no. 20, pp. 3087–3093, 1998. · ·

164. N. J. Robinson, C. M. Procter, E. L. Connolly, and M. L. Guerinot, "A ferric-chelate reductase for iron uptake from soils," Nature, vol. 397, no. 6721, pp. 694–697, 1999. · ·

165. N. Fusco, L. Micheletto, G. Dal Corso, L. Borgato, and A. Furini, "Identification of cadmium-regulated genes by cDNA-AFLP in the heavy metal accumulator Brassica juncea L," Journal of Experimental Botany, vol. 56, no. 421, pp. 3017–3027, 2005. · ·

166. J. E. Van De Mortel, H. Schat, P. D. Moerland et al., "Expression differences for genes involved in lignin, glutathione and sulphate metabolism in response to cadmium in Arabidopsis thaliana and the related Zn/Cd-hyperaccumulator Thlaspi caerulescens," Plant, Cell and Environment, vol. 31, no. 3, pp. 301–324, 2008. · ·

167. K. B. Singh, R. C. Foley, and L. Oñate-Sánchez, "Transcription factors in plant defense and stress responses," Current Opinion in Plant Biology, vol. 5, no. 5, pp. 430–436, 2002. · ·

168. G. Noctor and C. H. Foyer, "Ascorbate and glutathione: keeping active oxygen under control,"Annual Review of Plant Biology, vol. 49, pp. 249–279, 1998.

169. S. L. Singla-Pareek, M. K. Reddy, and S. K. Sopory, "Genetic engineering of the glyoxalase pathway in tobacco leads to enhanced salinity tolerance," Proceedings of the National Academy of Sciences of the United States of America, vol. 100, no. 25, pp. 14672–14677, 2003. · ·

170. Q. Shi and Z. Zhu, "Effects of exogenous salicylic acid on manganese toxicity, element contents and antioxidative system in cucumber," Environmental and Experimental Botany, vol. 63, no. 1–3, pp. 317–326, 2008. · ·

171. Z. S. Zhou, K. Guo, A. A. Elbaz, and Z. M. Yang, "Salicylic acid alleviates mercury toxicity by preventing oxidative stress in roots of Medicago sativa," Environmental and Experimental Botany, vol. 65, no. 1, pp. 27–34, 2009. · ·

172. J. L. Freeman, M. W. Persans, K. Nieman et al., "Increased glutathione biosynthesis plays a role in nickel tolerance in Thlaspi nickel hyperaccumulators W inside box sign," Plant Cell, vol. 16, no. 8, pp. 2176–2191, 2004. · ·

173. A. Metwally, I. Finkemeier, M. Georgi, and K. J. Dietz, "Salicylic acid alleviates the cadmium toxicity in barley seedlings," Plant Physiology, vol. 132, no. 1, pp. 272–281, 2003. · ·

174. Y. S. Wang, J. Wang, Z. M. Yang et al., "Salicylic acid modulates aluminum-induced oxidative stress in roots of Cassia tora," Acta Botanica Sinica, vol. 46, no. 7, pp. 819–828, 2004.

175. R. Bassi and S. S. Sharma, "Changes in Proline Content Accompanying the Uptake of Zinc and Copper by Lemna minor,"

Annals of Botany, vol. 72, no. 2, pp. 151–154, 1993. · ·

176. R. Bassi and S. S. Sharma, "Proline accumulation in wheat seedlings exposed to zinc and copper,"Phytochemistry, vol. 33, no. 6, pp. 1339–1342, 1993.
177. V. V. Talanova, A. F. Titov, and N. P. Boeva, "Effect of increasing concentrations of lead and cadmium on cucumber seedlings," Biologia Plantarum, vol. 43, no. 3, pp. 441–444, 2000.
178. S. Siripornadulsil, S. Traina, D. P. S. Verma, and R. T. Sayre, "Molecular mechanisms of proline-mediated tolerance to toxic heavy metals in transgenic microalgae," Plant Cell, vol. 14, no. 11, pp. 2837–2847, 2002. · ·
179. G. Y. Huang, Y. S. Wang, C. C. Sun, J. D. Dong, and Z. X. Sun, "The effect of multiple heavy metals on ascorbate, glutathione and related enzymes in two mangrove plant seedlings (Kandelia candel and Bruguiera gymnorrhiza)," Oceanological and Hydrobiological Studies, vol. 39, no. 1, pp. 11–25, 2010. · ·
180. S. S. Hussain, M. Ali, M. Ahmad, and K. H. M. Siddique, "Polyamines: natural and engineered abiotic and biotic stress tolerance in plants," Biotechnology Advances, vol. 29, no. 3, pp. 300–311, 2011. ·
181. T. Kusano, T. Berberich, C. Tateda, and Y. Takahashi, "Polyamines: Essential factors for growth and survival," Planta, vol. 228, no. 3, pp. 367–381, 2008. · ·
182. C. C. Lin and C. H. Kao, "Excess copper induces an accumulation of putrescine in rice leaves,"Botanical Bulletin of Academia Sinica, vol. 40, no. 3, pp. 213–218, 1999.
183. H. C. Ha, N. S. Sirisoma, P. Kuppusamy, J. L. Zweier, P. M. Woster, and R. A. Casero, "The natural polyamine spermine functions directly as a free radical scavenger," Proceedings of the National Academy of Sciences of the United States of America, vol. 95, no. 19, pp. 11140–11145, 1998. · ·
184. M. D. Groppa and M. P. Benavides, "Polyamines and abiotic stress: recent advances," Amino Acids, vol. 34, no. 1, pp. 35–45, 2008. · ·
185. Y. Y. Leshem and P. J. C. Kuiper, "Is there a GAS (general adaptation syndrome) response to various types of environmental stress?" Biologia Plantarum, vol. 38, no. 1, pp. 1–18, 1996.

186. M. N. V. Prasad and K. Strzalka, "Impact of heavy metals on photosynthesis," in Heavy Metal Stress in Plants, M. N. V. Prasad and J. Hagemeyer, Eds., pp. 117–138, Springer, Berlin, Germany, 1999.
187. M. D. Groppa, M. S. Zawoznik, M. L. Tomaro, and M. P. Benavides, "Inhibition of root growth and polyamine metabolism in sunflower (Helianthus annuus) seedlings under cadmium and copper stress," Biological Trace Element Research, vol. 126, no. 1–3, pp. 246–256, 2008. · ·
188. M. D. Groppa, E. P. Rosales, M. F. Iannone, and M. P. Benavides, "Nitric oxide, polyamines and Cd-induced phytotoxicity in wheat roots," Phytochemistry, vol. 69, no. 14, pp. 2609–2615, 2008. · ·
189. V. Prabhavathi and M. V. Rajam, "Mannitol-accumulating transgenic eggplants exhibit enhanced resistance to fungal wilts," Plant Science, vol. 173, no. 1, pp. 50–54, 2007. · ·
190. X. P. Wen, X. M. Pang, N. Matsuda, et al., "Inhibition of root growth and polyamine metabolism in sunflower (Helianthus annuus) seedlings under cadmium and copper stress over-expression of the apple spermidine synthase gene in pear confers multiple abiotic stress tolerance by altering polyamine titers," Transgenic Research, vol. 17, no. 2, pp. 251–263, 2008.
191. C. Franchin, T. Fossati, E. Pasquini et al., "High concentrations of zinc and copper induce differential polyamine responses in micropropagated white poplar (Populus alba)," Physiologia Plantarum, vol. 130, no. 1, pp. 77–90, 2007. · ·
192. L. Lamattina, C. García-Mata, M. Graziano, and G. Pagnussat, "Nitric oxide: the versatility of an extensive signal molecule," Annual Review of Plant Biology, vol. 54, pp. 109–136, 2003. · ·
193. L. A. Del Río, F. J. Corpas, and J. B. Barroso, "Nitric oxide and nitric oxide synthase activity in plants," Phytochemistry, vol. 65, no. 7, pp. 783–792, 2004. · ·
194. M. C. Palmieri, S. Sell, X. Huang et al., "Nitric oxide-responsive genes and promoters inArabidopsis thaliana: a bioinformatics approach," Journal of Experimental Botany, vol. 59, no. 2, pp. 177–186, 2008. · ·

195. Q. Shi, F. Ding, X. Wang, and M. Wei, "Exogenous nitric oxide protect cucumber roots against oxidative stress induced by salt stress," Plant Physiology and Biochemistry, vol. 45, no. 8, pp. 542–550, 2007. · ·
196. C. Zheng, D. Jiang, F. Liu et al., "Exogenous nitric oxide improves seed germination in wheat against mitochondrial oxidative damage induced by high salinity," Environmental and Experimental Botany, vol. 67, no. 1, pp. 222–227, 2009. · ·
197. M. H. Siddiqui, M. H. Al-Whaibi, and M. O. Basalah, "Role of nitric oxide in tolerance of plants to abiotic stress," Protoplasma, vol. 248, no. 3, pp. 447–455, 2011. ·
198. P. A. Karplus, M. J. Daniels, and J. R. Herriot, "Atomic structure of ferredoxin-NADPH C reductase, prototype for a structurally novel flavoenzyme family," Science, vol. 251, no. 4989, pp. 60–66, 1999.
199. J. Xu, W. Wang, J. Sun et al., "Involvement of auxin and nitric oxide in plant Cd-stress responses," Plant and Soil, vol. 346, no. 1, pp. 107–119, 2011. ·
200. C. Xiumin, Z. Yikai, C. Xiuling, J. Hong, and W. Xiaobin, "Effects of exogenous nitric oxide protects tomato plants under copper stress," in Proceedings of the 3rd International Conference on Bioinformatics and Biomedical Engineering (ICBBE ‹09), pp. 1–7, Beijing, China, June 2009. · ·
201. J. Xu, H. Yin, and X. Li, "Protective effects of proline against cadmium toxicity in micropropagated hyperaccumulator, Solanum nigrum L," Plant Cell Reports, vol. 28, no. 2, pp. 325–333, 2009. · ·
202. N. V. Laspina, M. D. Groppa, M. L. Tomaro, and M. P. Benavides, "Nitric oxide protects sunflower leaves against Cd-induced oxidative stress," Plant Science, vol. 169, no. 2, pp. 323–330, 2005. · ·
203. K. Apel and H. Hirt, "Reactive oxygen species: metabolism, oxidative stress, and signal transduction," Annual Review of Plant Biology, vol. 55, pp. 373–399, 2004. · ·
204. R. Mittler, S. Vanderauwera, M. Gollery, and F. Van Breusegem, "Reactive oxygen gene network of plants," Trends in Plant Science, vol. 9, no. 10, pp. 490–498, 2004. · ·

205. A. Mhamdi, G. Queval, S. Chaouch, S. Vanderauwera, F. Van Breusegem, and G. Noctor, "Catalase function in plants: a focus on Arabidopsis mutants as stress-mimic models," Journal of Experimental Botany, vol. 61, no. 15, pp. 4197–4220, 2010. · ·

206. S. Takahashi and N. Murata, "How do environmental stresses accelerate photoinhibition?"Trends in Plant Science, vol. 13, no. 4, pp. 178–182, 2008. · ·

207. T. Baszynski, L. Wajda, M. Król, D. Wolinska, Z. Krupa, and A. Tukendorf, "Photosynthetic activities of cadmium treated tomato plants," Plant Physiology, vol. 98, no. 3, pp. 365–370, 1980. ·

208. N. Mohanty and P. Mohanty, "Cation effects on primary processes of photosynthesis," inAdvances in Frontier Areas of Plant Biochemistry, R. Singh and S. K. Sawheny, Eds., pp. 1–18, Prentice-Hall, Delhi, India, 1988.

209. N. Atal, P. P. Saradhi, and P. Mohanty, "Inhibition of the chloroplast photochemical reactions by treatment of wheat seedlings with low concentrations of cadmium: analysis of electron transport activities and changes in fluorescence yield," Plant and Cell Physiology, vol. 32, no. 7, pp. 943–951, 1991.

210. P. Faller, K. Kienzler, and A. Krieger-Liszkay, "Mechanism of Cd^{2+} toxicity: Cd^{2+} inhibits photoactivation of Photosystem II by competitive binding to the essential Ca^{2+} site," Biochimica et Biophysica Acta, vol. 1706, no. 1-2, pp. 158–164, 2005. · ·

211. A. Krantev, R. Yordanova, T. Janda, G. Szalai, and L. Popova, "Treatment with salicylic acid decreases the effect of cadmium on photosynthesis in maize plants," Journal of Plant Physiology, vol. 165, no. 9, pp. 920–931, 2008. · ·

212. A. Siedlecka and T. Baszy Aski, "Inhibition of electron flow around photosystem I in chloroplasts of Cd-treated maize plants is due to Cd-induced iron deficiency," Physiologia Plantarum, vol. 87, no. 2, pp. 199–202, 1993.

213. A. Siedlecka, G. Samuelsson, P. Gardenstrom, L. A. Kleczkowski, and Z. Krupa, "The activatory model of plant response to moderate cadmium stress-relationship between carbonic anhydrase and Rubisco," in Photosynthesis: Mechanisms and Effects, G. Garab, Ed., pp. 2677–2680, Kluwer Academic Publishers, Dordrecht, The Netherlands, 1998.

214. F. J. Corpas, J. B. Barroso, and L. A. Del Río, "Peroxisomes as a source of reactive oxygen species and nitric oxide signal molecules in plant cells," Trends in Plant Science, vol. 6, no. 4, pp. 145–150, 2001. · ·

215. L. A. Del Río, L. M. Sandalio, F. J. Corpas, J. M. Palma, and J. B. Barroso, "Reactive oxygen species and reactive nitrogen species in peroxisomes. Production, scavenging, and role in cell signaling," Plant Physiology, vol. 141, no. 2, pp. 330–335, 2006. · ·

216. M. Rodríguez-Serrano, M. C. Romero-Puertas, I. Sparkes, C. Hawes, L. A. del Río, and L. M. Sandalio, "Peroxisome dynamics in Arabidopsis plants under oxidative stress induced by cadmium," Free Radical Biology and Medicine, vol. 47, no. 11, pp. 1632–1639, 2009. · ·

217. M. C. Romero-Puertas, M. Rodríguez-Serrano, F. J. Corpas, M. Gómez, L. A. Del Río, and L. M. Sandalio, "Cadmium-induced subcellular accumulation of O_2.- and H_2O_2 in pea leaves," Plant, Cell and Environment, vol. 27, no. 9, pp. 1122–1134, 2004. · ·

218. O. Blokhina and K. Fagerstedt, "Oxidative stress and antioxidant defenses in plants," in Oxidative Stress, Disease and Cancer, K. K. Singh, Ed., pp. 151–199, Imperial College Press, London, UK, 2006.

219. E. Heyno, C. Klose, and A. Krieger-Liszkay, "Origin of cadmium-induced reactive oxygen species production: mitochondrial electron transfer versus plasma membrane NADPH oxidase," New Phytologist, vol. 179, no. 3, pp. 687–699, 2008. · ·

220. Y. S. Wang, J. Wang, Z. M. Yang et al., "Salicylic acid modulates aluminum-induced oxidative stress in roots of Cassia tora," Acta Botanica Sinica, vol. 46, no. 7, pp. 819–828, 2004.

221. F. Hao, X. Wang, and J. Chen, "Involvement of plasma-membrane NADPH oxidase in nickel-induced oxidative stress in roots of wheat seedlings," Plant Science, vol. 170, no. 1, pp. 151–158, 2006. · ·

222. M. Rodríguez-Serrano, M. C. Romero-Puertas, A. Zabalza et al., "Cadmium effect on oxidative metabolism of pea (Pisum sativum L.) Roots. Imaging of reactive oxygen species and nitric oxide accumulation in vivo," Plant, Cell and Environment, vol. 29, no. 8, pp. 1532–1544, 2006. · ·

223. B. Pourrut, G. Perchet, J. Silvestre, M. Cecchi, M. Guiresse, and E. Pinelli, "Potential role of NADPH-oxidase in early steps of lead-induced oxidative burst in Vicia faba roots," Journal of Plant Physiology, vol. 165, no. 6, pp. 571–579, 2008. · ·

224. T. Remans, K. Opdenakker, K. Smeets, D. Mathijsen, J. Vangronsveld, and A. Cuypers, "Metal-specific and NADPH oxidase dependent changes in lipoxygenase and NADPH oxidase gene expression in Arabidopsis thaliana exposed to cadmium or excess copper," Functional Plant Biology, vol. 37, no. 6, pp. 532–544, 2010. · ·

225. B. Halliwell, "Reactive species and antioxidants. Redox biology is a fundamental theme of aerobic life," Plant Physiology, vol. 141, no. 2, pp. 312–322, 2006. · ·

226. C. H. Foyer and G. Noctor, "Redox homeostasis and antioxidant signaling: a metabolic interface between stress perception and physiological responses," Plant Cell, vol. 17, no. 7, pp. 1866–1875, 2005. · ·

227. T. S. Gechev, F. Van Breusegem, J. M. Stone, I. Denev, and C. Laloi, "Reactive oxygen species as signals that modulate plant stress responses and programmed cell death," BioEssays, vol. 28, no. 11, pp. 1091–1101, 2006. · ·

228. J. Espartero, I. Sanchez-Aguayo, and J. M. Pardo, "Molecular characterization of glyoxalase-I from a higher plant; Upregulation by stress," Plant Molecular Biology, vol. 29, no. 6, pp. 1223–1233, 1995.

229. J. P. Richard, "Mechanism for the formation of methylglyoxal from triosephosphates,"Biochemical Society Transactions, vol. 21, no. 2, pp. 549–553, 1993.

230. S. K. Yadav, S. L. Singla-Pareek, M. K. Reddy, and S. K. Sopory, "Methylglyoxal detoxification by glyoxalase system: a survival strategy during environmental stresses," Physiology and Molecular Biology of Plants, vol. 11, no. 1, pp. 1–11, 2005.

231. D. L. Pompliano, A. Peyman, and J. R. Knowles, "Stabilization of a reaction intermediate as a catalytic device: definition of the functional role of the flexible loop in triosephosphate isomerase," Biochemistry, vol. 29, no. 13, pp. 3186–3194, 1990.

232. Veena, V. S. Reddy, and S. K. Sopory, "Glyoxalase I from Brassica

juncea: molecular cloning, regulation and its over-expression confer tolerance in transgenic tobacco under stress," Plant Journal, vol. 17, no. 4, pp. 385–395, 1999. · ·

233. Z. Y. Chen, R. L. Brown, K. E. Damann, and T. E. Cleveland, "Identification of a maize kernel stress-related protein and its effect on aflatoxin accumulation," Phytopathology, vol. 94, no. 9, pp. 938–945, 2004.
234. M. N. Akhter Banu, M. A. Hoque, M. Watanabe-Sugimoto et al., "Proline and glycinebetaine ameliorated NaCl stress via scavenging of hydrogen peroxide and methylglyoxal but not superoxide or nitric oxide in tobacco cultured cells," Bioscience, Biotechnology and Biochemistry, vol. 74, no. 10, pp. 2043–2049, 2010. · ·
235. S. Ray, S. Dutta, J. Halder, and M. Ray, "Inhibition of electron flow through complex I of the mitochondrial respiratory chain of Ehrlich ascites carcinoma cells by methylglyoxal," Biochemical Journal, vol. 303, no. 1, pp. 69–72, 1994.
236. L. Wu and B. H. J. Juurlink, "Increased methylglyoxal and oxidative stress in hypertensive rat vascular smooth muscle cells," Hypertension, vol. 39, no. 3, pp. 809–814, 2002. · ·
237. F. W. R. Chaplen, "Incidence and potential implications of the toxic metabolite methylglyoxal in cell culture: a review," Cytotechnology, vol. 26, no. 3, pp. 173–183, 1998. · ·
238. K. M. Desai, T. Chang, H. Wang et al., "Oxidative stress and aging: is methylglyoxal the hidden enemy?" Canadian Journal of Physiology and Pharmacology, vol. 88, no. 3, pp. 273–284, 2010. · ·
239. S. Neill, R. Desikan, and J. Hancock, "Hydrogen peroxide signalling," Current Opinion in Plant Biology, vol. 5, no. 5, pp. 388–395, 2002. · ·
240. P. J. Thornalley, "The glyoxalase system: new developments towards functional characterization of a metabolic pathway fundamental to biological life," Biochemical Journal, vol. 269, no. 1, pp. 1–11, 1990.
241. M. Saxena, R. Bisht, S. D. Roy, S. K. Sopory, and N. Bhalla-Sarin, "Cloning and characterization of a mitochondrial glyoxalase II from Brassica juncea that is upregulated by NaCl, Zn, and

ABA,"Biochemical and Biophysical Research Communications, vol. 336, no. 3, pp. 813–819, 2005. · ·

242. M. A. Iannelli, F. Pietrini, L. Fiore, L. Petrilli, and A. Massacci, "Antioxidant response to cadmium in Phragmites australis plants," Plant Physiology and Biochemistry, vol. 40, no. 11, pp. 977–982, 2002. · ·
243. R. Boominathan and P. M. Doran, "Cadmium tolerance and antioxidative defenses in hairy roots of the cadmium hyperaccumulator, Thlaspi caerulescens," Biotechnology and Bioengineering, vol. 83, no. 2, pp. 158–167, 2003. · ·
244. É. Darkó, H. Ambrus, É. Stefanovits-Bányai, J. Fodor, F. Bakos, and B. Barnabás, "Aluminum toxicity, aluminum tolerance and oxidative stress in an Al-sensitive wheat genotype and in Al-tolerant lines developed by in vitro microspore selection," Plant Science, vol. 166, no. 3, pp. 583–591, 2004.
245. N. Singh, L. Q. Ma, M. Srivastava, and B. Rathinasabapathi, "Metabolic adaptations to arsenic-induced oxidative stress in Pteris vittata L and Pteris ensiformis L," Plant Science, vol. 170, no. 2, pp. 274–282, 2006. · ·
246. X. Jin, X. Yang, E. Islam, D. Liu, and Q. Mahmood, "Effects of cadmium on ultrastructure and antioxidative defense system in hyperaccumulator and non-hyperaccumulator ecotypes of Sedum alfredii Hance," Journal of Hazardous Materials, vol. 156, no. 1–3, pp. 387–397, 2008. · ·
247. X. Zeng, L. Q. Ma, R. Qiu, and Y. Tang, "Responses of non-protein thiols to Cd exposure in Cd hyperaccumulator Arabis paniculata Franch," Environmental and Experimental Botany, vol. 66, no. 2, pp. 242–248, 2009. · ·
248. D. K. Gupta, H. G. Huang, X. E. Yang, B. H. N. Razafindrabe, and M. Inouhe, "The detoxification of lead in Sedum alfredii H. is not related to phytochelatins but the glutathione," Journal of Hazardous Materials, vol. 177, no. 1–3, pp. 437–444, 2010. · ·
249. Q. Sun, Z. H. Ye, X. R. Wang, and M. H. Wong, "Increase of glutathione in mine population ofSedum alfredii: a Zn hyperaccumulator and Pb accumulator," Phytochemistry, vol. 66, no. 21, pp. 2549–2556, 2005. · ·

250. A. Koprivova, S. Kopriva, D. Jäger, B. Will, L. Jouanin, and H. Rennenberg, "Evaluation of transgenic poplars over-expressing enzymes of glutathione synthesis for phytoremediation of cadmium," Plant Biology, vol. 4, no. 6, pp. 664–670, 2002. · ·
251. R. A. Larson, "The antioxidants of higher plants," Phytochemistry, vol. 27, no. 4, pp. 969–978, 1988.
252. C. Herschbach and H. Rennenberg, "Significance of phloem-translocated organic sulfur compounds for the regulation of sulfur nutrition," Progress in Botany, vol. 62, pp. 177–193, and 2001.
253. R. Blum, A. Beck, A. Korte et al., "Function of phytochelatin synthase in catabolism of glutathione-conjugates," Plant Journal, vol. 49, no. 4, pp. 740–749, 2007. · ·
254. G. Noctor, "Metabolic signalling in defence and stress: the central roles of soluble redox couples,"Plant, Cell and Environment, vol. 29, no. 3, pp. 409–425, 2006. · ·
255. A. S. Gupta, R. G. Alscher, and D. McCune, "Response of photosynthesis and cellular antioxidants to ozone in Populus leaves," Plant Physiology, vol. 96, no. 2, pp. 650–655, 1991.
256. M. J. May and C. J. Leaver, "Oxidative stimulation of glutathione synthesis in Arabidopsis thaliana suspension cultures," Plant Physiology, vol. 103, no. 2, pp. 621–627, 1993.
257. K. A. Marrs, "The functions and regulation of glutathione S-transferases in plants," Annual Review of Plant Physiology and Plant Molecular Biology, vol. 47, no. 1, pp. 127–158, 1996.
258. M. J. May, T. Vernoux, C. Leaver, M. Van Montagu, and D. Inzé, "Glutathione homeostasis in plants: implications for environmental sensing and plant development," Journal of Experimental Botany, vol. 49, no. 321, pp. 649–667, 1998.
259. S. Qadir, M. I. Qureshi, S. Javed, and M. Z. Abdin, "Genotypic variation in phytoremediation potential of Brassica juncea cultivars exposed to Cd stress," Plant Science, vol. 167, no. 5, pp. 1171–1181, 2004. · ·
260. N. A. Anjum, S. Umar, A. Ahmad, M. Iqbal, and N. A. Khan, "Ontogenic variation in response ofBrassica campestris L. to cadmium toxicity," Journal of Plant Interactions, vol. 3, no. 3, pp. 189–198, 2008. · ·

261. A. Schützendübel, P. Nikolova, C. Rudolf, and A. Polle, "Cadmium and H_2O_2-induced oxidative stress in Populus x canescens roots," Plant Physiology and Biochemistry, vol. 40, no. 6–8, pp. 577–584, 2002. · ·

262. E. Grill, E. L. Winnacker, and M. H. Zenk, "Phytochelatins: the principal heavy-metal complexing peptides of higher plants," Science, vol. 230, no. 4726, pp. 674–676, 1985.

263. F. Pietrini, M. A. Iannelli, S. Pasqualini, and A. Massacci, "Interaction of cadmium with glutathione and photosynthesis in developing leaves and chloroplasts of Phragmites australis (Cav.) Trin. Ex steudel," Plant Physiology, vol. 133, no. 2, pp. 829–837, 2003. · ·

264. Z. S. Zhou, S. J. Wang, and Z. M. Yang, "Biological detection and analysis of mercury toxicity to alfalfa (Medicago sativa) plants," Chemosphere, vol. 70, no. 8, pp. 1500–1509, 2008. · ·

265. G. Y. Huang, Y. S. Wang, C. C. Sun, J. D. Dong, and Z. X. Sun, "The effect of multiple heavy metals on ascorbate, glutathione and related enzymes in two mangrove plant seedlings (Kandelia candel and Bruguiera gymnorrhiza)," Oceanological and Hydrobiological Studies, vol. 39, no. 1, pp. 11–25, 2010. · ·

266. J. Hartley-Whitaker, G. Ainsworth, and A. A. Meharg, "Copper- and arsenate-induced oxidative stress in Holcus lanatus L. clones with differential sensitivity," Plant, Cell and Environment, vol. 24, no. 7, pp. 713–722, 2001. · ·

267. N. Verbruggen, C. Hermans, and H. Schat, "Mechanisms to cope with arsenic or cadmium excess in plants," Current Opinion in Plant Biology, vol. 12, no. 3, pp. 364–372, 2009. · ·

268. S. Mishra, S. Srivastava, R. D. Tripathi, and P. K. Trivedi, "Thiol metabolism and antioxidant systems complement each other during arsenate detoxification in Ceratophyllum demersum L,"Aquatic Toxicology, vol. 86, no. 2, pp. 205–215, 2008. · ·

269. G. Innocenti, C. Pucciariello, M. Le Gleuher et al., "Glutathione synthesis is regulated by nitric oxide in Medicago truncatula roots," Planta, vol. 225, no. 6, pp. 1597–1602, 2007. · ·

270. J. B. Barroso, F. J. Corpas, A. Carreras et al., "Localization of S-nitrosoglutathione and expression of S-nitrosoglutathione

reductase in pea plants under cadmium stress," Journal of Experimental Botany, vol. 57, no. 8, pp. 1785–1793, 2006. · ·

271. J. Xiong, G. Fu, L. Tao, and C. Zhu, "Roles of nitric oxide in alleviating heavy metal toxicity in plants," Archives of Biochemistry and Biophysics, vol. 497, no. 1-2, pp. 13–20, 2010. · ·
272. D. P. Dixon and R. Edwards, "Glutathione transferases," in The Arabidopsis Book, vol. 8, The American Society of Plant Biologists, Austin, Tex, USA, 2010. ·
273. M. Z. Hossain and M. Fujita, "Purification of a phi-type glutathione S-transferase from pumpkin flowers, and molecular cloning of its cDNA," Bioscience, Biotechnology and Biochemistry, vol. 66, no. 10, pp. 2068–2076, 2002.
274. M. Fujita and M. Z. Hossain, "Molecular cloning of cDNAs for three tau-type glutathione S-transferases in pumpkin (Cucurbita maxima) and their expression properties," Physiologia Plantarum, vol. 117, no. 1, pp. 85–92, 2003. · ·
275. M. Fujita and M. Z. Hossain, "Modulation of pumpkin glutathione S-transferases by aldehydes and related compounds," Plant and Cell Physiology, vol. 44, no. 5, pp. 481–490, 2003. · ·
276. M. Z. Hossain, M. D. Hossain, and M. Fujita, "Induction of pumpkin glutathione S-transferases by different stresses and its possible mechanisms," Biologia Plantarum, vol. 50, no. 2, pp. 210–218, 2006. · ·
277. M. Z. Hossain, J. A. T. da Silva, and M. Fujita, "Differential roles of glutathione S-transferase in oxidative stress modulation," in Floriculture, Ornamental and Plant Biotechnology. Advances and Topical Issues, J. A. T. da Silva, Ed., pp. 108–116, Global Science Books Ltd, London, UK, 2006.
278. E. Skórzyńska-Polit, M. Drążkiewicz, and Z. Krupa, "The activity of the antioxidative system in cadmium-treated Arabidopsis thaliana," Biologia Plantarum, vol. 47, no. 1, pp. 71–78, 2004. · ·
279. J. W. Gronwald and K. L. Plaisance, "Isolation and characterization of glutathione S-transferase isozymes from sorghum," Plant Physiology, vol. 117, no. 3, pp. 877–892, 1998.

280. J. D. Hayes, J. U. Flanagan, and I. R. Jowsey, "Glutathione transferases," Annual Review of Pharmacology and Toxicology, vol. 45, pp. 51–88, 2005. · ·

281. Y. Hu, Y. Ge, C. Zhang, T. Ju, and W. Cheng, "Cadmium toxicity and translocation in rice seedlings are reduced by hydrogen peroxide pretreatment," Plant Growth Regulation, vol. 59, no. 1, pp. 51–61, 2009. · ·

282. L. Halusková, K. Valentovicová, J. Huttová, I. Mistrík, and L. Tamás, "Effect of abiotic stresses on glutathione peroxidase and glutathione S-transferase activity in barley root tips," Plant Physiology and Biochemistry, vol. 47, no. 11-12, pp. 1069–1074, 2009. ·

283. M. M. Rohman, M. S. Uddin, and M. Fujita, "Up-regulation of onion bulb glutathione S (GSTs) by abiotic stresses: a comparative study between two differently sensitive GSTs to their physiological inhibitors," Plant OMICS, vol. 3, no. 1, pp. 28–34, 2010.

284. R. Szollosi, I. S. Varga, L. Erdei, and E. Mihalik, "Cadmium-induced oxidative stress and antioxidative mechanisms in germinating Indian mustard (Brassica juncea L.) Seeds,"Ecotoxicology and Environmental Safety, vol. 72, no. 5, pp. 1337–1342, 2009. · ·

285. C. H. ZHANG andY. GE, "Response of glutathione and glutathione S-transferase in rice seedlings exposed to cadmium stress," Rice Science, vol. 15, no. 1, pp. 73–76, 2008. · ·

286. E. Gajewska and M. Skłodowska, "Differential biochemical responses of wheat shoots and roots to nickel stress: antioxidative reactions and proline accumulation," Plant Growth Regulation, vol. 54, no. 2, pp. 179–188, 2008. · ·

287. A. M. Reddy, S. G. Kumar, G. Jyothsnakumari, S. Thimmanaik, and C. Sudhakar, "Lead induced changes in antioxidant metabolism of horse gram (Macrotyloma uniflorum (Lam.) Verdc.) And Bengal gram (Cicer arietinum L.)," Chemosphere, vol. 60, no. 1, pp. 97–104, 2005.

288. C. Gullner, M. Uotila, and T. Kömives, "Responses of glutathione and glutathione S-transferase to cadmium and mercury exposure in pedunculate oak (Quercus robur) leaf discs," Botanica Acta, vol. 111, no. 1, pp. 62–65, 1998.

289. I. Cakman and W. J. Horst, "Effect of aluminum on lipid peroxidation, superoxide dismutase, catalase, and peroxidase activities in root tips of soybean (Glycine max L.)," Physiologia Plantarum, vol. 83, no. 3, pp. 463–468, 1991.
290. K. A. Marrs and V. Walbot, "Expression and RNA splicing of the maize glutathione S-transferase Bronze2 gene is regulated by cadmium and other stresses," Plant Physiology, vol. 113, no. 1, pp. 93–102, 1997.
291. P. Dixit, P. K. Mukherjee, V. Ramachandran, and S. Eapen, "Glutathione transferase from Trichoderma virens enhances cadmium tolerance without enhancing its accumulation in transgenic Nicotiana tabacum," PLoS ONE, vol. 6, no. 1, Article ID e16360, 2011. ·
292. C. H. Foyer and G. Noctor, "Ascorbate and glutathione: the heart of the redox hub," Plant Physiology, vol. 155, no. 1, pp. 2–18, 2011. ·
293. C. H. Foyer and G. Noctor, "Oxidant and antioxidant signalling in plants: a re-evaluation of the concept of oxidative stress in a physiological context," Plant, Cell and Environment, vol. 28, no. 8, pp. 1056–1071, 2005. · ·
294. A. Cuypers, J. Vangronsveld, and H. Clijsters, "Peroxidases in roots and primary leaves ofPhaseolus vulgaris Copper and Zinc phytotoxicity: a comparison," Journal of Plant Physiology, vol. 159, no. 8, pp. 869–876, 2002.
295. R. A. Fatima and M. Ahmad, "Certain antioxidant enzymes of Allium cepa as biomarkers for the detection of toxic heavy metals in wastewater," Science of the Total Environment, vol. 346, no. 1–3, pp. 256–273, 2005. · ·
296. B. Semane, A. Cuypers, K. Smeets et al., "Cadmium responses in Arabidopsis thaliana: glutathione metabolism and antioxidative defence system," Physiologia Plantarum, vol. 129, no. 3, pp. 519–528, 2007. · ·
297. S. Srivastava, A. K. Srivastava, P. Suprasanna, and S. F. D›souza, "Comparative antioxidant profiling of tolerant and sensitive varieties of Brassica juncea L. to arsenate and arsenite exposure," Bulletin of Environmental Contamination and Toxicology, vol. 84, no. 3, pp. 342–346, 2010. · ·

298. S. Srivastava, S. Mishra, R. D. Tripathi, S. Dwivedi, P. K. Trivedi, and P. K. Tandon, "Phytochelatins and antioxidant systems respond differentially during arsenite and arsenate stress in Hydrilla verticillata (L.f.) Royle," Environmental Science and Technology, vol. 41, no. 8, pp. 2930–2936, 2007. · ·

299. M. Gupta, P. Sharma, N. B. Sarin, and A. K. Sinha, "Differential response of arsenic stress in two varieties of Brassica juncea L," Chemosphere, vol. 74, no. 9, pp. 1201–1208, 2009. · ·

300. H. Hartikainen, T. L. Kue, and V. Piironem, "Selenium as an antioxidant and prooxidant in rye grass," Plant and Soil, vol. 225, no. 1-2, pp. 193–200, 2000.

301. M. M. A. Boojar and F. Goodarzi, "The copper tolerance strategies and the role of antioxidative enzymes in three plant species grown on copper mine," Chemosphere, vol. 67, no. 11, pp. 2138–2147, 2007. · ·

302. P. Aravind and M. N. V. Prasad, "Cadmium-Zinc interactions in a hydroponic system usingCeratophyllum demersum L.: adaptive ecophysiology, biochemistry and molecular toxicology,"Brazilian Journal of Plant Physiology, vol. 17, no. 1, pp. 3–20, 2005.

303. K. Asada, "The water-water cycle in chloroplasts: scavenging of active oxygens and dissipation of excess photons," Annual Review of Plant Biology, vol. 50, pp. 601–639, 1999.

304. Z. Chen, T. E. Young, J. Ling, S. C. Chang, and D. R. Gallie, "Increasing vitamin C content of plants through enhanced ascorbate recycling," Proceedings of the National Academy of Sciences of the United States of America, vol. 100, no. 6, pp. 3525–3530, 2003. · ·

305. F. Chen, F. Wang, F. Wu, W. Mao, G. Zhang, and M. Zhou, "Modulation of exogenous glutathione in antioxidant defense system against Cd stress in the two barley genotypes differing in Cd tolerance," Plant Physiology and Biochemistry, vol. 48, no. 8, pp. 663–672, 2010. · ·

306. A. Paradiso, R. Berardino, M. C. De Pinto et al., "Increase in ascorbate-glutathione metabolism as local and precocious systemic responses induced by cadmium in durum wheat plants," Plant and Cell Physiology, vol. 49, no. 3, pp. 362–374, 2008. · ·

307. R. Maheshwari and R. S. Dubey, "Nickel-induced oxidative stress and the role of antioxidant defence in rice seedlings," Plant Growth Regulation, vol. 59, no. 1, pp. 37–49, 2009. · ·
308. L. M. Cervilla, B. Blasco, J. J. Ríos, L. Romero, and J. M. Ruiz, "Oxidative stress and antioxidants in tomato (Solanum lycopersicum) plants subjected to boron toxicity," Annals of Botany, vol. 100, no. 4, pp. 747–756, 2007. · ·
309. L. Yin, S. Wang, A. E. Eltayeb et al., "Overexpression of dehydroascorbate reductase, but not monodehydroascorbate reductase, confers tolerance to Aluminum stress in transgenic Tobacco,"Planta, vol. 231, no. 3, pp. 609–621, 2010. · ·
310. F. Y. Zhao, W. Liu, and S. Y. Zhang, "Different responses of plant growth and antioxidant system to the combination of cadmium and heat stress in transgenic and non-transgenic rice," Journal of Integrative Plant Biology, vol. 51, no. 10, pp. 942–950, 2009. · ·
311. C. H. Foyer and B. Halliwell, "The presence of glutathione and glutathione reductase in chloroplasts: a proposed role in ascorbic acid metabolism," Planta, vol. 133, no. 1, pp. 21–25, 1976. · ·
312. H. R. Lascano, L. M. Casano, M. N. Melchiorre, and V. S. Trippi, "Biochemical and molecular characterisation of wheat chloroplastic glutathione reductase," Biologia Plantarum, vol. 44, no. 4, pp. 509–516, 2001. · ·
313. D. Contour-Ansel, M. L. Torres-Franklin, M. H. Cruz De Carvalho, A. D›Arcy-Lameta, and Y. Zuily-Fodil, "Glutathione reductase in leaves of cowpea: cloning of two cDNAs, expression and enzymatic activity under progressive drought stress, desiccation and abscisic acid treatment,"Annals of Botany, vol. 98, no. 6, pp. 1279–1287, 2006. · ·
314. M. C. Romero-Puertas, F. J. Corpas, L. M. Sandalio et al., "Glutathione reductase from pea leaves: response to abiotic stress and characterization of the peroxisomal isozyme," New Phytologist, vol. 170, no. 1, pp. 43–52, 2006. · ·
315. N. A. Anjum, S. Umar, M. Iqbal, and N. A. Khan, "Cadmium causes oxidative stress in moongbean [Vigna radiata (L.) Wilczek] by affecting antioxidant enzyme systems and ascorbate-glutathione cycle metabolism," Russian Journal of Plant Physiology, vol. 58, no. 1, pp. 92–99, 2011.

316. S. Srivastava, S. Mishra, R. D. Tripathi, S. Dwivedi, and D. K. Gupta, "Copper-induced oxidative stress and responses of antioxidants and phytochelatins in Hydrilla verticillata (L.f.) Royle,"Aquatic Toxicology, vol. 80, no. 4, pp. 405–415, 2006. · ·

317. M. Israr and S. V. Sahi, "Antioxidative responses to mercury in the cell cultures of Sesbania drummondii," Plant Physiology and Biochemistry, vol. 44, no. 10, pp. 590–595, 2006. · ·

318. M. Russo, C. Sgherri, R. Izzo, and F. Navari-Izzo, "Brassica napus subjected to copper excess: phospholipases C and D and glutathione system in signalling," Environmental and Experimental Botany, vol. 62, no. 3, pp. 238–246, 2008. · ·

319. R. K. Tewari, P. Kumar, and P. N. Sharma, "Antioxidant responses to enhanced generation of superoxide anion radical and hydrogen peroxide in the copper-stressed mulberry plants," Planta, vol. 223, no. 6, pp. 1145–1153, 2006. · ·

320. S. Verma and R. S. Dubey, "Lead toxicity induces lipid peroxidation and alters the activities of antioxidant enzymes in growing rice plants," Plant Science, vol. 164, no. 4, pp. 645–655, 2003. · ·

321. M. D. Hossain, M. M. Rohman, and M. Fujita, "Comparative investigation of glutathione S-transferases, glyoxalase I and allinase activities in different vegetable crops," Journal of Crop Science and Biotechnology, vol. 10, no. 1, pp. 21–28, 2007.

322. M. Tuomainen, V. Ahonen, S. O. Karenlampi, et al., "Characterization of the glyoxalase 1 geneTcGLX1 in the metal hyperaccumulator plant Thlaspi caerulescens," Planta, vol. 233, no. 6, pp. 1173–1184, 2011.

323. H. El-Shabrawi, B. Kumar, T. Kaul, M. K. Reddy, S. L. Singla-Pareek, and S. K. Sopory, "Redox homeostasis, antioxidant defense, and methylglyoxal detoxification as markers for salt tolerance in Pokkali rice," Protoplasma, vol. 245, no. 1, pp. 85–96, 2010. · ·

324. M. K. Maiti, S. Krishnasamy, H. A. Owen, and C. A. Makaroff, "Molecular characterization of glyoxalase II from Arabidopsis thaliana," Plant Molecular Biology, vol. 35, no. 4, pp. 471–481, 1997. · ·

325. T. M. Zang, D. A. Hollman, P. A. Crawford, M. W. Crowder, and C. A. Makaroff, "Arabidopsisglyoxalase II contains a zinc/iron binuclear metal center that is essential for substrate binding and

catalysis," Journal of Biological Chemistry, vol. 276, no. 7, pp. 4788–4795, 2001. ·

326. S. K. Yadav, S. L. Singla-Pareek, and S. K. Sopory, "An overview on the role of methylglyoxal and glyoxalases in plants," Drug Metabolism and Drug Interactions, vol. 23, no. 1-2, pp. 51–68, 2008.
327. M. A. Hoque, M. N. A. Banu, Y. Nakamura, Y. Shimoishi, and Y. Murata, "Proline and glycinebetaine enhance antioxidant defense and methylglyoxal detoxification systems and reduce NaCl-induced damage in cultured tobacco cells," Journal of Plant Physiology, vol. 165, no. 8, pp. 813–824, 2008. · ·
328. M. A. Hoque, Banu M. N. A., E. Okuma, et al., "Exogenous proline and glycinebetaine ingresses NaCl-induced ascorbate-glutathione cycle enzyme activities and proline improves salt tolerance more than glycinebetaine in tobacco Bright yellow-2 suspension-cultured cells," Journal of Plant Physiology, vol. 164, no. 5, pp. 553–561, 2007.
329. Y. Y. Chao, Y. T. Hsu, and C. H. Kao, "Involvement of glutathione in heat shock- and hydrogen peroxide-induced cadmium tolerance of rice (Oryza sativa L.) seedlings," Plant and Soil, vol. 318, no. 1-2, pp. 37–45, 2009. · ·
330. C. F. Tang, Y. G. Liu, G. M. Zeng et al., "Effects of exogenous spermidine on antioxidant system responses of Typha latifolia L. under Cd^{2+} stress," Journal of Integrative Plant Biology, vol. 47, no. 4, pp. 428–434, 2005. · ·
331. A. Levine, R. Tenhaken, R. Dixon, and C. Lamb, "H_2O_2 from the oxidative burst orchestrates the plant hypersensitive disease resistance response," Cell, vol. 79, no. 4, pp. 583–593, 1994. · ·
332. T. K. Prasad, M. D. Anderson, B. A. Martin, and C. R. Stewart, "Evidence for chilling-induced oxidative stress in maize seedlings and a regulatory role for hydrogen peroxide," Plant Cell, vol. 6, no. 1, pp. 65–74, 1994. · ·
333. F. J. Xu, C. W. Jin, W. J. Liu, Y. S. Zhang, and X. Y. Lin, "Pretreatment with H_2O_2 alleviates aluminum-induced oxidative stress in wheat seedlings," Journal of Integrative Plant Biology, vol. 53, no. 1, pp. 44–53, 2011. ·

334. R. K. Tewari, E. J. Hahn, and K. Y. Paek, "Modulation of copper toxicity-induced oxidative damage by nitric oxide supply in the adventitious roots of Panax ginseng," Plant Cell Reports, vol. 27, no. 1, pp. 171–181, 2008. · ·

335. N. Ahsan, J. Renaut, and S. Komatsu, "Recent developments in the application of proteomics to the analysis of plant responses to heavy metals," Proteomics, vol. 9, no. 10, pp. 2602–2621, 2009. · ·

336. K. Kosová, P. Vítámvás, I. T. Prášil, and J. Renaut, "Plant proteome changes under abiotic stress—contribution of proteomics studies to understanding plant stress response," Journal of Proteomics, vol. 74, no. 8, pp. 1301–1322, 2011. ·

337. K. Lee, D. W. Bae, S. H. Kim et al., "Comparative proteomic analysis of the short-term responses of rice roots and leaves to cadmium," Journal of Plant Physiology, vol. 167, no. 3, pp. 161–168, 2010. · ·

338. Y. Wang, Y. Qian, H. Hu, Y. Xu, and H. Zhang, "Comparative proteomic analysis of Cd-responsive proteins in wheat roots," Acta Physiologiae Plantarum, pp. 1–9, 2010. · ·

339. N. Ahsan, D. G. Lee, I. Alam et al., "Comparative proteomic study of arsenic-induced differentially expressed proteins in rice roots reveals glutathione plays a central role during As stress," Proteomics, vol. 8, no. 17, pp. 3561–3576, 2008. · ·

340. R. A. Ingle, J. A. C. Smith, and L. J. Sweetlove, "Responses to nickel in the proteome of thehyperaccumulator plant Alyssum lesbiacum," BioMetals, vol. 18, no. 6, pp. 627–641, 2005. · ·

341. M. Labra, E. Gianazza, R. Waitt et al., "Zea mays L. protein changes in response to potassium dichromate treatments," Chemosphere, vol. 62, no. 8, pp. 1234–1244, 2006. · ·

342. M. H. Tuomainen, N. Nunan, S. J. Lehesranta et al., "Multivariate analysis of protein profiles of metal hyperaccumulator Thlaspi caerulescens accessions," Proteomics, vol. 6, no. 12, pp. 3696–3706, 2006. · ·

343. N. Ahsan, D. G. Lee, S. H. Lee et al., "Excess copper induced physiological and proteomic changes in germinating rice seeds," Chemosphere, vol. 67, no. 6, pp. 1182–1193, 2007. · ·

344. Y. Zhen, J. L. Qi, S. S. Wang et al., "Comparative proteome analysis of differentially expressed proteins induced by Al toxicity in soybean," Physiologia Plantarum, vol. 131, no. 4, pp. 542–554, 2007. · ·
345. Q. Yang, Y. Wang, J. Zhang, W. Shi, C. Qian, and X. Peng, "Identification of aluminum-responsive proteins in rice roots by a proteomic approach: cysteine synthase as a key player in Al response," Proteomics, vol. 7, no. 5, pp. 737–749, 2007. · ·
346. C. Ortega-Villasante, L. E. Hernández, R. Rellán-Álvarez, F. F. Del Campo, and R. O. Carpena-Ruiz, "Rapid alteration of cellular redox homeostasis upon exposure to cadmium and mercury in alfalfa seedlings," New Phytologist, vol. 176, no. 1, pp. 96–107, 2007. · ·
347. P. Mohanpuria, N. K. Rana, and S. K. Yadav, "Cadmium induced oxidative stress influence on glutathione metabolic genes of Camellia sinensis (L.) O. Kuntze," Environmental Toxicology, vol. 22, no. 4, pp. 368–374, 2007. · ·
348. C. Xiang and D. J. Oliver, "Glutathione metabolic genes coordinately respond to heavy metals and jasmonic acid in Arabidopsis," Plant Cell, vol. 10, no. 9, pp. 1539–1550, 1998. · ·
349. I. Ogawa, H. Nakanishi, S. Mori, and N. K. Nishizawa, "Time course analysis of gene regulation under cadmium stress in rice," Plant and Soil, vol. 325, no. 1, pp. 97–108, 2009. · ·
350. P. D. B. Adamis, D. S. Gomes, M. L. C. C. Pinto, A. D. Panek, and E. C. A. Eleutherio, "The role of glutathione transferases in cadmium stress," Toxicology Letters, vol. 154, no. 1-2, pp. 81–88, 2004. · ·
351. S. K. Panda and H. Matsumoto, "Changes in antioxidant gene expression and induction of oxidative stress in pea (Pisum sativum L.) Under Al stress," BioMetals, vol. 23, no. 4, pp. 753–762, 2010. · ·
352. H. Luo, H. Li, X. Zhang, and J. Fu, "Antioxidant responses and gene expression in perennial ryegrass (Lolium perenne L.) Under cadmium stress," Ecotoxicology, vol. 20, no. 4, pp. 770–778, 2011. ·

353. K. Smeets, J. Ruytinx, B. Semane et al., "Cadmium-induced transcriptional and enzymatic alterations related to oxidative stress," Environmental and Experimental Botany, vol. 63, no. 1–3, pp. 1–8, 2008. · ·

354. P. Goupil, D. Souguir, E. Ferjani, O. Faure, A. Hitmi, and G. Ledoigt, "Expression of stress-related genes in tomato plants exposed to arsenic and chromium in nutrient solution," Journal of Plant Physiology, vol. 166, no. 13, pp. 1446–1452, 2009. · ·

355. S. Heiss, A. Wachter, J. Bogs, C. Cobbett, and T. Rausch, "Phytochelatin synthase (PCS) protein is induced in Brassica juncea leaves after prolonged Cd exposure," Journal of Experimental Botany, vol. 54, no. 389, pp. 1833–1839, 2003. · ·

356. D. Gonzalez-Mendoza, A. Q. Moreno, and O. Zapata-Perez, "Coordinated responses of phytochelatin synthase and metallothionein genes in black mangrove, Avicennia germinans, exposed to cadmium and copper," Aquatic Toxicology, vol. 83, no. 4, pp. 306–314, 2007. · ·

357. K. Akashi, N. Nishimura, Y. Ishida, and A. Yokota, "Potent hydroxyl radical-scavenging activity of drought-induced type-2 metallothionein in wild watermelon," Biochemical and Biophysical Research Communications, vol. 323, no. 1, pp. 72–78, 2004. · ·

358. H. Zhang, W. Xu, J. Guo, Z. He, and M. Ma, "Coordinated responses of phytochelatins and metallothioneins to heavy metals in garlic seedlings," Plant Science, vol. 169, no. 6, pp. 1059–1065, 2005. · ·

359. D. Duressa, K. Soliman, and D. Chen, "Identification of aluminum responsive genes in Al-tolerant soybean line PI 416937," International Journal of Plant Genomics, vol. 2010, Article ID 164862, 13 pages, 2010. ·

360. F. Lin, J. Xu, J. Shi, H. Li, and B. Li, "Molecular cloning and characterization of a novel glyoxalase I gene TaGly I in wheat (Triticum aestivum L.)," Molecular Biology Reports, vol. 37, no. 2, pp. 729–735, 2010. · ·

361. L. Lanfranco, A. Bolchi, E. C. Ros, S. Ottonello, and P. Bonfante, "Differential expression of a metallothionein gene during the presymbiotic versus the symbiotic phase of an arbuscular

mycorrhizal fungus," Plant Physiology, vol. 130, no. 1, pp. 58–67, 2002. · ·

362. F. Ouziad, U. Hildebrandt, E. Schmelzer, and H. Bothe, "Differential gene expressions in arbuscular mycorrhizal-colonized tomato grown under heavy metal stress," Journal of Plant Physiology, vol. 162, no. 6, pp. 634–649, 2005. · ·
363. S. Perotto and E. Martino, "Molecular and cellular mechanisms of heavy metal tolerance in mycorrhizal fungi: what perspectives for bioremediation?" Minerva Biotecnologica, vol. 13, no. 1, pp. 55–63, 2001.
364. U. Galli, H. Schuepp, and C. Brunold, "Heavy metal binding by mycorrhizal fungi," Physiologia Plantarum, vol. 92, no. 2, pp. 364–368, 1994.
365. Y. Guo, E. George, and H. Marschner, "Contribution of an arbuscular mycorrhizal fungus to the uptake of cadmium and nickel in bean and maize plants," Plant and Soil, vol. 184, no. 2, pp. 195–205, 1996.
366. U. Hildebrandt, M. Regvar, and H. Bothe, "Arbuscular mycorrhiza and heavy metal tolerance,"Phytochemistry, vol. 68, no. 1, pp. 139–146, 2007.
367. K. Turnau, P. Ryszka, V. Gianinazzi-Pearson, and D. Van Tuinen, "Identification of arbuscular mycorrhizal fungi in soils and roots of plants colonizing zinc wastes in southern Poland,"Mycorrhiza, vol. 10, no. 4, pp. 169–174, 2001. · ·
368. N. Weyens, S. Truyens, E. Saenen et al., "Endophytes and their potential to deal with co-contamination of organic contaminants (toluene) and toxic metals (nickel) during phytoremediation," International Journal of Phytoremediation, vol. 13, no. 3, pp. 244–255, 2011. ·
369. C. L. Rugh, "Genetically engineered phytoremediation: one man›s trash is another man›s transgene," Trends in Biotechnology, vol. 22, no. 10, pp. 496–498, 2004. · ·
370. Y. P. Tong, R. Kneer, and Y. G. Zhu, "Vacuolar compartmentalization: a second-generation approach to engineering plants for phytoremediation," Trends in Plant Science, vol. 9, no. 1, pp. 7–9, 2004. · ·

371. S. L. Doty, "Enhancing phytoremediation through the use of transgenics and endophytes," New Phytologist, vol. 179, no. 2, pp. 318–333, 2008. · ·
372. J. R. Domínguez-Solís, M. C. López, F. J. Ager, M. D. Ynsa, L. C. Romero, and C. Gotor, "Increased cysteine availability is essential for cadmium tolerance and accumulation inArabidopsis thaliana," Plant Biotechnology Journal, vol. 2, no. 6, pp. 469–476, 2004. · ·
373. Y. L. Zhu, E. A. H. Pilon-Smits, L. Jouanin, and N. Terry, "Overexpression of glutathione synthetase in Indian mustard enhances cadmium accumulation and tolerance," Plant Physiology, vol. 119, no. 1, pp. 73–79, 1999.
374. Y. L. Zhu, E. A. H. Pilon-Smits, A. S. Tarun, S. U. Weber, L. Jouanin, and N. Terry, "Cadmium tolerance and accumulation in Indian mustard is enhanced by overexpressing -glutamylcysteine synthetase," Plant Physiology, vol. 121, no. 4, pp. 1169–1177, 1999.
375. G. Noctor, M. Strohm, L. Jouanin, K. J. Kunert, C. H. Foyer, and H. Rennenberg, "Synthesis of glutathione in leaves of transgenic poplar overexpressing -glutamylcysteine synthetase," Plant Physiology, vol. 112, no. 3, pp. 1071–1078, 1996.
376. G. Creissen, P. Broadbent, R. Stevens, A. R. Wellburn, and P. Mullineaux, "Manipulation of glutathione metabolism in transgenic plants," Biochemical Society Transactions, vol. 24, no. 2, pp. 465–469, 1996.
377. S. Reisinger, M. Schiavon, N. Terry, and E. A. H. Pilon-Smits, "Heavy metal tolerance and accumulation in Indian mustard (Brassica juncea L.) Expressing bacterial -glutamylcysteine synthetase or glutathione synthetase," International Journal of Phytoremediation, vol. 10, no. 5, pp. 440–454, 2008.
378. L. A. Ivanova, D. A. Ronzhina, L. A. Ivanov, L. V. Stroukova, A. D. Peuke, and H. Rennenberg, "Over-expression of gsh1 in the cytosol affects the photosynthetic apparatus and improves the performance of transgenic poplars on heavy metal-contaminated soil," Plant Biology, vol. 13, no. 4, pp. 649–659, 2011. ·
379. Y. Li, A. C. P. Heaton, L. Carreira, and R. B. Meagher, "Enhanced tolerance to and accumulation of mercury, but not arsenic, in plants overexpressing two enzymes required for thiol peptide

synthesis," Physiologia Plantarum, vol. 128, no. 1, pp. 48–57, 2006. · ·

380. M. Pomponi, V. Censi, V. di Girolamo et al., "Overexpression of Arabidopsis phytochelatin synthase in tobacco plants enhances Cd^{2+} tolerance and accumulation but not translocation to the shoot," Planta, vol. 223, no. 2, pp. 180–190, 2006. ·
381. K. Gasic and S. S. Korban, "Expression of Arabidopsis phytochelatin synthase in Indian mustard (Brassica juncea) plants enhances tolerance for Cd and Zn," Planta, vol. 225, no. 5, pp. 1277–1285, 2007.
382. J. L. Couselo, J. Navarro-Avino, and A. Ballester, "Expression of the phytochelatin synthaseTaPCS1 in transgenic aspen, insight into the problems and qualities in phytoremediation of Pb,"International Journal of Phytoremediation, vol. 12, no. 4, pp. 358–370, 2010. · ·
383. S. Wojas, S. Clemens, A. Sklodowska, and D. M. Antosiewicz, "Arsenic response of AtPCS1-andCePCS-expressing plants—effects of external As(V) concentration on As-accumulation pattern and NPT metabolism," Journal of Plant Physiology, vol. 167, no. 3, pp. 169–175, 2010. ·
384. G.-Y. Liu, Y.-X. Zhang, and T.-Y. Chai, "Phytochelatin synthase of Thlaspi caerulescens enhanced tolerance and accumulation of heavy metals when expressed in yeast and tobacco," Plant Cell Reports, vol. 30, no. 6, pp. 1067–1076, 2011. ·
385. Y. Li, O. P. Dhankher, L. Carreira et al., "Overexpression of phytochelatin synthase in Arabidopsisleads to enhanced arsenic tolerance and cadmium hypersensitivity," Plant and Cell Physiology, vol. 45, no. 12, pp. 1787–1797, 2004.
386. J. Guo, X. Dai, W. Xu, and M. Ma, "Overexpressing GSH1 and AsPCS1 simultaneously increases the tolerance and accumulation of cadmium and arsenic in Arabidopsis thaliana," Chemosphere, vol. 72, no. 7, pp. 1020–1026, 2008. · ·
387. S. Misra and L. Gedamu, "Heavy metal tolerant transgenic Brassica napus L. and Nicotiana tabacum L. plants," Theoretical and Applied Genetics, vol. 78, no. 2, pp. 161–168, 1989.
388. K. Sekhar, B. Priyanka, V. D. Reddy, and K. V. Rao, "Metallothionein 1 (CcMT1) of pigeon pea (Cajanus cajan, L.) Confers enhanced

tolerance to copper and cadmium in Escherichia coli andArabidopsis thaliana," Environmental and Experimental Botany, vol. 72, no. 2, pp. 131–139, 2011.

389. G. S. Sanghera, P. L. Kashyap, G. Singh, and J. A. T. da Silva, "Transgenics: fast track to plant stress amelioration," Transgenic Plant Journal, vol. 5, no. 1, pp. 1–26, 2011.

390. C.-J. Ruan and J. A. T. da Silva, "Metabolomics: creating new potentials for unraveling the mechanisms in response to salt and drought stress and for the biotechnological improvement of xero-halophytes," Critical Reviews in Biotechnology, vol. 31, no. 2, pp. 153–169, 2011.

391. S. S. Gill, N. A. Khan, N. K. Anjum, and N. Tuteja, "Amelioration of cadmium stress in crop plants by nutrients management: morphological, physiological and biochemical aspects," Plant Stress, vol. 5, no. 1, pp. 1–23, 2011.

5

Fatty Acid-Amino Acid Conjugates are Essential for Systemic Activation of Salicylic Acid-Induced Protein Kinase and Accumulation of Jasmonic Acid in Nicotiana Attenuata

Christian Hettenhausen[1], Maria Heinrich[2], Ian T Baldwin[2], and Jianqiang Wu[1*]

[1]Kunming Institute of Botany, Chinese Academy of Sciences, Kunming, 650201, China

[2]Max Planck Institute for Chemical Ecology, Hans-Knoell Str. 8, Jena, 07745, Germany

ABSTRACT

Background

Herbivory induces the activation of mitogen-activated protein kinases (MAPKs), the accumulation of jasmonates and defensive metabolites in damaged leaves and in distal undamaged leaves. Previous studies mainly focused on individual responses and a limited number of systemic leaves, and more research is needed for a better understanding of how different plant parts respond to herbivory. In the wild tobacco Nicotiana attenuata, FACs (fatty acid-amino acid conjugates) inManduca sexta oral secretions (OS) are the major elicitors that induce herbivory-specific signaling but their role in systemic signaling is largely unknown.

Results

Here, we show that simulated herbivory (adding M. sexta OS to fresh wounds) dramatically increased SIPK (salicylic acid-induced protein kinase) activity and jasmonic acid (JA) levels in damaged leaves and in certain (but not all) undamaged systemic leaves, whereas wounding alone had no detectable systemic effects; importantly, FACs and wounding are both required for activating these systemic responses. In contrast to the activation of SIPK and elevation of JA in specific systemic leaves, increases in the activity of an important anti-herbivore defense, trypsin proteinase inhibitor (TPI), were observed in all systemic leaves after simulated herbivory, suggesting that systemic TPI induction does not require SIPK activation and JA increases. Leaf ablation experiments demonstrated that within 10 minutes after simulated herbivory, a signal (or signals) was produced and transported out of the treated leaves, and subsequently activated systemic responses.

Conclusions

Our results reveal that N. attenuata specifically recognizes herbivore-derived FACs in damaged leaves and rapidly send out a long-distance signal to phylotactically connected leaves to activate MAPK and JA signaling, and we propose that FACs that penetrated into wounds rapidly induce the production of another long-distance signal(s) which travels to all systemic leaves and activates TPI defense.

BACKGROUND

Herbivores pose a major threat to plants. To cope with this challenge, plants have evolved sophisticated defense systems to perceive damage and herbivore-derived elicitors (the so-called herbivore-associated molecular patterns, HAMPs) [1] and activate a chain reaction of downstream signaling events, including rapid activation of mitogen-activated protein kinases (MAPKs) [2]-[4], biosynthesis of phytohormones, such as jasmonic acid (JA), JA-isoleucine conjugate (JA-Ile), and ethylene [5], and reshaping transcriptomes, proteomes, and metabolomes.

It is believed that systemic responses prevent insects from escaping plant defense by moving to undefended tissues. Systemic defense was first discovered in tomato (Lycopersicon esculentum): after wounding, a signal was found to move to other parts of the plants and induce the production of an important defensive compound, proteinase inhibitor I (PI-I) [6]. In a wild tobacco, Nicotiana attenuata, in addition to PIs, transcriptional and metabolomic analyses indicated that various genes and metabolites are also up-regulated in systemic undamaged leaves and roots [7]-[9]. MAPKs and the phytohormones JA and JA-Ile are all upstream signaling molecules, which play important roles in regulating plant resistance to herbivores [3],[4],[10]-[13]. Wounding or herbivory activates MAPKs within a few minutes [3],[4],[14],[15] and rapidly induces the biosynthesis of JA, with levels peaking within 1–2 h [16],[17].

In tomato, cultivated tobacco, forage and turf grasses, rapid MAPK activation was also detected in certain systemic leaves after wounding [18]-[20]; however, wounding or treatment of simulated herbivory (wounding and application of herbivore oral secretions to wounds) did not result in changes of MAPK activity in the adjacent systemic leaf in N. attenuata[3], suggesting that systemic activation of MAPKs might be species-specific or dependent on leaf positions. Recently, it was found that wounding rapidly induces JA accumulation in systemic leaves in Arabidopsis [21],[22]. In contrast, wound treatment did not induce the accumulation of systemic jasmonates in N. attenuata, but increased JA and JA-Ile levels were found in systemic leaves after simulated herbivore feeding [23],[24]. Therefore, in addition to a long-distance signal that induces the accumulation of defensive compounds such as PIs in systemic leaves, another (or the same) signal or several signals rapidly travel to distal leaves and activates MAPK signaling and JA biosynthesis. A prerequisite for obtaining deeper insight into the molecular mechanisms underlying systemic defense is a thorough description of the spatial and temporal herbivory-induced responses in local and systemic leaves.

The wild tobacco, N. attenuata, is a diploid annual plant that inhabits the deserts of western North America. N. attenuata has been intensively studied in the aspect of how it responds to herbivory of the specialist insect Manduca sexta[25]. Feeding of M. sexta elicits the production of plant defense metabolites not only in the local leaves but also in systemic leaves distal to the wound sites [26]. Previous research on Arabidopsis, tomato, and tobacco has suggested that MAPK and JA signaling are involved in systemic responses [18],[19],[22]-[24]; however, most studies only focused on a rather limited number of systemic leaves and examined the responses either on the signaling or metabolite level. Here we comprehensively investigated the changes in MAPK activity, accumulation of JA/JA-Ile, as well as the levels of trypsin protease inhibitors (TPI), a typical systemic defense in Solanaceae, in local and systemic leaves after wounding and simulated herbivore treatments. We found that a rapid mobile signal induces salicylic acid-induced protein kinase (SIPK) activation and JA/JA-Ile accumulation in certain, but not all, systemic leaves in N. attenuata, and the production of this signal is highly dependent on fatty acid-amino acid conjugates (FACs) in M. sexta oral secretions (OS) that are introduced into wounds during feeding; furthermore, neither wounding nor FACs alone can induce elevated SIPK activity and JA/JA-Ile levels in systemic leaves. Using TPI activity assay and leaf ablation approach, we demonstrate that the pattern of TPI induction is different from that of systemically induced SIPK and JA/JA-Ile, and we propose that another signal travels at a similar speed to almost all systemic leaves to activate TPI biosynthesis.

RESULTS

Simulated M. sexta Herbivory Treatment Induces a Specific Spatial and Temporal Pattern of JA Accumulation in Nicotiana attenuata Systemic Leaves

Given the central role of JA in regulating plant resistance to herbivores, we first examined whether simulated herbivore feeding induces systemic JA production. Because JA-Ile, the conjugate of JA and isoleucine, but not JA itself, functions as the active jasmonate hormone [27], the concentrations of JA-Ile were also determined. Slightly elongated plants (about 10 cm in height, Figure 1a) were wounded at node 0 [local leaf; hereafter leaf 0, and leaves X were used for naming the leaves at node X (X represents the node number)], which was the second fully expanded leaf, and 20 μl of 1/5-diluted M. sexta OS were applied to the wounds (W+OS) to simulate M. sexta herbivory. JA and JA-Ile levels in local and systemic leaves were determined using a HPLC-MS/MS method. In the treated leaves, JA and JA-Ile levels increased after 10 min, the levels were highest 1 h after the treatments, and decreased to almost the basal levels 2.5 h after induction (Figure 1b). In contrast, JA and JA-Ile levels in systemic leaves showed a very distinct pattern. JA accumulated almost exclusively in leaves +3, with the highest levels 90 min after elicitation, whereas the other systemic leaves contained only minor amounts (Figure 1c). Importantly, the JA levels in leaves +3 were remarkable high: at 90 min after W+OS treatment, JA contents reached up to 6 $\mu g\ g^{-1}$ fresh mass (FM) JA, which were more than twice as much as the highest JA levels detected in the local leaves. The systemic distribution of JA-Ile was similar to that of JA but the highest levels in leaves +3 did not exceed those detected in local leaves, although 90 min after W+OS treatment JA-Ile contents in leaves +3 were also 2-fold greater than those in local leaves (Figure 1c). To determine whether systemic JA accumulations were limited to younger leaves, we elicited leaves 0 with W+OS and quantified JA and JA-Ile levels in leaves −4 (4 positions older than the elicited leaf) to leaves +4 (the youngest leaf) 90 min after W+OS elicitation (Figure 1d). Again, leaves +3 accumulated high JA levels (about 6.5 $\mu g\ g^{-1}$

FM), but increased contents of JA could also be detected in leaves −3 -2, and +2, with 1, 0.45, and 0.4 μg g^{-1} FM, respectively (Figure 1d). Remarkably, these leaves accumulated relative high amounts of JA-Ile: leaves −3 contained 225 ng g^{-1} FM, as did leaves +3 (Figure 1d).

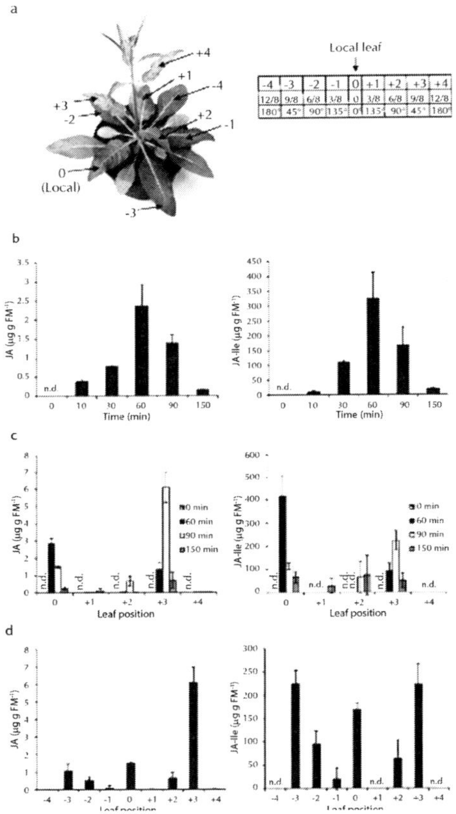

Figure 1: W + OS-induced JA accumulation in local and systemic leaves. The leaves undergoing source-sink transition, designated as the leaves 0, were wounded with a pattern wheel and 20 μl of 1/5 diluted M. sexta OS were immediately applied to wounds (W + OS). Samples were harvested at indicated times, and their JA and JA-Ile contents were analyzed. a Numbering of the leaf positions in a bolting N. attenuata plant and the additive angular distances assuming 3/8 phyllotaxis proceeding from the leaf 0. b JA and JA-Ile JA levels in treated local leaves 0. c JA and JA-Ile accumulation in local leaves 0 and in younger systemic leaves at different times after W + OS treatment. d JA and JA-Ile levels in local and systemic leaves 90 min after W + OS treatment. Values are mean ± SE; N = 5; n.d. = not detectable.

Thus, simulated M. sexta herbivory highly increases the accumulations of JA in local and systemic leaves, but systemically induced JA levels follow a pattern with very different increased levels in different leaves.

Both Wounding and FACs are required for Systemic JA Accumulation

In maize leaves, wounding induces JA accumulation only at the immediate site of damage, whereas insect elicitors also induce JA accumulation in distant tissues [28]. Previous research onN. attenuata revealed that after simulated herbivory, JA levels in distal leaves accumulate to less than 10% of the local maximum [9],[23],[24], and after wounding alone no increase in JA was detected [9]. To gain insight into the responses of systemic leaves to mechanical wounding, leaves 0 were wounded and 20 μl of water were applied (W+W), and JA and JA-Ile accumulations were determined in all leaves. Local JA levels increased 10 min after the treatment and reached high values between 30 and 90 min (about 550 ng g^{-1} FM), which were significantly smaller than those detected after W+OS elicitation; after 150 min, JA levels decreased back to almost basal levels (Figure 2a). JA-Ile followed the JA pattern but accumulated to the highest levels after 30 min (137 ng g^{-1} FM) (Figure 2a). In contrast to W+OS, W+W treatment did not lead to detectably increased JA and JA-Ile levels in any systemic leaves 90 min after treatment (Figure 2b) and other times examined (30, 60, and 150 min; these data are not shown). These findings suggest that the systemic JA accumulation is OS-dependent and wounding alone has non-detectable effect on systemic JA levels. FACs in M. sexta OS are known to be the elicitors for OS-specific plant responses, such as MAPK activation [3], JA burst, and accumulation of defense metabolites [17],[29]. Given that applying OS to wounds (W+OS) induced systemic JA accumulation, we next explored whether FACs were responsible for this systemic response.

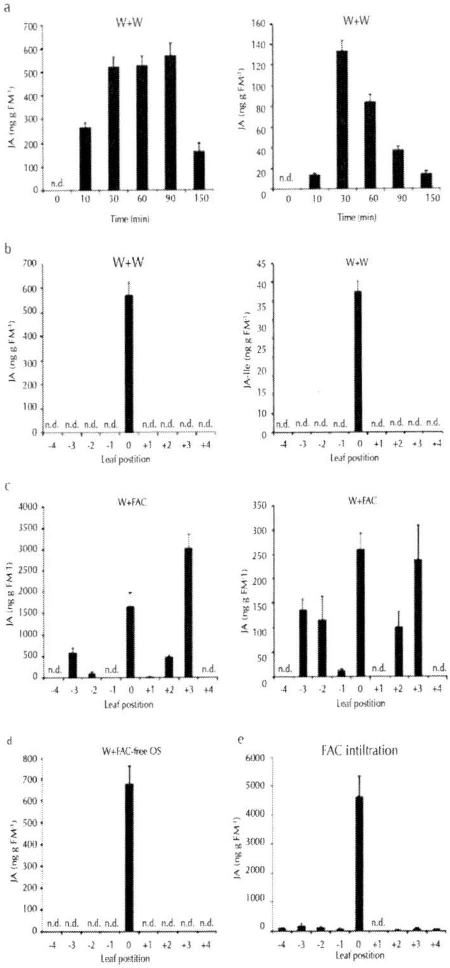

Figure 2: Wounding and FACs are both required for systemic JA accumulation. a JA levels in local leaves 0 after wounding. The leaves 0 were wounded with a pattern wheel and 20 μl of water were immediately applied to wounds (W+W), and the JA contents were analyzed. b JA and JA-Ile contents in local and systemic leaves 90 min after W+W treatment. c and d JA and JA-Ile values in different leaves 90 min after applying FAC or FAC-free OS to wounds. Leaves 0 were wounded with a pattern wheel and 20 μl of FAC (27.6 ng/μl; W+FAC) or 20 μl FAC-free M. sexta OS (W+FAC-free OS) were immediately applied to wounds. e JA contents in different leaves 90 min after pressure infiltration of 100 μl FAC (27.6 ng/μl) into leaves 0. Values are mean±SE; N = 5; n.d. = not detectable.

Twenty microliter of N-linolenoyl-L-Glu, one of the most abundant FACs in M. sexta OS [17], at 27.6 ng/μl (similar to its concentration in 1/5 diluted OS), were applied to freshly wounded N. attenuata leaves 0 (W+FAC) and leaf samples were harvested 90 min after the treatment when systemic hormone levels attained the highest values (Figure 1c); in parallel, we applied FAC-free OS to wounds (W+FAC-free OS). The W+FAC-induced JA and JA-Ile levels in local and systemic leaves largely resembled those after W+OS treatment (Figure 2c), and similar to W+W, FAC-free OS highly induced JA in the local tissue but systemic JA remained below the detection limit, indicating that FACs are necessary to elicit systemic JA (Figure 2d). To exclude the possibility that FACs were transported from local to distal leaves through the plant vascular system and thus induced systemic JA accumulation, we pressure-inoculated 100 μl of a FAC solution (27.6 ng/μl) into leaves 0 and measured local and systemic (leaves +3) JA levels after 90 min. The solvent for the FAC (0.05% Tween 20) was inoculated as the control treatment, and it did not induce local or systemic JA (data not shown). In contrast, FAC-inoculated leaves accumulated 4.6 μg g^{-1} FM JA, almost 3 times as much as the induced JA levels after W+FAC treatments (Figure 2e), and importantly, FAC inoculation did not induce a strong accumulation of JA in systemic leaves, unlike what we saw in W+OS-treated plants (Figure 2e).

These data indicate that both FACs and wounding are required to induce a systemic signal that leads to JA accumulation in distal leaves.

Simulated Herbivory Treatment Induces MAPK Activity in Systemic Leaves

MAPKs are an essential part of the signaling cascade induced by wounding or herbivore attack. In tomato, wounding activates MAPKs both locally and systemically [18]. Wounding tobacco leaves with carborundum quickly increases the levels of WIPK (wound-induced protein kinase) transcripts in systemic leaves, and cutting tobacco stem activates WIPK systemically [19]. FACs are strong elicitors that amplify wounding-induced MAPK activation and potentiate the elicited JA burst in N. attenuate [3],[17]. To explore whether systemic JA accumulation was correlated with increased MAPK activity in these tissues, we performed a series of in-gel MAPK activity assays. The basal SIPK activity in uninduced plants was similarly very low in all leaves (Additional file 1). Following W+OS induction in leaves 0, SIPK activity increased locally and systemically and the distribution of SIPK activity levels in different leaves greatly resembled that of JA levels in these leaves (Figure 3a). Silencing SIPK highly compromises herbivory-induced JA accumulation [3], and these data suggest that SIPK activity might also be required for systemic JA induction: SIPK activity was the highest in the treated local leaves but about 50% less in leaves +3, and leaves −3 and −2 had activity levels slightly above those in controls (Additional file 1). To gain further insight into the regulation of systemic kinase activation, we performed a time course experiment and compared kinase activity in local and the systemic leaves +3. In treated leaves SIPK activity rapidly increased within 10 min, peaked at 30 min, and remained high until 90 min after induction (Figure 3b). In contrast, increase of SIPK activity in leaves +3 was not detected until 30 min, and was very transient with a maximum at 60 min after the treatment (Figure 3b). This delayed MAPK response and subsequent JA accumulation in systemic leaves might reflect the time that the mobile signal(s) needs to travel from the treated leaves to the distal ones.

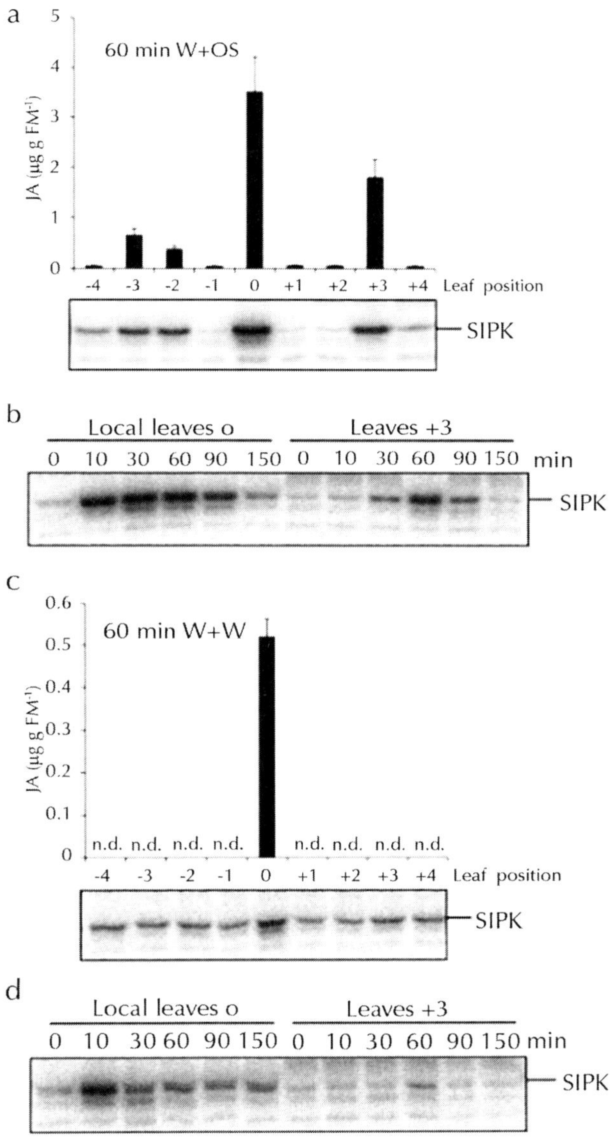

Figure 3: SIPK activation and JA accumulation show similar leaf distribution. The leaves 0 were wounded with a pattern wheel and 20 µl of 1/5 diluted M. sexta OS (a and b) or water (c and d) were immediately applied to wounds (W+OS). Local and systemic leaves (N=5) were harvested at indicated times and JA contents (mean±SE) were analyzed; SIPK activity was analyzed in pooled samples using an in-gel kinase assay.

Consistent with the lack of elevated JA levels, W+W did not increase systemic SIPK activity (Figure 3c). A more detailed analysis of the kinetic of SIPK activation also revealed no strong increase of SIPK activity in the leaves +3 (Figure 3d).

We conclude that after M. sexta herbivory, but not wounding, a mobile signal is rapidly propagated from damaged leaves to specific systemic leaves to induce MAPK signaling, and activation of MAPKs likely further triggers JA biosynthesis.

Systemic Induction of Trypsin Protease Inhibitors Does Not Require Increased MAPK Activity or JA Contents in Systemic Leaves

M. sexta attack increases the levels of TPI transcripts and activity in N. attenuata. This response is not limited to attacked leaves but spreads to systemic ones [30],[31]. TPI expression is dependent on JA signaling as COI1- and LOX3-silenced plants that are defective in JA perception and production, respectively, have very little TPI activity and do not accumulate TPI after W+OS elicitation [32],[33]. To investigate the pattern of TPI activity in local and systemic leaves and to reveal whether it is correlated with MAPK activation and induced JA/JA-Ile levels, we treated leaves 0 with W+W and W+OS and analyzed TPI activity in local and systemic leaves 3 days after elicitation. Without treatment, highest TPI levels were found in young leaves, and the older ones had very low TPI activity (Figure 4). Wounding elicited increased TPI activity levels in leaves 0 and in leaves +2 and +3, with values 2- to 3-fold of those in uninduced respective controls. In contrast, W+OS treatment induced TPI levels in almost all leaves (Figure 4). Similar to W+W treatment, highest values were detected in the local leaves 0 and in leaves +2 and +3, whose TPI activity levels were twice as much as those induced by wounding; remarkably, despite relatively high W+OS-induced JA-Ile levels in systemic leaves −3 and −2 (Figure 1d), these leaves exhibited only minor TPI activity (Figure 4).

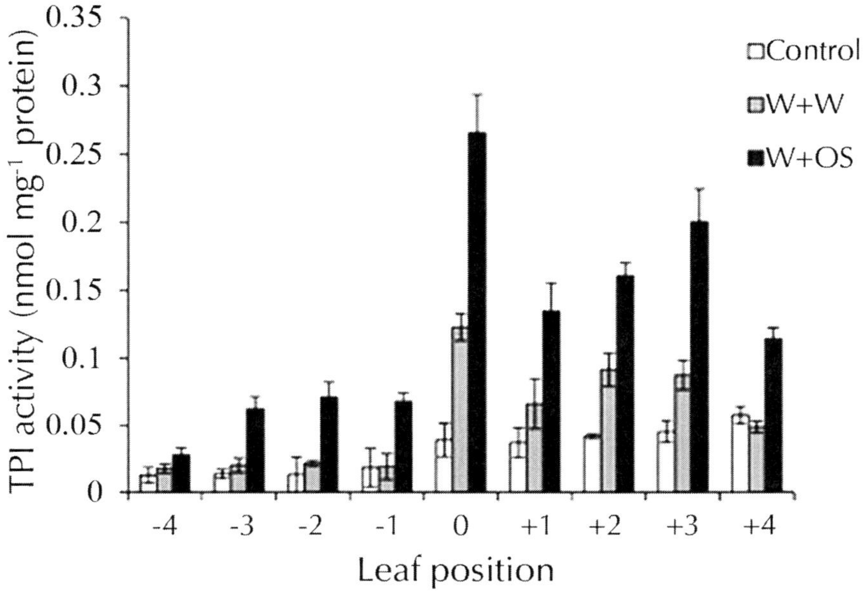

Figure 4: TPI activity in local and systemic leaves. Leaves 0 were wounded with a pattern wheel and 20 μl of water (W+W) or M. sexta OS (W+OS) were applied to the wounds. Treated leaves and systemic leaves were harvested 3 days after the treatment and TPI activity was analyzed. Values are mean ± SE; N = 5.

We supposed that older leaves might have decreased inducibility of TPI after elicitation of JA. To examine the inducibility of TPI in different leaves in response to jasmonate elicitation, methyl jasmonate was applied to leaves at all positions on individual plants and TPI activity was quantified after 3 days. TPI activity levels were lowest in the oldest leaves −4, increased in younger leaves, and the youngest leaves +4 had about 3.6 times more TPI activity than leaves −4 (Additional file2), confirming that JA-induced TPI levels decrease with increasing leaf age.

Therefore, unlike wounding, simulated M. sexta feeding induces increase of TPI activity in almost all leaves, although systemic TPI activity increases more strongly in younger leaves. Importantly, systemic leaves that have highly induced TPI activity do not necessarily have elevated MAPK activity and JA contents.

Rapid Mobile Long-Distance Signals Induce Systemic Defense Responses

The increased MAPK activity, JA levels, and TPI activity in systemic leaves after W+OS elicitation revealed that certain long-distance signals are propagated from local leaves to systemic ones to activate these responses. To estimate the time required for the TPI-inducing systemic signal to exit from the wounded leaf, we treated leaves 0 with W+OS and then removed them (by excising from petioles) at 0, 1, 5, and 10 min after the treatments; untreated plants and plants treated with W+OS whose local leaves were retained were used for comparisons. Three days after these initial treatments, TPI activity was measured in systemic leaves (Figure 5a). Immediately removing the treated leaves did not induce any changes of TPI activity, and similarly, excision of the damaged leaves in 1 or 5 min also induced very little systemic TPI (Figure 5a). However, when the local leaves were removed 10 min after the treatment, TPI activity levels in systemic leaves almost fully elevated to those in plants whose treated leaves were retained (Figure 5a). These results suggest that systemic TPI induction involves a signal that exits the wounded leaves between 5 to 10 min, and given that the petiole lengths are about 3 cm, the speed of the signal traveling out of the treated leaves is approximately 0.3 cm/min. These findings are consistent with an earlier study inN. attenuata where it was shown that removing a 3-mm-wide zone adjacent to the W+OS treatment site within 40 s did not prevent the induction of JA in the remaining leaf tissue [34].

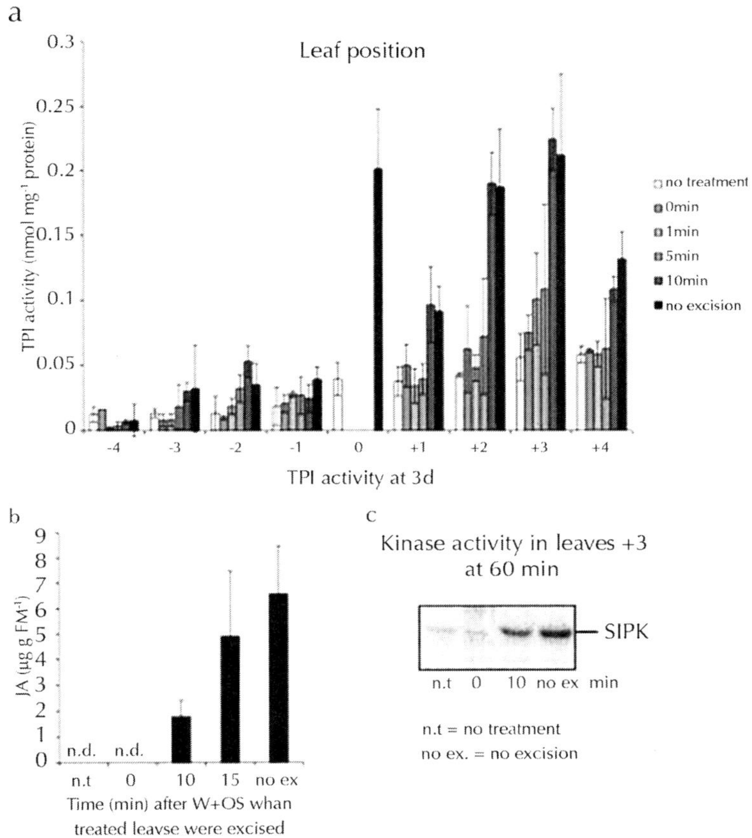

Figure 5: Systemic responses after W+OS elicitation and leaf excision. Local leaves 0 were W+OS-elicited, and these leaves including petioles were ablated at indicated times after treatment and the elicited systemic responses were determined. a TPI activity (mean ± SE, N =5) in different leaves, 3 d after elicitation of leaves 0, which were either not excised or ablated at different times [untreated plants ("no treatment") served as comparisons]. b JA accumulation (mean ± SE, N = 5) in leaves +3, 90 min after local leaves were elicited with W+OS and ablated at indicated times. c SIPK activity in systemic leaves +3, 60 min after the leaves 0 were treated with W+OS.

To investigate how fast the signal that triggers MAPK activation and JA accumulation travels out of herbivore-damaged leaves, we excised W+OS-elicited leaves at different times after the treatment and measured JA accumulation in leaves +3 after 90 min when JA contents reach the highest values. Leaf excision alone did not increase systemic JA levels (Figure 5b), and excising the local leaves 10 min after treatment resulted in 30% increased JA contents; excising the local leaves 15 min after the treatment almost fully elicited JA levels in leaves +3 (Figure 5b). Likewise, leaves +3 from plants whose local leaves were ablated 10 min after W+OS treatment showed similar SIPK activity levels as those whose local leaves were retained (Figure 5c). It seems that the speed of this signal is not very different from that of the signal activating systemic TPI.

DISCUSSION

Herbivore feeding induces plant defense responses not only in the local attacked leaves, but also in distal undamaged ones. How plants regulate these defense responses is still poorly understood. Here we demonstrate that M. sexta OS applied to wounds elicits systemic induction of MAPK activity and JA accumulation. Our results suggest that N. attenuata is able to recognize herbivore feeding by perceiving FACs penetrated into wounds and deploying specific responses in undamaged systemic leaves, including MAPK activation, JA accumulation, and later, increased TPI activity.

Herbivory But Not Wounding Elicits Early Systemic Responses

Studies on N. tabacum revealed 3/8 phyllotaxis for rosette stage plants and 5/13 phyllotaxis for the stem of elongated plants [35]-[37]. Accordingly, the systemic leaves analyzed in the present study were not directly vascular connected with the treated leaves but grew in specific angular distances resulting in different transvascular resistance levels [30],[38]. As the transvascular resistance is often higher than the axial resistance, especially when the stems are relatively short[38], the angular distances between the treated leaves and systemic leaves may significantly influence the systemic signaling. In tomato, the intensity of systemic TPI accumulation was found to correlate with the degree of vascular linkage between and within leaves [39],[40]. The same is true for salicylic acid transport in N. tabacum[41]. We detected highest JA levels in leaves +3, which have together with leaves −3 the smallest angular distance (45 degree) to the local leaves at node 0 (Figure 1a and d). Also leaves +2 and −2, with a shift of about 90 degrees to node 0, had significantly increased JA levels 90 min after W+OS (Figure 1d). In contrast, leaves +1 and −1 with about 135 degree, and +4 and −4 leaves with about 180 degree angles to the local leaves did not show increased JA levels even 150 min after elicitation (Figure 1c). Clearly, the angular distance between local and systemic leaves is important in determining the levels of JA in those leaves and the elicited JA contents decrease with increasing angles.

Several other studies conducted on N. attenuata revealed only minor systemic JA concentrations, which were about 5-10% of the locally induced JA levels [9],[23],[24],[42]. However, our comprehensive analysis indicated that systemic responses depend on leaf positions and the time after treatment. Furthermore, in Arabidopsis and Solanum nigrum, systemic JA levels also increase to only 10% of the local values [21],[22],[43], implying that systemic defense signaling might be species specific. It was found that N. attenuata systemic leaves −1 do not have elevated SIPK activity after simulated M. sexta herbivory treatment was applied to leaves +1 [3]; however, this comprehensive study pointed out that after W+OS treatment, some systemic leaves do have highly elevated SIPK activity, and the previously proposed model should be updated.

Two lines of evidence support the notion that highly elevated systemic JA levels are unlikely to be transported from the damaged leaves to the systemic ones, but JA is de novo synthesized in the systemic leaves: Firstly, W+OS-induced JA levels in leaves +3 even exceeded those in the local leaves. Secondly, our leaf ablation experiments revealed that 10 min after local induction, the systemic signal had left the treated leaves and at this time point W+W and W+OS treatment elicited similar amounts of JA in local leaves (Figure 1b and 2a) but only W+OS induced systemic JA accumulation. These findings are also supported by the studies in N. attenuata and Arabidopsis that JA-Ile and MeJA are de novo synthesized in systemic leaves, not transported from the wounded leaves [9],[22],[23].

In Arabidopsis, wounding is sufficient to elevate systemic JA levels [21],[22], but in Zea mays, Solanum nigrum, and N. attenuata, wounding alone induces JA accumulation only at the adjacent site of damage, whereas insect elicitors induce JA accumulation in distant tissues

[9],[28],[44]. Similarly, systemic MAPK activation after wounding has been reported in some plant species, including soybean, tomato, and tobacco [18],[19],[45]; but wounding alone failed to induce systemic MAPK activity in N. attenuata (this study). By adding FACs to wounds and by removing them from OS, we show that FACs are the elicitors of the systemic JA response; however, FACs themselves appeared not to be the systemic signal, given that 1) FACs are quickly degraded after entering plant tissue [44], and 2) inoculating FACs into local tissue did not elicit JA responses in distal leaves (Figure 2e). The reason why FACs require wounding for activating systemic JA accumulation remains unknown, but it is possible that wounds are necessary for efficient loading of FACs to mechanically broken vascular tissues. During infiltration, FACs may remain in the apoplast and could not be transported to systemic leaves. Radio-traceable FACs could be used for elucidating whether FACs can be transported to systemic leaves.

These findings strongly suggest that a rapid mobile signal, which is elicited by FACs penetrated into wounds, but not by wounding alone, activates SIPK, and thereafter, SIPK activates JA biosynthesis in systemic leaves.

Herbivory, but not Wounding, Strongly Activates the Late Systemic Response, TPI Accumulation

We found that unlike SIPK and JA, which were activated only in specific systemic leaves, simulated herbivory elicited the accumulation of TPI in all systemic leaves tested, but wounding only elevated TPI levels in systemic leaves +2 and +3. The different distribution between early (SIPK and JA) and late (TPI) responses argues that the signal that triggers systemic TPI accumulation is likely different from the one that activates SIPK and initiates JA biosynthesis, and systemically increased JA levels are not important for elevation of TPI activity. Alternatively, the systemic leaf +3 with very pronounced JA accumulation (and MAPK activity) could serve as a "hub" for jasmonate distribution throughout the plant by inducing leaves in close phyllotactic positions and other distal leaves. Moreover, it cannot be excluded that TPI protein itself is re-distributed within the whole plant and thus also accumulates in leaves without a previous JA induction. These possibilities should be examined further.

The biological significance of the specific spatial distribution of systemic SIPK and JA remains unknown. We hypothesize that certain defense responses, such as terpenoids, some of which are known to have a function as indirect defenses [46], downstream of these signaling factors are also specifically mounted in these systemic leaves, and furthermore, systemic activation of SIPK and JA may also induce transmissible signals to other parts of the plant to further propagate or strengthen systemic defenses (metabolites).

In tomato, grafting of mutants deficient in JA production and perception indicated that induction of systemic TPI requires both the biosynthesis of jasmonic acid at the site of wounding and the ability to perceive a jasmonate signal in systemic leaves, but JA biosynthesis in systemic leaves and JA perception in local ones are not important for systemic TPI induction [47], [48]. Our leaf ablation experiments showed that the signal released within 10 min after W+OS treatment from local leaves almost fully induced TPI activity in systemic leaves; furthermore, within 10 min, local JA levels only elevated to about 10% of the highest JA levels produced, and these were similar in W+W and W+OS-treated leaves (Figures 1b and 5b). Thus, these data suggest that JA levels induced in local leaves are not directly involved in controlling systemic TPI accumulation. We propose that a signal induced by FACs is transported to systemic leaves, and there together with JA signaling (but not JA biosynthesis), induces TPI production. This intriguing observation clearly deserves more attention.

The Nature of the Mobile Signals

Several studies suggest the involvement of hydraulic or electric signals in systemic signaling [22], [49]-[52]. Given that our treatments W+W and W+OS likely generate similar hydraulic pressures to the systemic tissues, the hypothesis that hydraulic pressure is the only mobile signal can be ruled out. In lima bean (Phaseolus lunatus), FACs, but not wounding alone, specifically induce changes of cell membrane polarization [53]. Recent data from Arabidopsis indicate that wounding activates surface potential changes and experimental current injection into leaves leads to activation of JA biosynthesis and transcriptome changes [54]. It would be valuable to examine the changes of surface potentials of N. attenuata in local and systemic leaves. In N. attenuata, the signal that induces systemic SIPK and JA accumulation exits the treated leaves at about 0.3 cm/min; in Arabidopsis, the speed of the mobile signal, which induces JA-Ile accumulation in systemic leaves, is about 2 cm/min [22]; in contrast, in Solanum nigrum, the mobile signal that elicits the systemic defensive compound, leucine aminopeptidase, needs much longer time – 90 to 240 min to exit the local leaves [44]. Elucidating the nature of the mobile signals in different species will also shed light on the large variations of the speeds of these signal transmissions.

In addition to TPIs, nicotine, and terpene-derived volatiles serve as important herbivory-inducible systemic defenses in N. attenuata [55]-[57]. Given that very likely different mobile signals induce systemic accumulation of JA (and activation of SIPK) and TPI, possibly other types of mobile signals are responsible for activating other systemic defenses; for example, recently, it was found that in N. attenuata JA perception and synthesis are important for wounding-induced putrescine methyltransferase transcript levels in roots and for the transport of de novo synthesized nicotine to leaves, implying that the regulation of root nicotine is modulated by a pathway different from the one that controls systemic TPI [58]. Transcriptome rearrangements and metabolite accumulations have also been observed in systemic leaves in other species, such as Arabidopsis, tomato, poplar, and soybean [6],[22],[59],[60]. The identities of the transmissible signals, whether they are similar or species-specific, and how they are transported and function, and importantly, the ecological function of systemic defense are all very interesting questions to explore.

CONCLUSIONS

This study comprehensively demonstrates how plants respond to leaf herbivory on multiple levels, including signaling and defensive metabolite accumulation in local and systemic leaves, and highlights the importance of insect-derived elicitors in plant systemic defenses.

METHODS

Plant Growth and Sample Treatments

Nicotiana attenuata Torr. Ex W. (originally collected from the DI ranch, Santa Clara, UT) (Solanaceae) seeds were from an inbred line maintained in the Baldwin laboratory, and the seeds can be distributed by I.T. Baldwin, Max Planck Institute for Chemical Ecology, upon request. Voucher specimens of N. attenuata can be accessed at the Cornell University Herbarium (1989, I.T. Baldwin).

Seed germination and plant cultivation followed Krügel et al. [61]. Seeds were germinated on Petri dishes to synchronize their germination, and the seedlings were transferred to soil after 10 days. Four- to 5-week-old plants were used for all experiments.

For collection of M. sexta oral secretions (OS), M. sexta larvae were reared on N. attenuata plants until the third to fifth instars. OS were collected on ice as described in Roda et al.[62] and stored under nitrogen at −20°C. For simulated herbivory treatment, leaves at position 0 were wounded with a pattern wheel and 1/5 diluted OS were immediately rubbed onto each wounded leaf (W+OS); for wounding treatment, leaves were wounded with a pattern wheel, and 20 μl of water were rubbed onto each leaf (W+W). MeJA (methyl jasmonate) was dissolved in heat-liquefied lanolin at a concentration of 7.5 mg/ml; 20 μl of the resulting lanolin paste was applied to the base of the leaves, and pure lanolin was applied as a control. FAC (N-linolenoyl-L-Glu) was synthesized in-house [17], which was dissolved in 0.05% Tween 20 at a concentration of 27.6 ng/μl (similar to that in 1/5 diluted OS). FAC-free OS was prepared by passing OS four times through spin columns filled with Amberlite IRA-400 resin (Sigma-Aldrich) [17]. Twenty microliters of each test solution were applied to each leaf. After specific times, leaves were excised, immediately frozen in liquid nitrogen, and stored at −80°C until use.

Analysis of JA and JA-Ile Concentrations

One milliliter of ethyl acetate spiked with 200 ng of D_2-JA and 40 ng of $^{13}C_6$-JA-Ile, the internal standards for JA and JA-Ile, respectively, was added to each briefly crushed leaf sample (~150 mg). Samples were then ground on a FastPrep homogenizer (Thermo Electron). After being centrifuged at 13,000 g for 10 min at 4°C, supernatants were transferred to fresh tubes and evaporated to dryness on a vacuum concentrator (Eppendorf). Each residue was resuspended in 0.5 ml of 70% methanol (v/v) and centrifuged at 13,000 g for 15 min at 4°C to remove particles. The supernatants were analyzed on a HPLC-MS/MS (LCMS8040, Shimadzu).

In-Gel Kinase Activity Assay

Leaf tissue pooled from 4 replicate leaves was crushed in liquid nitrogen, and 200 μl of protein extraction buffer [100 mM HEPES, pH 7.5, 5 mM EDTA, 5 mM EGTA, 10 mM Na_3VO_4, 10 mM NaF, 50 mM -glycerolphosphate, 1 mM phenylmethylsulfonyl fluoride, 10% glycerol, and EDTA-free proteinase inhibitor cocktail (Roche Diagnostics)] was added to ~100 mg of tissue. Leaf tissue was then completely suspended by vortexing. After being centrifuged at 4°C at maximum speed for 20 min, supernatants were transferred to fresh tubes. Protein concentrations were measured using a Bio-Rad protein assay kit with bovine serum albumin as a standard. In-gel MAPK activity assays were done following Zhang & Klessig [63] using myelin basic protein (MBP) as the substrate. Gel images were obtained on an FLA-3000 phosphor imager system (Fujifilm).

Analyses of TPI Activity

TPI activity was analyzed with a radial diffusion assay described by van Dam et al.[31].

Availability of Supporting Data

The data sets supporting the results of this article are included within the article and its additional files.

ACKNOWLEDGEMENTS

We thank Dr. Mario Kallenbach for synthesizing FACs. This work was funded by the Strategic Priority Research Program of the Chinese Academy of Sciences, Grant No. XDB11050200, the Max Planck Society, a grant from the Yunnan Talent Recruitment Program (Grant No. 2012HA016), and the 1000 Young Talent Recruitment.

REFERENCES

1. Mithofer A, Boland W: Recognition of herbivory-associated molecular patterns. Plant Physiol 2008, 146(3):825-831.
2. Hettenhausen C, Baldwin IT, Wu JQ: Nicotiana attenuata MPK4 suppresses a novel jasmonic acid (JA) signaling-independent defense pathway against the specialist insect Manduca sexta, but is not required for the resistance to the generalist Spodoptera littoralis. New Phytol 2013, 199(3):787-799.
3. Wu JQ, Hettenhausen C, Meldau S, Baldwin IT: Herbivory rapidly activates MAPK signaling in attacked and unattacked leaf regions but not between leaves of Nicotiana attenuata. Plant Cell 2007, 19(3):1096-1122.
4. Kandoth PK, Ranf S, Pancholi SS, Jayanty S, Walla MD, Miller W, Howe GA, Lincoln DE, Stratmann JW: Tomato MAPKs LeMPK1, LeMPK2, and LeMPK3 function in the systemin-mediated defense response against herbivorous insects. Proc Natl Acad Sci U S A 2007, 104(29):12205-12210.
5. Schmelz EA, Engelberth J, Alborn HT, Tumlinson JH, Teal PEA: Phytohormone-based activity mapping of insect herbivore-produced elicitors. Proc Natl Acad Sci U S A 2009, 106(2):653-657.
6. Green TR, Ryan CA: Wound-induced proteinase inhibitor in plant leaves - possible defense mechanism against insects. Science 1972, 175(4023):776.
7. Heidel AJ, Baldwin IT: Microarray analysis of salicylic acid- and jasmonic acid-signalling in responses of Nicotiana attenuata to attack by insects from multiple feeding guilds. Plant Cell Environ 2004, 27(11):1362-1373.

8. Kim SG, Yon F, Gaquerel E, Gulati J, Baldwin IT: Tissue Specific Diurnal Rhythms of Metabolites and Their Regulation during Herbivore Attack in a Native Tobacco, Nicotiana attenuata. Plos One 2011, 6(10):e26214
9. von Dahl CC, Baldwin IT: Methyl jasmonate and cis-jasmone do not dispose of the herbivore-induced jasmonate burst in Nicotiana attenuata. Physiol Plant 2004, 120(3):474-481.
10. Li Q, Xie QG, Smith-Becker J, Navarre DA, Kaloshian I: Mi-1-mediated aphid resistance involves salicylic acid and mitogen-activated protein kinase signaling cascades. Mol Plant Microbe In 2006, 19(6):655-664.
11. Howe GA, Jander G: Plant immunity to insect herbivores. Annu Rev Plant Biol 2008, 59:41-66.
12. Wasternack C: Jasmonates: an update on biosynthesis, signal transduction and action in plant stress response, growth and development. Ann Bot-London 2007, 100(4):681-697.
13. Kessler A, Halitschke R, Baldwin IT: Silencing the jasmonate cascade: induced plant defenses and insect populations. Science 2004, 305(5684):665-668.
14. Seo S, Okamoto M, Seto H, Ishizuka K, Sano H, Ohashi Y: Tobacco MAP kinase: a possible mediator in wound signal transduction pathways. Science 1995, 270(5244):1988-1992.
15. Ichimura K, Mizoguchi T, Yoshida R, Yuasa T, Shinozaki K: Various abiotic stresses rapidly activate Arabidopsis MAP kinases ATMPK4 and ATMPK6. Plant J 2000, 24(5):655-665.
16. Park JH, Halitschke R, Kim HB, Baldwin IT, Feldmann KA, Feyereisen R: A knock-out mutation in allene oxide synthase results in male sterility and defective wound signal transduction in Arabidopsis due to a block in jasmonic acid biosynthesis. Plant J 2002, 31(1):1-12.
17. Halitschke R, Schittko U, Pohnert G, Boland W, Baldwin IT: Molecular interactions between the specialist herbivore Manduca sexta (Lepidoptera, Sphingidae) and its natural host Nicotiana attenuata. III. Fatty acid-amino acid conjugates in herbivore oral secretions are necessary and sufficient for herbivore-specific plant responses. Plant Physiol 2001, 125(2):711-717.

18. Stratmann JW, Ryan CA: Myelin basic protein kinase activity in tomato leaves is induced systemically by wounding and increases in response to systemin and oligosaccharide elicitors. Proc Natl Acad Sci U S A 1997, 94(20):11085-11089.

19. Seo S, Sano H, Ohashi Y: Jasmonate-based wound signal transduction requires activation of WIPK, a tobacco mitogen-activated protein kinase. Plant Cell 1999, 11(2):289-298.

20. Dombrowski JE, Hind SR, Martin RC, Stratmann JW: Wounding systemically activates a mitogen-activated protein kinase in forage and turf grasses. Plant Sci 2011, 180(5):686-693.

21. Glauser G, Grata E, Dubugnon L, Rudaz S, Farmer EE, Wolfender JL: Spatial and temporal dynamics of jasmonate synthesis and accumulation in Arabidopsis in response to wounding. J Biol Chem 2008, 283(24):16400-16407.

22. Koo AJK, Gao XL, Jones AD, Howe GA: A rapid wound signal activates the systemic synthesis of bioactive jasmonates in Arabidopsis. Plant J 2009, 59(6):974-986.

23. Wang L, Allmann S, Wu J, Baldwin IT: Comparisons of LIPOXYGENASE3- and JASMONATE-RESISTANT4/6-silenced plants reveal that jasmonic acid and jasmonic acid-amino acid conjugates play different roles in herbivore resistance of Nicotiana attenuata. Plant Physiol 2008, 146(3):904-915.

24. Stitz M, Gase K, Baldwin IT, Gaquerel E: Ectopic expression of AtJMT in Nicotiana attenuata: creating a metabolic sink has tissue-specific consequences for the jasmonate metabolic network and silences downstream gene expression. Plant Physiol 2011, 157(1):341-354.

25. Wu JQ, Baldwin IT: New insights into plant responses to the attack from insect herbivores. Annu Rev Genet 2010, 44:1-24.

26. Baldwin IT: An ecologically motivated analysis of plant-herbivore interactions in native tobacco. Plant Physiol 2001, 127(4):1449-1458.

27. Browse J: Jasmonate passes muster: a receptor and targets for the defense hormone. Annu Rev Plant Biol 2009, 60:183-205.

28. Engelberth J, Seidl-Adams I, Schultz JC, Tumlinson JH: Insect elicitors and exposure to green leafy volatiles differentially upregulate major octadecanoids and transcripts of 12-oxo

phytodienoic acid reductases in Zea mays. Mol Plant Microbe In 2007, 20(6):707-716.

29. Halitschke R, Gase K, Hui DQ, Schmidt DD, Baldwin IT: Molecular interactions between the specialist herbivore Manduca sexta (Lepidoptera, Sphingidae) and its natural host Nicotiana attenuata. VI. Microarray analysis reveals that most herbivore-specific transcriptional changes are mediated by fatty acid-amino acid conjugates. Plant Physiol 2003, 131(4):1894-1902.
30. Schittko U, Baldwin IT: Constraints to herbivore-induced systemic responses: bidirectional signaling along orthostichies in Nicotiana attenuata. J Chem Ecol 2003, 29(3):763-770.
31. van Dam NM, Horn M, Mares M, Baldwin IT: Ontogeny constrains systemic protease inhibitor response in Nicotiana attenuata. J Chem Ecol 2001, 27(3):547-568.
32. Halitschke R, Baldwin IT: Antisense LOX expression increases herbivore performance by decreasing defense responses and inhibiting growth-related transcriptional reorganization in Nicotiana attenuata. Plant J 2003, 36(6):794-807.
33. Paschold A, Halitschke R, Baldwin IT: Co(i)-ordinating defenses: NaCOI1 mediates herbivore-induced resistance in Nicotiana attenuata and reveals the role of herbivore movement in avoiding defenses. Plant J 2007, 51(1):79-91.
34. Schittko U, Preston CA, Baldwin IT: Eating the evidence? Manduca sexta larvae can not disrupt specific jasmonate induction in Nicotiana attenuata by rapid consumption. Planta 2000, 210(2):343-346.
35. Allard HA: Some aspects of the phyllotaxy of tobacco. J Agric Res 1942, 64:0049-0055.
36. Shirova M, Lister GR DNC, Krotov G: Translocation of C14 in tobacco at different stages of development following assimilation of C14 O2 by a single leaf. Can J Bot 1961, 39:855-864.
37. Jones H, Martin RV, PORTER HK: Translocation of 14carbon in tobacco following assimilation of 14carbon dioxide by a single leaf. Ann Bot London 1959, 23:493-510.
38. Fiscus EL, Parsons LR, Alberte RS: Phyllotaxy and water relations in tobacco. Planta 1973, 112(4):285-292.

39. Rhodes JD, Thain JF, Wildon DC: Evidence for physically distinct systemic signalling pathways in the wounded tomato plant. Ann Bot-London 1999, 84(1):109-116.
40. Orians CM, Pomerleau J, Ricco R: Vascular architecture generates fine scale variation in systemic induction of proteinase inhibitors in tomato. J Chem Ecol 2000, 26(2):471-485.
41. Shulaev V, Leon J, Raskin I: Is salicylic-acid a translocated signal of systemic acquired-resistance in tobacco. Plant Cell 1995, 7(10):1691-1701.
42. Pluskota WE, Qu N, Maitrejean M, Boland W, Baldwin IT: Jasmonates and its mimics differentially elicit systemic defence responses in Nicotiana attenuata. J Exp Bot 2007, 58(15–16):4071-4082.
43. VanDoorn A, Bonaventure G, Schmidt DD, Baldwin IT: Regulation of jasmonate metabolism and activation of systemic signaling in Solanum nigrum: COI1 and JAR4 play overlapping yet distinct roles. New Phytol 2011, 190(3):640-652.
44. VanDoorn A, Kallenbach M, Borquez AA, Baldwin IT, Bonaventure G: Rapid modification of the insect elicitor N-linolenoyl-glutamate via a lipoxygenase-mediated mechanism on Nicotiana attenuata leaves. Bmc Plant Biol 2010, 10:164-174
45. Lee S, Hirt H, Lee Y: Phosphatidic acid activates a wound-activated MAPK in Glycine max. Plant J 2001, 26(5):479-486.
46. Halitschke R, Kessler A, Kahl J, Lorenz A, Baldwin IT: Ecophysiological comparison of direct and indirect defenses in Nicotiana attenuata. Oecologia 2000, 124(3):408-417.
47. Li L, Li CY, Lee GI, Howe GA: Distinct roles for jasmonate synthesis and action in the systemic wound response of tomato. Proc Natl Acad Sci U S A 2002, 99(9):6416-6421.
48. Li CY, Schilmiller AL, Liu GH, Lee GI, Jayanty S, Sageman C, Vrebalov J, Giovannoni JJ, Yagi K, Kobayashi Y, Howe GA: Role of beta-oxidation in jasmonate biosynthesis and systemic wound signaling in tomato. Plant Cell 2005, 17(3):971-986.
49. Felle HH, Zimmermann MR: Systemic signalling in barley through action potentials. Planta 2007, 226(1):203-214.
50. Malone M: Rapid, long-distance signal transmission in higher plants. Adv Bot Res 1996, 22:163-228.

51. Rhodes JD, Thain JF, Wildon DC: The pathway for systemic electrical signal conduction in the wounded tomato plant. Planta 1996, 200(1):50-57.

52. Zimmermann MR, Maischak H, Mithofer A, Boland W, Felle HH: System potentials, a novel electrical long-distance apoplastic signal in plants, induced by wounding. Plant Physiol 2009, 149(3):1593-1600.

53. Maffei M, Bossi S, Spiteller D, Mithofer A, Boland W: Effects of feeding Spodoptera littoralis on lima bean leaves. I. Membrane potentials, intracellular calcium variations, oral secretions, and regurgitate components. Plant Physiol 2004, 134(4):1752-1762.

54. Mousavi SA, Chauvin A, Pascaud F, Kellenberger S, Farmer EE: GLUTAMATE RECEPTOR-LIKE genes mediate leaf-to-leaf wound signalling. Nature 2013, 500(7463):422-426.

55. Kessler A, Baldwin IT: Defensive function of herbivore-induced plant volatile emissions in nature. Science 2001, 291(5511):2141-2144.

56. Steppuhn A, Gase K, Krock B, Halitschke R, Baldwin IT: Nicotine's defensive function in nature. PLoS Biol 2004, 2(8):E217.

57. Schuman MC, Barthel K, Baldwin IT: Herbivory-induced volatiles function as defenses increasing fitness of the native plant Nicotiana attenuata in nature. eLife 2012, 1:e00007.

58. Fragoso V, Rothe E, Baldwin IT, Kim SG: Root jasmonic acid synthesis and perception regulate folivore-induced shoot metabolites and increase Nicotiana attenuata resistance. New Phytol 2014, 202(4):1335-1345.

59. Philippe RN, Ralph SG, Mansfield SD, Bohlmann J: Transcriptome profiles of hybrid poplar (Populus trichocarpa x deltoides) reveal rapid changes in undamaged, systemic sink leaves after simulated feeding by forest tent caterpillar (Malacosoma disstria). New Phytol 2010, 188(3):787-802.

60. Rostas M, Eggert K: Ontogenetic and spatio-temporal patterns of induced volatiles in Glycine max in the light of the optimal defence hypothesis. Chemoecology 2008, 18(1):29-38.

61. Krugel T, Lim M, Gase K, Halitschke R, Baldwin IT: Agrobacterium-mediated transformation of Nicotiana attenuata, a model

ecological expression system. Chemoecology 2002, 12(4):177-183.

62. Roda A, Halitschke R, Steppuhn A, Baldwin IT: Individual variability in herbivore-specific elicitors from the plant's perspective. Mol Ecol 2004, 13(8):2421-2433.

63. Zhang S, Klessig DF: Salicylic acid activates a 48-kD MAP kinase in tobacco. Plant Cell 1997, 9(5):809-824.

16

Greater Efficiency of Photosynthetic Carbon Fixation Due to Single Amino-Acid Substitution

Judith Katharina Paulus[1], Daniel Schlieper[1], and Georg Groth[1]

[1]Cluster of Excellence on Plant Sciences (CEPLAS), Institute of Biochemical Plant Physiology, Heinrich Heine University, Universitaetsstr. 1, 40225 Düsseldorf, Germany

ABSTRACT

The C4-photosynthetic carbon cycle is an elaborated addition to the classical C3-photosynthetic pathway, which improves solar conversion efficiency. The key enzyme in this pathway, phosphoenolpyruvate carboxylase, has evolved from an ancestral non-photosynthetic C3 phosphoenolpyruvate carboxylase. During evolution, C4 phosphoenolpyruvate carboxylase has increased its kinetic efficiency and reduced its sensitivity towards the feedback inhibitors malate andaspartate. An open question is the molecular basis of the shift in inhibitor tolerance. Here we show that a single-point mutation is sufficient to account for the drastic differences between the inhibitor tolerances of C3 and C4 phosphoenolpyruvate carboxylases. We solved high-resolution X-ray crystal structures of a C3 phosphoenolpyruvate carboxylase and a closely related C4 phosphoenolpyruvate carboxylase. The comparison of both structures revealed that Arg884 supports tight inhibitor binding in the C3-type enzyme. In the C4 phosphoenolpyruvate carboxylase isoform, this arginine is replaced by glycine. The substitution reduces inhibitor affinity and enables the enzyme to participate in the C4 photosynthesis pathway.

INTRODUCTION

Based on the type of CO_2 assimilation, plants can be divided into three photosynthetic types: the C3-type, the C4-type and the Crassulacean Acid Metabolism[1]. In the classical C3-photosynthetic pathway, the primary CO_2 fixation is catalysed by the enzyme ribulose-1, 5-bisphosphate carboxylase/oxygenase (RuBisCO) resulting in the formation of the three-carbon compound 3-phosphoglycerate. However, in high temperature conditions RuBisCO is prone to energy loss by a process called photorespiration[2]. In C4 plants, this energy loss is minimized by an additional CO_2concentrating mechanism. This new mechanism evolved to adapt to stress factors such as heat, high light and salinity in combination with low CO_2 availability in recent geological history[3].

The relevant key enzyme of this pathway, phosphoenolpyruvate (PEP) carboxylase (PEPC), catalyses the HCO_3-dependent carboxylation of PEP to form the four-carbon molecule oxaloacetate[1]. Subsequent to the carboxylation reaction, oxaloacetate is reduced to malate or transaminated to aspartate. Both C4 molecules form a reservoir pool for the malic enzyme or PEP carboxykinase. These enzymes generate a high CO_2 concentration at the active site of RuBisCO. Thereby RuBisCO's oxygenase activity is reduced and the photosynthetic efficiency is increased in terms of use of water, nitrogen and other mineral nutrients for the production of valuable biomass[3]. For the CO_2 concentration mechanism, it is necessary to spatially separate the primary CO_2 fixation by PEPC and the CO_2 release to RuBisCO. Most C4 plants realize this by a characteristic anatomical feature, the Kranz anatomy, which spatially separates RuBisCO in the bundle-sheath cells from the initial site of CO_2 assimilation in the mesophyll cells[4]. Other mechanisms of compartmentation of the photosynthetic enzymes within cells have also been reported[5].

Another crucial step in the evolution of the C4 pathway is the recruitment of enzymes such as PEPC and the malic enzyme, which are required for initial CO_2 fixation and CO_2 release, respectively[6]. The predecessors for these C4 enzymes are enzymes from C3 plants and are involved in non-photosynthetic metabolic processes. However, the C4-type enzymes have distinctly different kinetic and regulatory properties. For instance, C4 PEPC shows tenfold larger substrate saturation constants for PEP[7] than the C3 PEPC and higher tolerance towards feedback inhibition by the C4-dicarboxylic acids malate and aspartate[8]. Previous studies imply that the acquisition of this enhanced tolerance towards feedback inhibition is an essential achievement in the evolution of C4 PEPC from the C3 ancestor[9].

A prime example of the evolution of C4 photosynthesis is found in the genus *Flaveria* (yellowtops) in the Asteraceae family. It includes species that perform C3 photosynthesis (for example, *F. pringlei*), C4 photosynthesis (for example, *F. trinervia*) as well as C3–C4 intermediate metabolism (for example, *F. pubescens*) [10]. All species within the genus are closely related and therefore ideal model organisms to pinpoint the key mutations that led from C3 to C4 photosynthesis[11]. In the case of PEPC, the C4 isoform from *F. trinervia* (encoded by the *ppcA* gene) and it's corresponding non-photosynthetic C3 isoform, the orthologous *ppcA* gene of *F. pringlei*[12], share 94% amino-acid (aa) identity. But along with differences in location and function, they show distinct differences in their kinetic and regulatory properties[8]. The C3 *ppcA* gene of *F. pringlei* is assumed to be similar to the PEPC that was ancestral to the C3 and the C4 PEPCs in the genus *Flaveria*[13].

The differences in kinetic efficiency have been addressed by reciprocal domain swapping and site-specific mutagenesis experiments. The increased PEP saturation kinetics of the C4 isoform depends on a single aa (Ser774)[14]. Less is known about the regulatory motif conferring the high malate/aspartate tolerance of C4 PEPC, which is imperative for the C4 cycle. Mutagenesis experiments indicate that the malate tolerance is largely mediated by the carboxy-terminal region of C4 PEPC (aa 645–966)[8], and residues Arg641, Lys829, Arg888 and Asn964 (*Flaveria*numbering) together have been identified as the malate-binding motif in the crystal structure of a C4-type PEPC from maize[15]. Mutagenesis of residues Lys829 and Arg888 was shown to completely disrupt the feedback inhibitor-binding site and results in enzymes with highly reduced malate sensitivity[16]. However, as this malate-binding motif is also found in the C3-type ortholog, these residues cannot account for the different feedback inhibitor sensitivity of C3- and C4-type PEPCs. Despite intensive studies[17, 18], no specific residue or motif was identified to account for the increased malate/ aspartate tolerance of the photosynthetic C4 PEPC in comparison with the C3 PEPC isoform.

As sequence analysis and mutagenesis studies failed to elucidate the molecular basis for malate/aspartate tolerance, we determined the crystal structures of PEPC isoforms from the C4 plant *F. trinervia* (2.5 Å) as well as from the C3 plant *F. pringlei* (2.7 Å) in their inhibited T-conformation. Our structures help to define the molecular adaptation that occurred when the housekeeping C3 isoform mutated to the photosynthetic C4 PEPC.

X-Ray Crystallography

Crystal structures of PEPC from *Zea mays* (maize), a representative C4 isoform, and from*Escherichia coli*, a non-photosynthetic C3 PEPC catalysing typical housekeeping functions, are known[17, 19]. However, the low level of sequence similarity (40%) [17] in these enzymes impeded the correlation of structural differences to C4-related features. Furthermore, the fact that the C4 structure was crystallized in the activated R-state and the C3 structure in the inhibited T-state may have resulted in the high level of backbone variation in multiple regions of the monomers (Fig. 1a). In contrast, any differences detected in the substrate site or the inhibitor-binding pocket of the PEPC isoforms from *F. pringlei* and *F. trinervia* can be attributed to a C3/C4-specific function. We crystallized PEPC from *F. pringlei* and *F. trinervia* with the inhibitor aspartate. We chose aspartatebecause malate and aspartate are equivalent feedback inhibitors and the addition of malateimpeded crystal growth. The crystallographic data and the refinement statistics are shown in Table 1. The Ramachandran plot of the refined C3 PEPC structure showed that the backbone conformation of 97.2% of the residues lies in the favoured region, 2.5% in the allowed region and 0.3% in the outlier region. The C4 PEPC structure has 97.9% favoured residues, 1.8% allowed residues and 0.2% outliers.

a

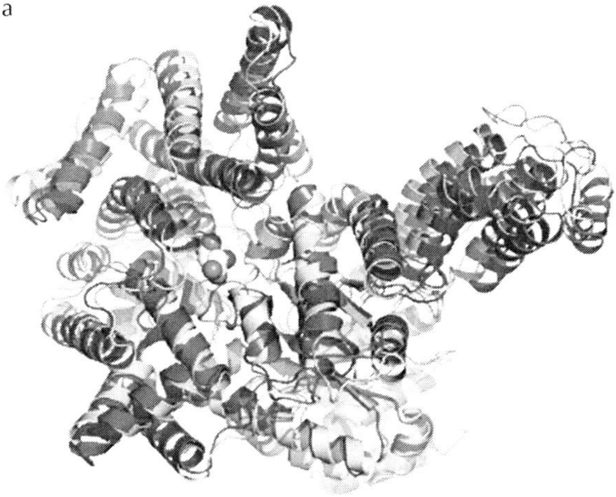

b

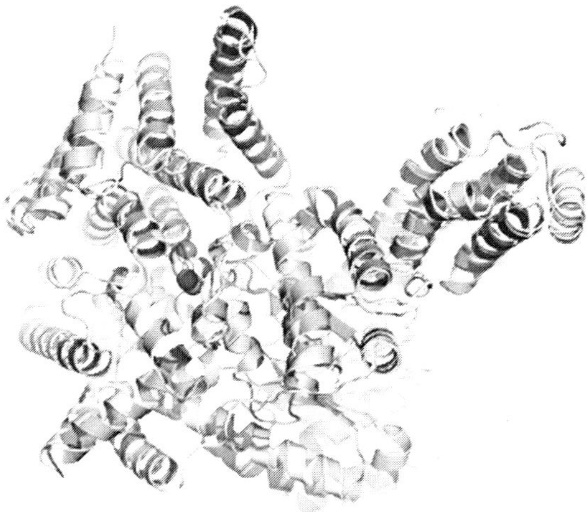

Figure 1: Structural comparison of C3 PEPC and C4 PEPC monomers. (a) Overlay of PEPC from *E. coli* (C3-type, grey) [17] and *Z. mays* (C4-type, blue) [17]. The calculated root-mean-square deviaton (r.m.s.d.) of both monomers is 2.0 Å. (b) Superposition of PEPC from *F. pringlei*(C3-type, green) and *F. trinervia* (C4-type, yellow). The *Flaveria* PEPCs have a high overall topologi-

cal similarity with a r.m.s.d. of 0.4 Å. Structures are shown as ribbons with the aspartate/malate-binding site on top. The bound inhibitors are shown as spheres representing van der Waals radii.

Table 1: Data collection and refinement statistics

PEPC crystal	C3 (*F. pringlei*)	$C3_{truncated}$	C4 (*F. trinervia*)
r.m.s.d., root-mean-square deviation.			
A single crystal was used for each data set.			
*Highest resolution shell is shown in parenthesis.			
†Signal-to-noise ratio of merged intensities.			
‡Protein B factors are calculated as the sum of the contribution from the TLS parameters and the residual B factors.			

The structural superposition of the PEPCs from *F. pringlei* and *F. trinervia* (Fig. 1b) confirms the high level of structural similarity, which is also reflected in the low backbone root-mean-square deviation of 0.4 Å. A closer look at the inhibitor-binding site reveals that residue 884, which is located close to the feedback inhibitor-binding site, differs in the C3 and the C4 isoforms (Fig. 2and corresponding stereo views in Fig. 3). The C3 PEPC isoform from *F. pringlei* has an argininethat is involved in inhibitor binding and functions as a molecular clamp tightening the inhibitory aspartate molecule in the binding pocket. In the C4 PEPC isoform from *F. trinervia* this arginine is replaced by glycine. The substitution reduces the number of protein–inhibitor interactions and provides more steric space for the inhibitor to dissociate from the binding pocket (Fig. 4). This important change is caused by a single-point mutation changing codon AGA to GGA. Thus, the structural comparison between the two *Flaveria* PEPC isoforms provides a unique way to analyse the function of these enzymes.

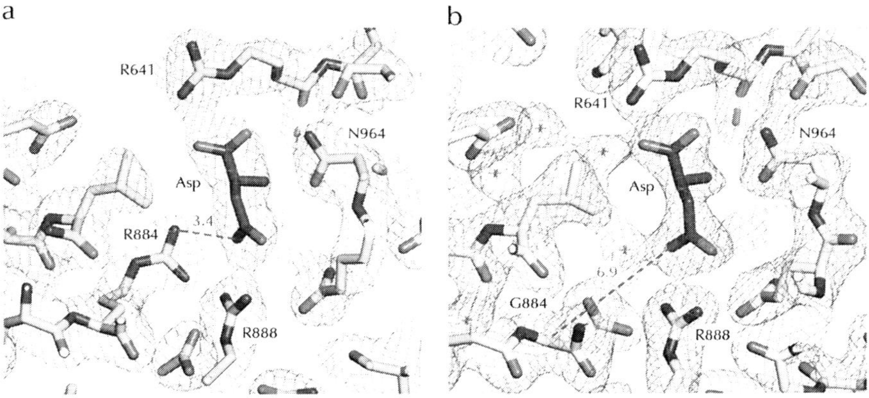

Figure 2: Comparison of the inhibitor-binding sites.(a) C3 PEPC from *F. pringlei*. (b) C4 PEPC from *F. trinervia*. Arg884 in C3 PEPC functions as a clamp by providing an additional hydrogen bond for inhibitor binding, while Gly884 in C4 PEPC is more than 6 Å away from the inhibitor. Structures and $2F_o-F_c$ electron densities at 1.2 of the malate/aspartate-binding pockets.

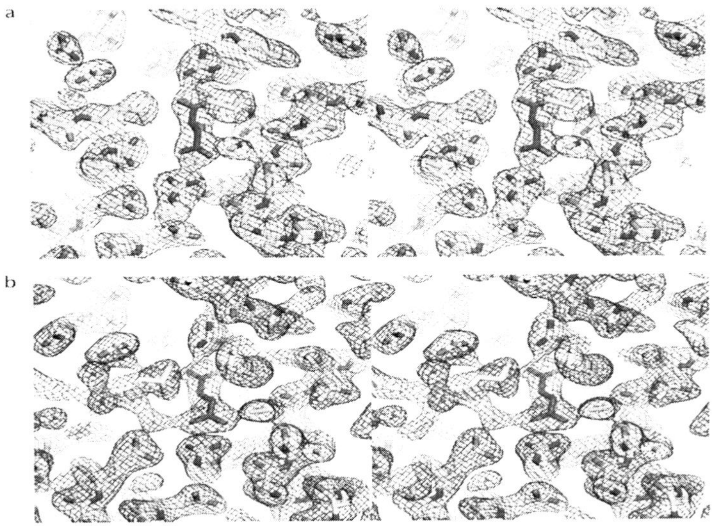

Figure 3: Electron density maps and final model structures.(a) C3 PEPC from *F. pringlei,* (b) C4 PEPC from *F. trinervia*. Shown are the refined $2F_o-F_c$ electron density maps around the inhibitor-binding sites. The maps are contoured at 1.2, where represents the root mean square of the density fluctuation of

the unit cell. The model structures are shown as sticks. Carbon atoms are coloured green, yellow and purple for *F. pringlei* PEPC, *F. trinveria* PEPC and the inhibitor Asp, respectively. Nitrogen atoms are coloured blue, oxygens red and sulphurs yellow.

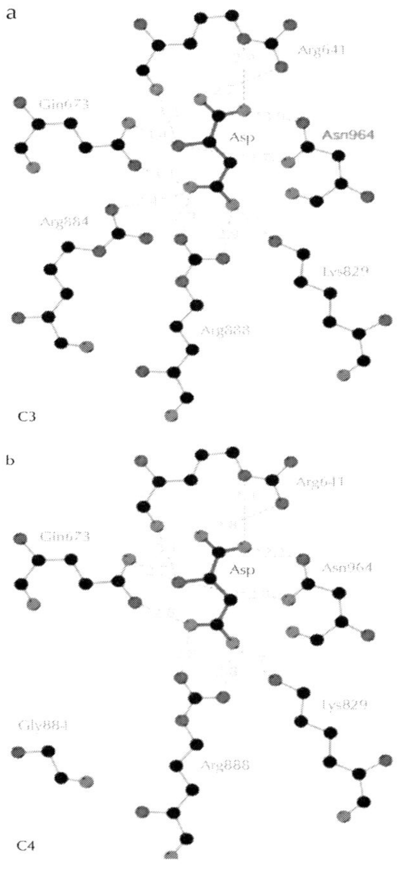

Figure 4: Side by side comparison of the feedback inhibitor-binding site. (a) Inhibitor-binding site of C3 PEPC (*F. pringlei*). (b) Inhibitor-binding site of C4 PEPC (*F. trinervia*). The distances between the inhibitor aspartate and amino acids of the binding pocket are given in Å. The estimated coordinate errors are 0.24 Å (*F. pringlei* C3 PEPC) and 0.29 Å (*F. trinervia* C4 PEPC). Residues Arg641, Lys829, Arg888 and Asn964 have been identified as the malate-binding motif[15]. In C3 PEPC, Arg884 provides an additional hydrogen bond for inhibitor binding. In C4 PEPC, this residue is replaced by glycine, which is 6.9 Å away from the inhibitor molecule.

Mutagenesis and Enzyme Kinetics

To substantiate these findings, we assessed the interactions in the malate/aspartate-binding site. We generated a series of mutations and examined their effect on the catalytic function of PEPC. When the key residue arginine 884 in C3 PEPC from *F. pringlei* was replaced by glycine as found in C4 PEPC, the malate tolerance of the engineered enzyme was increased by a factor of ~3. This increase corresponds to the IC_{50} value obtained with C4 wild-type PEPC from *F. trinervia*. The reverse test is the mutation from glycine to arginine in *F. trinervia* C4 PEPC. This mutation resulted in an increased malate inhibition to the level of wild-type C3 PEPC from *F. pringlei* (Fig. 5). Thus, the single aa in position 884 defines the malate-binding characteristics of the enzyme. This finding suggests that the substitution R884G is an important event in the evolution of the PEPC in the genus *Flaveria*. Sequence alignments with other PEPCs show that the arginine is conserved in typical C3 PEPCs while a glycine is located in the PEPCs from prominent C4 crop plants (Fig. 6). These findings provide strong evidence that the substitution R884G is a key event in the evolution of a C4 PEPC in many lineages of C4 plants.

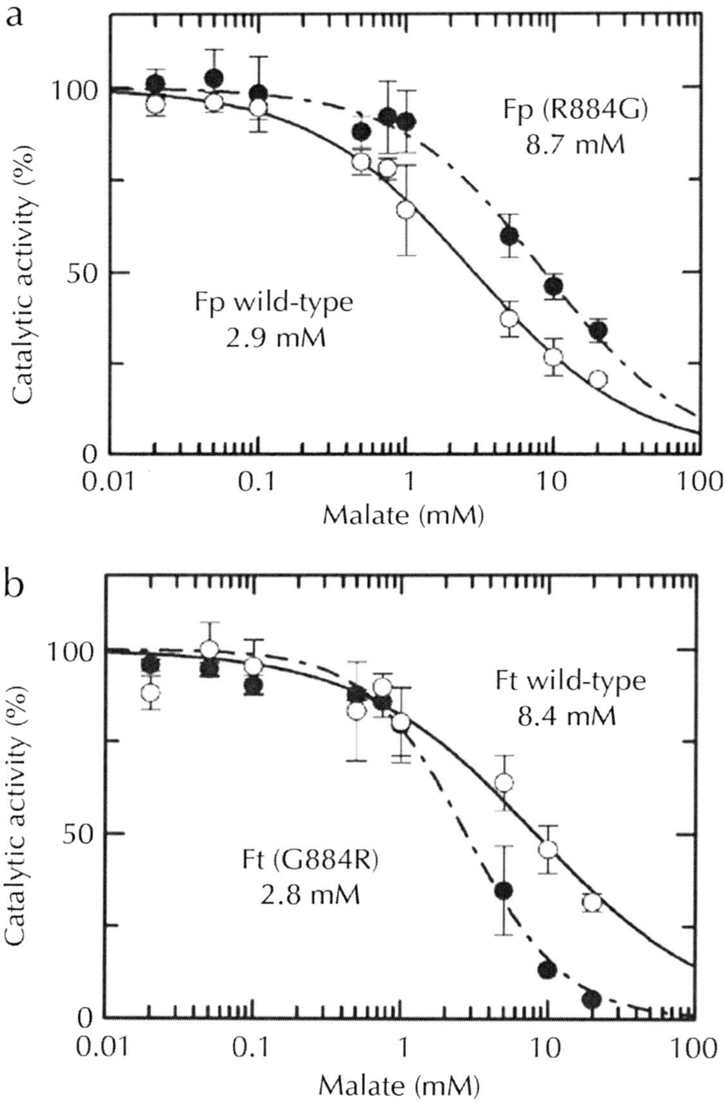

Figure 5: Determination of IC50 values for wild-type and mutant PEPC from *F. pringlei* (C3) and *F. trinervia* (C4). (a) The mean activity values of *F. pringlei* wild-type C3 PEPC (open circles) and *F. pringlei* R884G mutant (solid circles) are plotted against the concentration of the feedback inhibitor malate. The calculated IC_{50}values are 2.9 mM for the wild-type (solid curve) and 8.7 mM for the R884G mutant C3 PEPC (dashed curve). (b) Plot of mean activity

values of *F. trinervia* wild-type C4 PEPC (open circles) and *F. trinervia*G884R mutant (closed circles). IC_{50} values calculated from the experimental data are 8.4 mM for wild-type and 2.8 mM for the mutant C4 PEPC. S.d. was calculated from three independent experiments. Activities were measured in triplicates for each experiment.

Ft	770	PWIF**S**WTQTR (...)DLLEGDPYLKQGIRLRDPYIT	893
So	767	PWIF**S**WTQTR (...)DLLEGDPYLKQGIRLRDPYIT	890
Sb	767	PWIF**S**WTQTR (...)DLLEGDPYLKQGIRLRDPYIT	890
Zm	776	PWIF**S**WTQTR (...)DLLEGDPYLKQGIRLRDPYIT	899
Fp	770	PWIF**A**WTQTR (...)DLLEGDPYLKQRIRLRDPYIT	893
Os	769	PWIF**A**WTQTR (...)DLLEGDPYLKQRIRLRDPYIT	892
Nt	770	PWIF**A**WTQTR (...)DLLEGDPYLKQRIRLRDPYIT	893
At	773	PWIF**A**WTQTR(...)DLLEGDPYLKQRIRLRDPYIT	896

Figure 6: Sequence alignment of C3 and C4 determining regions in various PEPCs. Amino acids that determine C3/C4-specific function are highlighted. In the substrate-binding centre, Ala774 (*Flaveria* numbering) mediates C3 specificity, while Ser774 determines the increased kinetic efficiency of C4 PEPC (black boxes). Malate binding in the inhibitory site is controlled by Arg884 in the C3 enzyme, while the increased tolerance of the C4 PEPC towards the feedback inhibitor is mediated by Gly884 (grey boxes). The sequences are: *Ft* (*F. trinervia* UniProt No. P30694), *So* (*Saccharum officinarum* Q9FS96), *Sb* (*Sorghum bicolor* P15804), *Zm* (*Zea mays* P04711), *Fp* (*F. pringlei* GenBank CAA88829.1), *Os* (*Oryza sativa* subsp. *indica* Q84XH0), *Nt* (*Nicotiana tabacum* P27154) and *At* (*Arabidopsis thaliana* Q84VW9).

DISCUSSION

The crystal structures of a C3-type and a C4-type PEPC obtained from two closely related species allowed us to pinpoint the molecular switch regulating feedback inhibition in a key enzyme of the C4-photosynthetic pathway. The underlying mechanism of the adapted feedback inhibitor sensitivity in C4 PEPC isoforms has not previously been identified by molecular genetics and bioinformatics. Our structural and biochemical studies reveal that the increased inhibitor tolerance is caused by a single aa substitution emphasizing that even small changes like point mutations can cause dramatic differences in kinetic and regulatory characteristics of enzymes. While residue Arg884 in the feedback inhibitor-binding pocket of C3 PEPC provides an additional hydrogen bond to the inhibitor and promotes stronger binding, the corresponding residue Gly884 in C4 PEPCreduces the steric restriction of the binding site and provides fewer contacts with the inhibitor. Therefore, the C4 enzyme is more tolerant towards feedback inhibition than the C3 enzyme.

Our findings on the importance of residue 884 for malate sensitivity are not limited to the C3 and C4 species from the genus *Flaveria*. Alignments with other protein sequences from typical C3 and C4 PEPCs show that the arginine is conserved in C3 PEPCs, while PEPCs from typical C4 crop plants such as *Z. mays, Saccharum officinarium* (sugar cane) and *Sorgum bicolor* have a glycineat this position (Fig. 6). The results show that the sensitivity of PEPC towards the feedback inhibitors malate and aspartate is accomplished by a single aa substitution at position 884. In the same way, increased PEP saturation kinetics of the C4 isoform is controlled by a single aa switch at position 774 (ref. 14). Nevertheless, enzymes with non-optimized feedback inhibitor-binding sites and with incomplete malate tolerance may be found in plants where high malate concentrations are circumvented by other means. For example, the PEPC of the C4-like plant *Flaveria brownii* carries the C4-specific aa in position 774 while an arginine remains in position 884. *F. brownii* is considered a C4-like species because the intercellular enzyme compartmentation is not as strict as in a true C4 plant and a substantial amount of CO_2 still enters through the C3 cycle[20]. The *F. brownii* PEPC therefore functions in C4-like photosynthesis but shows only moderate tolerance towards feedback inhibition[21]. The full potential of malate tolerance in such enzymes would be obtained by the glycine–arginine substitution at position 884 as realized in the C4 PEPC of *F. trinvervia*, though kinetic data to test this hypothesis are not available.

Previous attempts to pinpoint the molecular origin of the different kinetic and regulatory characteristics of C4 PEPC suggest that multiple residues are involved in the transfer of a C3- to a C4-type enzyme[22, 23]. Based on phylogenetic analysis of PEPC in grasses (Poaceae), Christin *et al.*[24] identified 21 codons that have undergone positive selection in C4 species. However, all but two of the aas concerned are identical in the C3 plant *F. pringlei* and the C4 plant *F. trinervia*. One of the C4-adaptive aa is Ser774 (*Flaveria* numbering), which defines the PEP saturation kinetics[14]. The other aa is Asn659, which is—as revealed by our structural data—localized on the surface of the protein and can hardly account for the differences in inhibitor sensitivity between *F. pringlei* PEPC and *F. trinervia* PEPC.

In summary, it is obvious that during evolution the main kinetic and regulatory changes of the C4 enzyme are the result of two point mutations rather than extensive modular changes such as recombination or exon shuffling. This notion is consistent with the idea that the evolution of new enzymatic activity is often achieved by the duplication of an existing gene that is capable of catalysing the chemical reaction required[22].

Within the next 50 years, the world's population will increase by a third[25] and a corresponding increase of crop yields is needed. A concerted effort of research and development is necessary to drastically increase crop yields—similar to the 'green revolution'[26]. One approach is to turn C3 plants into C4 plants and, as a result, to increase the photosynthesis efficiency[27]. The best candidates for this change are the C3 plants rice and wheat, two of the most important crops of the world. Attempts were made to introduce C4 enzymes into C3 plants[28, 29]. However, the key trigger that changes plants in anatomy as well as enzyme function and regulation has not yet been found.

With our data, we complement the knowledge about PEPC, the key enzyme of C4 photosynthesis. A single mutation in the active site (774)[14] and a single mutation at the inhibitory site (884) are sufficient to switch the function from C3 to C4 activity. Now, physiological experiments could test if instead of introducing the maize PEPC into rice[28, 29], the rice PEPC itself could be engineered to function as a C4 enzyme.

METHODS

Recombinant PEPC

The coding sequences of PEPC from *F. trinervia* (ppcA, EMBL-Bank X61304) and *F. pringlei* (ppcA, EMBL-Bank Z48966) were cloned in pETEV16b (Novagene), together with the coding sequence of an amino-terminal His10-tag and a TEV cleavage site. Plasmids were transformed in*E. coli* BL21(DE3) (*F. trinervia*) or BL21-Gold(DE3) (*F. pringlei*) (Agilent Technologies). Point mutations and truncations were introduced into the constructs as described in the QuikChange Site-Directed Mutagenesis Kit (Agilent Technologies) using Pwo polymerase. Constructs are listed in Table 2. Cultures were grown in 2YT media (5 g l^{-1} sodium chloride, 10 g l^{-1} yeast extract and 16 g l^{-1} peptone) with ampicillin at 37 °C to OD_{600} 0.8. Production of recombinant proteins was induced with 0.5 mM IPTG and cells were grown for 18 h at 16 °C. After harvesting, cells were suspended in lysis buffer (50 mM Tris, 10 mM $MgCl_2$, 150 mM NaCl, DNAse, lysozyme, pH 7.5) and disrupted by sonification. Proteins were purified by nickel affinity chromatography, concentrated by Amicon Ultra-15 centrifugal filter units (50 kDa cutoff, Millipore) and buffered in concentration buffer (50 mM Tris, 10 mM $MgCl_2$, 1 mM DTT, +/− 10 mM aspartate, pH 7.5) using PD10 columns (GE Healthcare).

Table 2: Constructs in pETEV16b

Construct	Source	Modification
C3	*F. pringlei*	—
C3(R884G)	*F. pringlei*	Point mutation GGA to AGA on aa position 884
$C3_{truncated}$	*F. pringlei*	Deletion of TEV site and linker region, deletion of the first five aa in the coding region
C4	*F. trinervia*	—
C4(G884R)	*F. trinervia*	Point mutation AGA to GGA on aa position 884

Crystallization

Crystallization trials were performed in micro batch technique. PEPC at a concentration range of 8–10 mg ml^{-1} in concentration buffer was mixed with precipitation solution at a 1:1 ratio and covered with mineral oil. Precipitation solutions were 0.1 M tri-sodium citrate pH 5.6, 0.2 M ammonium sulphate and 10% (w/v) PEG 4000 (*F. pringlei*) or 15% (w/v) PEG 4000 (*F. trinervia*). After 2 days at 21 °C, crystals of a typical size of 0.5 × 0.3 × 0.3 mm^3 were obtained and transferred to cryoprotection buffer containing 32% precipitant, 48% concentration buffer and 20% ethylene glycol and swiftly cryo-cooled in liquid nitrogen.

Structure Determination

X-ray diffraction data were collected at the European Synchrotron Radiation Facility (ESRF), Grenoble, at beamlines BM14 and ID23-2, and at EMBL/DESY, Hamburg, at beamlines DORIS X12, X13 and PETRA P14. The C3 PEPC dataset from *F. pringlei* was measured at 0.8726 Å wavelength (ID23-2), $C3_{truncated}$ at 1.1268 Å (P14) and the C4 PEPC dataset from *F. trinervia* at 0.9763 Å (BM14). Data of C4 PEPC with the inhibitor aspartate were integrated with Mosflm[30] and further processed with tools of the CCP4 suite[31], Pointless[32] and Scala[32] in particular. Initial phases of C4 PEPC (*F. trinervia*) were obtained by molecular replacement with *Z. mays* C4 PEPC (PDB Code: 1JQO, 73% sequence identity) using Phaser[33]. Model building was initially done automatically with ARP/wARP[34]. This step efficiently reduced model bias. Several rounds of manual building followed using Coot[35] and refinement by Refmac5[36] using non-crystallographic restraints and partially describing the atomic displacement parameters in groups of translation, libration and screw-motion (TLS). The structure comprises two monomers within the asymmetric unit. Crystallographic symmetry completes the tetramer. Additional electron density was found at the inhibitor-binding site. However, to avoid model bias the aspartate molecule was not fitted in the structure until model building was almost complete. The structures were validated using the tools provided by Coot, CCP4 and Rampage[37].

The C3 PEPC structure from *F. pringlei* was obtained by essentially the same workflow. Model building started using a 3.2 Å data set collected at ERSF ID23-2. Molecular replacement was done using the C4 PEPC structure from *F. trinervia*. As the region comprising the His_{10}-tag, the TEV protease recognition site and the first five aas of C4 PEPC were disordered in this structure, we deleted the TEV site and the first five aas (construct $C3_{truncated}$). Crystals from this new construct $C3_{truncated}$ were obtained under identical conditions and diffracted to 2.7 Å (EMBL/DESY PETRA P14). These data were integrated with XDS[38] and the model was built as described including ARP/wARP autobuilding to reduce model bias.

Figures of the atomic structures were generated with Pymol (Schrödinger). Schematic diagrams of protein–ligand interactions were made using LigPlot[39].

Inhibitor Studies

The IC_{50} of malate inhibition was determined by a coupled spectrometric assay as described[8]. In this assay, conversion of PEP to oxaloacetate is monitored via the reduction of oxaloacetate to malate by malate dehydrogenase. The reaction velocity is measured as a decrease in OD_{340} resulting from the oxidation of NADH in this reaction. The inhibition of the PEPC by malate was measured in the presence of PEP ($2 \times K_m$), at malate concentrations from 0–20 mM. The half maximal inhibitory concentrations (IC_{50}) of wild-type and mutant enzymes were calculated from the experimental data using the Grafit software package (Erithacus Software).

ACKNOWLEDGEMENTS

The ppcA complementary DNAs were kindly provided by Peter Westhoff. Christian Niehus cloned the truncated version of C3 PEPC. We thank the staff of ESRF (Grenoble) and EMBL/DESY (Hamburg) for their kind support, especially Hassan Belrhali (ESRF BM14), David Flot (ESRF ID23-2), Matthew Groves (EMBL/DESY DORIS X12/X13), Gleb Bourenkov and Thomas Schneider (EMBL/DESY PETRA P14). We thank Elisabeth Stratmann for editorial assistance. This study was supported by a grant from Deutsche Forschungsgemeinschaft through graduate programme IRTG 1525 iGRAD-Plant.

REFERENCES

1. Sage, R. F. The evolution of C4 photosynthesis. *New Phytol.* 161, 341–370 (2004).
2. Peterhänsel, C. & Maurino, V. G. Photorespiration redesigned. *Plant Physiol.* 155, 49(2011).
3. Sage, R. F., Sage, T. L. & Kocacinar, F. Photorespiration and the evolution of C4 photosynthesis. *Annu. Rev. Plant Physiol.* 63, 17.1–17.29 (2012).
4. Stitt, M. & Heldt, H. W. Generation and maintenance of concentration gradients between the mesophyll and bundle sheath in maize leaves. *Biochim. Biophys. Acta Bioenerg.* 808,400–414 (1985).
5. Voznesenskaya, E. V., Franceschi, V. R., Kiirats, O., Freitag, H. & Edwards, G. E.. Kranz anatomy is not essential for terrestrial C4 plant photosynthesis. *Nature* 414, 543–546 (2001).
6. Hibberd, J. M. & Covshoff, S. The regulation of gene expression required for C4 photosynthesis. *Annu. Rev. Plant Physiol.* 61, 181–207 (2010).
7. Ting, I. P. & Osmond, C. B. Multiple forms of plant phosphoenolpyruvate carboxylase associated with different metabolic pathways. *Plant Physiol.* 51, 448–453 (1973).
8. Jacobs, B., Engelmann, S., Westhoff, P. & Gowik, U. Evolution of C4 phosphoenolpyruvate carboxylase in *Flaveria*: determinants for high tolerance towards the inhibitor L-malate. *Plant Cell. Environ.* 31, 793–803 (2008).
9. Bläsing, O. E., Ernst, K., Streubel, M., Westhoff, P. & Svensson, P. The non-photosynthetic phosphoenolpyruvate carboxylases of the C4 dicot *Flaveria trinervia* - implications for the evolution of C4 photosynthesis. *Planta* 215, 448–456 (2002).
10. Monson, R. K., d. Moore, B., Ku, M. S. B. & Edwards, G. E.. Cofunction of C3-photosynthetic and C4-photosynthetic

pathways in C3, C4 and C3-C4 intermediate *Flaveria*species. *Planta* 168, 493–502 (1986).

11. Brown, N. J., Parsley, K. & Hibberd, J. M. The future of C4 research - maize, *Flaveria* or Cleome? *Trends Plant Sci.* 10, 215–221 (2005).
12. Westhoff, P. & Gowik, U. Evolution of C4 phosphoenolpyruvate carboxylase. Genes and proteins: a case study with the genus *Flaveria*. *Ann. Bot.* 93, 13–23 (2004).
13. Svensson, P., Bläsing, O. E. & Westhoff, P. Evolution of C4 phosphoenolpyruvate carboxylase. *Arch. Biochem. Biophys.* 414, 180–188 (2003).
14. Bläsing, O. E., Westhoff, P. & Svensson, P. Evolution of C4 phosphoenolpyruvate carboxylase in *Flaveria*–a conserved serine residue in the carboxyterminal part of the enzyme is a major determinant for C4-specific characteristics. *J. Biol. Chem.* 275, 27917–27923 (2000).
15. Izui, K., Matsumura, H., Furumoto, T. & Kai, Y. Phosphoenolpyruvate carboxylase: a new era of structural biology. *Annu. Rev. Plant Physiol.* 55, 69–84 (2004).
16. Endo, T. *et al.* Maize C4-form phosphoenolpyruvate carboxylase engineered to be functional in C3 plants: mutations for diminished sensitivity to feedback inhibitors and for increased substrate affinity. *J. Exp. Bot.* 59, 1811–1818 (2008).
17. Matsumura, H. *et al.* Crystal structures of C4 form maize and quaternary complex of *E. coli*phosphoenolpyruvate carboxylases. *Structure* 10, 1721–1730 (2002).
18. Dharmarajan, L., Kraszewski, J. L., Mukhopadhyay, B. & Dunten, P. W. Structure of an archaeal-type phosphoenolpyruvate carboxylase sensitive to inhibition by aspartate. *Proteins*79, 1820–1829 (2011).
19. Kai, Y. *et al.* Three-dimensional structure of phosphoenolpyruvate carboxylase: a proposed mechanism for allosteric inhibition. *Proc. Natl Acad. Sci. USA* 96, 823–828 (1999).
20. Cheng, S. H., Moore, B. D., Edwards, G. E. & Ku, M. S. Photosynthesis in *Flaveria brownii*, a C4-like species: leaf anatomy, characteristics of CO2 exchange, compartmentation of photosynthetic enzymes, and metabolism of CO2. *Plant Physiol.* 87, 867–873 (1988).

21. Engelmann, S., Bläsing, O. E., Gowik, U., Svensson, P. & Westhoff, P.. Molecular evolution of C4 phosphoenolpyruvate carboxylase in the genus *Flaveria*—a gradual increase from C3 to C4 characteristics. *Planta* 217, 717–725 (2003).
22. Lynch, M. Genomics. Gene Duplication and Evolution. *Science* 297, 945–947 (2002).
23. Besnard, G., Muasya, A. M., Russier, F., Roalson, E. H., Salamin, N. & Christin, P. A. Phylogenomics of C4 photosynthesis in sedges (Cyperaceae): multiple appearances and genetic convergence. *Mol. Biol. Evol.* 26, 1909–1919 (2009).
24. Christin, P. A., Salamin, N., Savolainen, V., Duvall, M. R. & Besnard, G. C4 photosynthesis evolved in grasses via parallel adaptive genetic changes. *Curr. Biol.* 17, 1241–1247 (2007).
25. Bloom, D. E.. 7 billion and counting. *Science* 333, 562–569 (2011).
26. Khush, G. S. Green revolution: the way forward. *Nat. Rev. Genet.* 2, 815–822 (2001).
27. Surridge, C. Agricultural biotech: the rice squad. *Nature* 416, 576–578 (2002).
28. Ku, M. S. *et al.* Introduction of genes encoding C4 photosynthesis enzymes into rice plants: physiological consequences. *Novartis Found Symp.* 236, 100–116 (2001).
29. Ku, M. S. *et al.* High-level expression of maize phosphoenolpyruvate carboxylase in transgenic rice plants. *Nat. Biotechnol.* 1, 76–80 (1999).
30. Leslie, A. G. W. & Powell, H. R. Processing diffraction data with Mosflm. *Evol. Methods Macromol. Crystallogr.* 245, 41–51 (2007).
31. Collaborative Computational Project. Number 4. The CCP4 Suite: programs for protein crystallography. *Acta Cryst.* D50, 760–763 (1994).
32. Evans, P. R. An introduction to data reduction: space-group determination, scaling and intensity statistics. *Acta Cryst.* D67, 282–292 (2011).
33. McCoy, A. J. *et al.* Phaser crystallographic software. *J. Appl. Cryst.* 40, 658–674 (2007).

34. Langer, G., Cohen, S. X., Lamzin, V. S. & Perrakis, A. Automated macromolecular model building for X-ray crystallography using ARP/wARP version 7. *Nat. Protoc.* 3, 1171–1179(2008).
35. Emsley, P., Lohkamp, B., Scott, W. & Cowtan, K. Features and development of Coot. *Acta Cryst.* D66, 486–501 (2010).
36. Murshudov, G. N. *et al.* REFMAC5 for the refinement of macromolecular crystal structures.*Acta Cryst.* D67, 355–367 (2011).
37. Lovell, S. C. *et al.* Structure validation by Cα geometry: ϕ/ψ and Cβ deviation. *Proteins: Struct. Funct. Genet.* 50, 437–450 (2003).
38. Kabsch, W. XDS. *Acta Cryst.* D66, 125–132 (2010).
39. Wallace, A. C., Laskowski, R. A. & Thornton, J. M. LIGPLOT: a program to generate schematic diagrams of protein-ligand interactions. *Protein Eng.* 8, 127–134 (1995).

CHAPTER 7

Different Accumulation of Free Amino Acids during Short- and Long-Term Osmotic Stress in Wheat

Zita Kovács[1], Livia Simon-Sarkadi[1], Ildikó Vashegyi[2, 3], and Gábor Kocsy[2]

[1]Department of Applied Biotechnology and Food Science, Budapest University of Technology and Economics, P.O. Box 91, 1521 Budapest, Hungary

[2]Agricultural Institute, Centre for Agricultural Research, Hungarian Academy of Sciences, P.O. Box 19, 2462 Martonvásár, Hungary

[3]Research Institute of Chemical and Process Engineering, Faculty of Information Technology, University of Pannonia, P.O. Box 158, 8200 Veszprém, Hungary

ABSTRACT

The effect of wheat chromosome 5A on free amino acid accumulation induced by osmotic stress was compared in chromosome 5A substitution lines with different freezing tolerance. Treatment with 15% polyethylene glycol (PEG) resulted in greater total free amino acid content even after 3 days compared to the controls. The ratio of amino acids belonging to various amino acid families differed after 3-week treatment in the control and PEG-treated plants only in the case of the freezing-sensitive substitution line. There was a transient increase with a maximum after 3 days in the amounts of several amino acids, after which their concentrations exhibited a more gradual increase. During the first days of osmotic stress the Glu, Gln, Asp, Asn, Thr, Ser, Leu, and His concentrations were greater in the tolerant substitution line than in the sensitive one, while the opposite relationship was observed at the end of the PEG treatment. The coordinated changes in the levels of individual amino acids indicated that they are involved in both the short- and long-term responses to osmotic stress. The alterations differed in the two chromosome 5A substitution lines, depending on the stress tolerance of the chromosome donor genotype.

INTRODUCTION

Abiotic stresses result in severe injury and yield loss in crops. A common consequence of drought and low temperature is the induction of osmotic stress. The accumulation of compatible solutes, such as certain amino acids especially Pro, prevents the development of severe osmotic stress. Amino acids also have several other roles in plants, for example they regulate ion transport and stomatal opening and affect the synthesis and activity of enzymes, gene expression, and redox homeostasis, helping the plants to cope with the harmful effects of osmotic stress [1].

Osmotic stress, which can be induced by limited water supply or osmotic agents, alters both the amino acid pattern and the concentrations of individual amino acids. Osmotic stress induced by polyethylene glycol resulted in an increase in the concentrations of Asp, Glu, Asn, Thr, Ser, Ala, and Pro during the first day of treatment in the roots of maize [2]. After dehydration the amino acid contents changed to differing extents in the desiccation-tolerant young leaves and sensitive old leaves of the resurrection plant Sporobolus stapfianus [3], indicating that these changes depended on the level of stress tolerance.

Proline plays a very important role as an osmoprotectant in the adaptation to osmotic stress [4]. The accumulation induced by osmotic stress is based on increased synthesis and decreased degradation [5]. Osmotic stress resulted in a greater increase in Pro content in a drought-tolerant rice genotype than in a sensitive one, due to the greater expression of the gene encoding pyrroline-5-carboxylate synthase and activity of this enzyme involved in the Pro synthesis [6]. Further evidence for the protective role of Pro was found in transgenic plants, where its overproduction increased tolerance to osmotic stress [5, 7].

Besides Pro, the involvement of other amino acids in the response to osmotic stress was also demonstrated. Arginine was found to function as a compatible solute, improving stress tolerance in yeast [8]. Under hyperosmotic conditions the transcription of genes encoding the enzymes involved in Arg biosynthesis increased, while those involved in degradation decreased. Both exogenous Arg and the overproduction of two enzymes required for its synthesis increased tolerance. In addition, osmotic stress resulted in the increased expression of asparagine synthase genes in sunflower [9] and in wheat [10]. The overexpression of glutamine synthase increased tolerance to osmotic stress in rice [11]. These observations show that the changes in amino acid concentrations induced by osmotic stress may be due to the altered expression of genes encoding the enzymes involved in their metabolism.

Chromosome 5A of wheat was found to affect the cold-induced accumulation of certain free amino acids [12]. Since this chromosome has an important role in the stress response and controls the expression of several stress-related genes [13], it seemed likely that it was involved in controlling the effect of osmotic stress on free amino acid levels. This hypothesis was tested by comparing the changes induced by short- and long-term osmotic stress in the free amino acid composition and in the concentration of individual amino acids in wheat chromosome 5A substitution lines with different freezing tolerance. Osmotic stress also occurs during freezing, so these genotypes were expected to give different responses.

MATERIALS AND METHODS

Plant Material and Treatment

A specific genetic system consisting of the moderately freezing-sensitive wheat (Triticum aestivum L. ssp.aestivum) cv. Chinese Spring (CS) and two chromosome 5A substitution lines the freezing-tolerant Chinese Spring (Cheyenne 5A) (CS(Ch5A)) and the freezing-sensitive Chinese Spring (Triticum aestivum L. ssp.spelta 5A) (CS(Tsp5A)) was used in the experiments [14]. The seeds were obtained from the Martonvásár Cereal Gene Bank (Agricultural Institute, Agricultural Research Centre, Hungarian Academy of Sciences, Martonvásár, Hungary). According to previous studies in which the osmotic stress tolerance of the used genotypes was also checked, CS and CS (Tsp5A) are sensitive and CS (Ch5A) is tolerant [15, 16]. It was assumed thus difference in the tolerance may also affect the accumulation of amino acids during osmotic stress.

Seedlings were grown in hydroponic culture using half-strength modified Hoagland nutrient solution [17] in a plant growth chamber (Conviron PGV-36, Controlled Environments Ltd., Canada) at 18/15°C day/night temperature and 70/75% relative humidity with 16 h illumination at 270 µmoL photons $m^{-2} s^{-1}$for 12 days prior to the stress period. Treatment with 15% polyethylene glycol (PEG) lasted for 21 days, under cultivation conditions similar to those described above. Control plants were cultivated without any supplementation. Samples were taken at the beginning of the treatment and after 1, 3, 7, and 21 days. The experiments were repeated three times and in each experiment three parallels were investigated.

Determination of Free Amino Acid Content

Shoot samples of 300–600 mg fresh weight were crushed in liquid nitrogen and extracted with 2 mL cold 10% trichloroacetic acid for 1 hour with gentle agitation on a shaker (C. Gerhardt GmbH & Co. KG, Germany) at room temperature. Each sample was filtered through a 0.2 µm pore membrane filter (Sartorius AG, Germany). The biochemical analysis was carried out on an automatic amino acid analyser (Ingos

Ltd., Czech Republic) equipped with an Ionex Ostion LCP5020 cation-exchange column (2 2 × 0 . 3 7 cm). Free amino acids were separated by stepwise gradient elution using a Li^+-citric buffer system (Ingos Ltd., Czech Republic). Colorimetric detection was accomplished at 570 nm and 440 nm (for Pro) after postcolumn derivatisation with ninhydrin reagent.

Statistical Analysis

The free amino acid data were evaluated using analysis of variance and the significant differences were calculated.

RESULTS

Effect of Osmotic Stress on Plant Growth

Application of PEG in 15% concentration resulted only in mild osmotic stress as it was shown by growth data (Figure 1). There were only very slight changes in fresh (Figure 1(a)) and dry (Figure 1(b)) weight of the shoots during the first week of the treatment. The growth of the plants sped up during the subsequent two weeks. The fresh and dry weight data did not differ significantly between the treated and control plants and the three genotypes at the end of the osmotic stress. The plants had no visible injuries at the end of the 3-week osmotic stress.

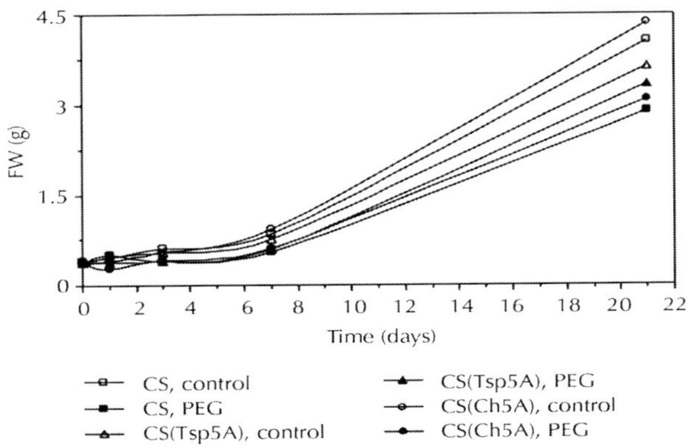

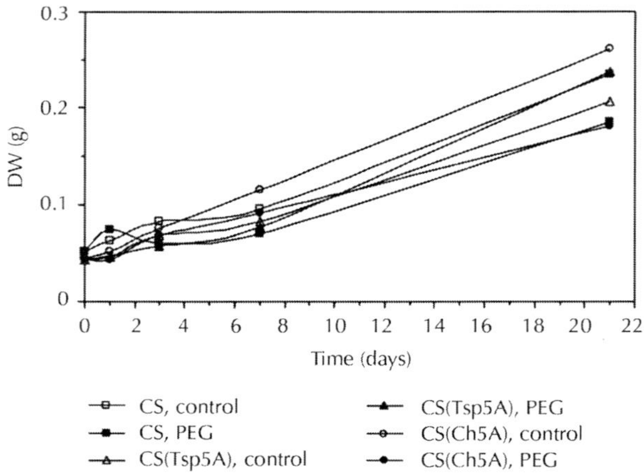

Figure 1: Effect of osmotic stress on fresh ((a); significant difference (SD): 0.612) and dry weight ((b); SD: 0.065). The time course of changes was compared under control and stress (15% PEG) conditions in the freezing-sensitive CS and CS (Tsp5A) and the freezing-tolerant CS (Ch5A) wheat genotypes. The standard deviations of the data were below 10% in each case (not shown on the graphs because of the better transparency).

Effect of Osmotic Stress on Plant Amino Acid Composition

The total free amino acid concentration increased already after 3 days osmotic stress in all the genotypes, but a great difference between control and treated plants was observed after 3 weeks, especially in the case of CS and CS(Tsp5A) (Figure 2). Comparing the two substitution lines following PEG treatment, the total free amino acid content was relatively greater after 3 d in the freezing-tolerant CS (Ch5A) and after 21 d in the freezing-sensitive CS (Tsp5A). The ratio of amino acids belonging to the various amino acid families changed similarly during the 3-week treatment in the control and PEG-treated plants except for the greater difference in the ratio of glutamate and aspartate family in CS(Tsp5A) (Figure 3).

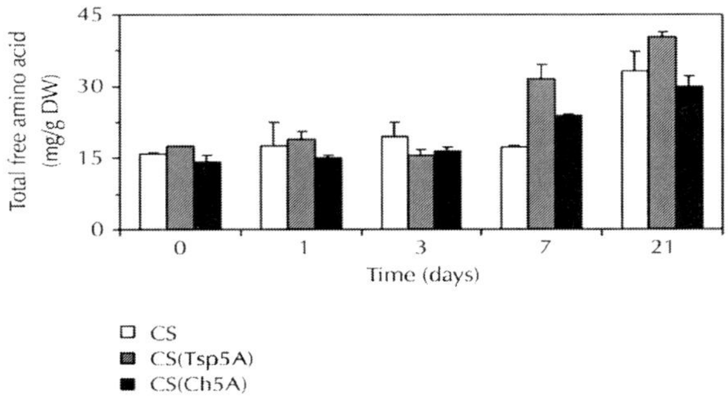

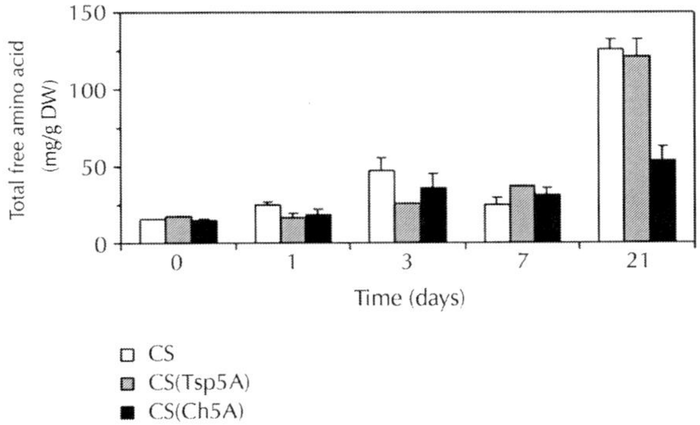

Figure 2: Effect of osmotic stress on total free amino acid content. The time course of changes was compared under control ((a); SD: 4.47) and stress (15% PEG; (b); SD: 12.68) conditions in the freezing-sensitive CS and CS (Tsp5A) and the freezing-tolerant CS (Ch5A) wheat genotypes.

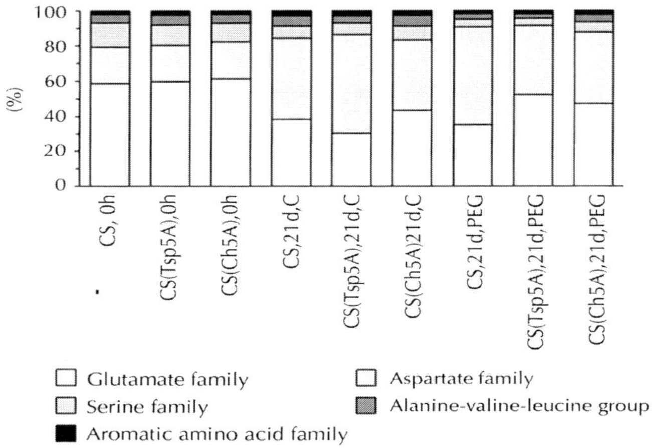

Figure 3: Changes in the ratio of amino acid families during osmotic stress. A comparison was made of the freezing-sensitive CS and CS (Tsp5A) and the freezing-tolerant CS (Ch5A) wheat genotypes, grown for three weeks under control (C) or stress conditions (15% PEG). The experiment was repeated with three parallels three times with similar results.

In the case of the individual amino acids little or no change was observed during the 3 weeks of treatment in the control plants, while the level of most of the 19 amino acids examined was increased by osmotic stress (Figures 4–8). Their concentrations were significantly greater in the PEG-treated plants than in control ones at the most sampling points. The levels of several amino acids were relatively greater after 3 d of PEG treatment in CS (Ch5A) and after 3 weeks in CS (Tsp5A).

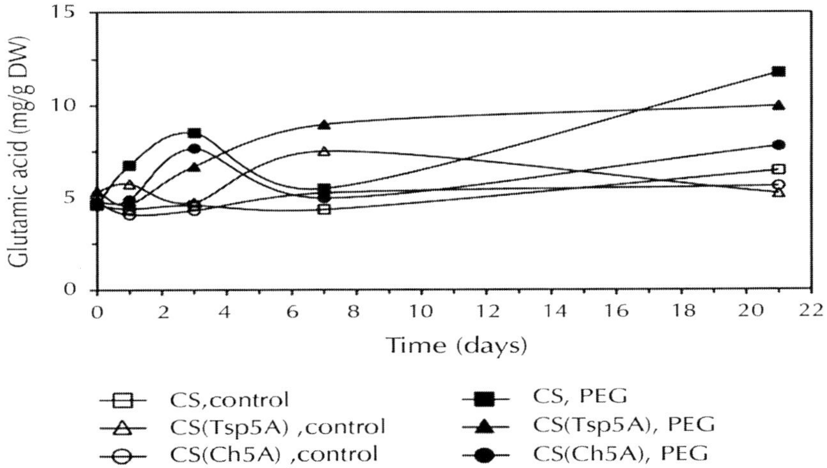
Glutamic acid (mg/g DW)
15
10
5
0
0 2 4 6 8 10 12 14 16 18 20 22
Time (days)
CS,control
CS(Tsp5A) ,control
CS(Ch5A) ,control
CS, PEG
CS(Tsp5A), PEG
CS(Ch5A), PEG

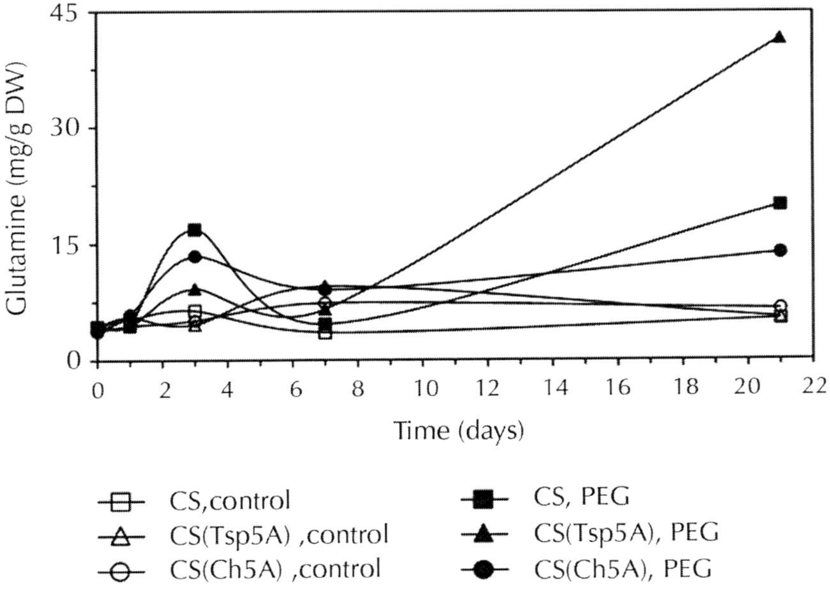
Glutamine (mg/g DW)
45
30
15
0
0 2 4 6 8 10 12 14 16 18 20 22
Time (days)
CS,control
CS(Tsp5A) ,control
CS(Ch5A) ,control
CS, PEG
CS(Tsp5A), PEG
CS(Ch5A), PEG

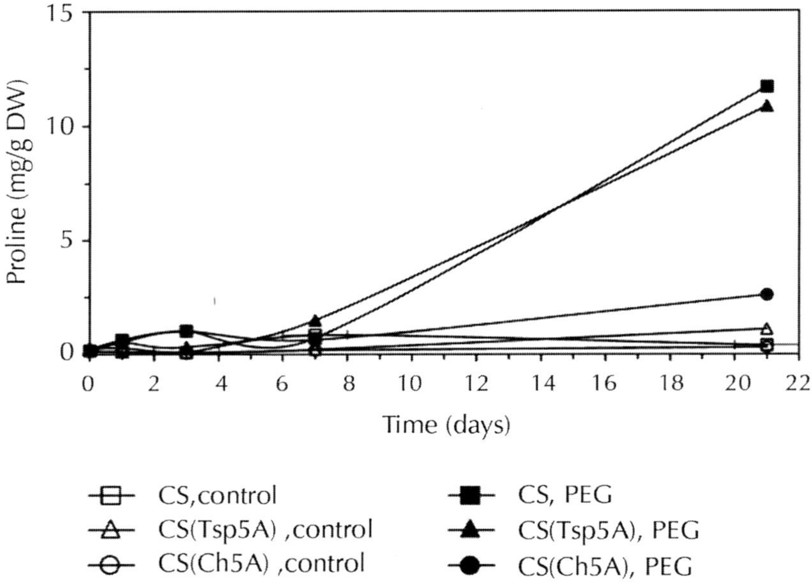
15
10
5
0
Proline (mg/g DW)
0 2 4 6 8 10 12 14 16 18 20 22
Time (days)
CS,control
CS(Tsp5A) ,control
CS(Ch5A) ,control
CS, PEG
CS(Tsp5A), PEG
CS(Ch5A), PEG

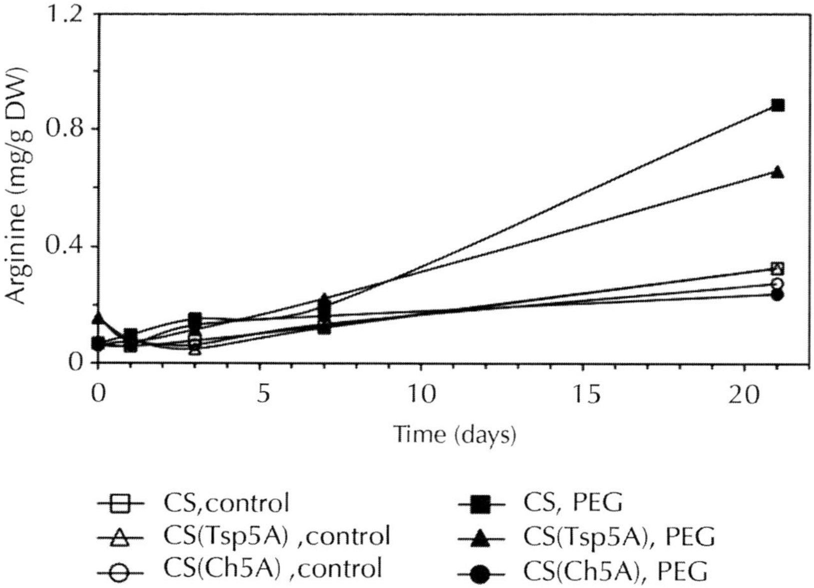
1.2
0.8
0.4
0
Arginine (mg/g DW)
0 5 10 15 20
Time (days)
CS,control
CS(Tsp5A) ,control
CS(Ch5A) ,control
CS, PEG
CS(Tsp5A), PEG
CS(Ch5A), PEG

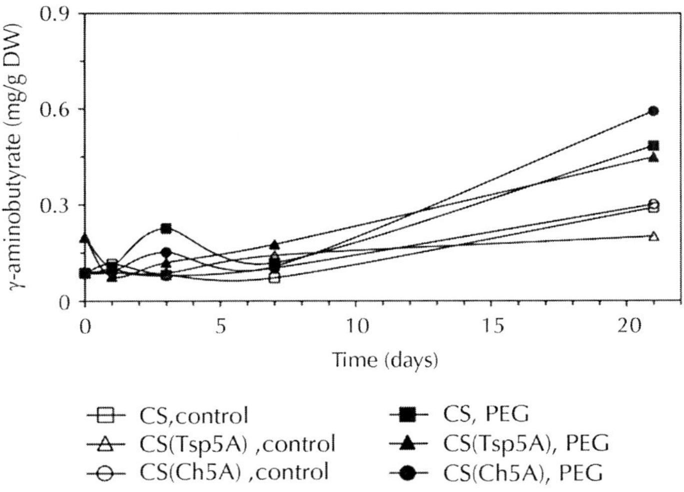

Figure 4: Effect of osmotic stress on the concentration of amino acids belonging to the glutamate family. The time course of changes was compared under control and stress (15% PEG) conditions in the freezing-sensitive CS and CS (Tsp5A) and the freezing-tolerant CS (Ch5A) wheat genotypes. The standard deviations of the data were below 10% in each case (not shown on the graphs because of the better transparency). (A): Glu (SD: 1.66), (b): Gln (SD: 5.10), (c): Pro (SD: 1.67), (d): Arg (SD: 0.070), and (e): γ-aminobutyrate (SD: 0.128).

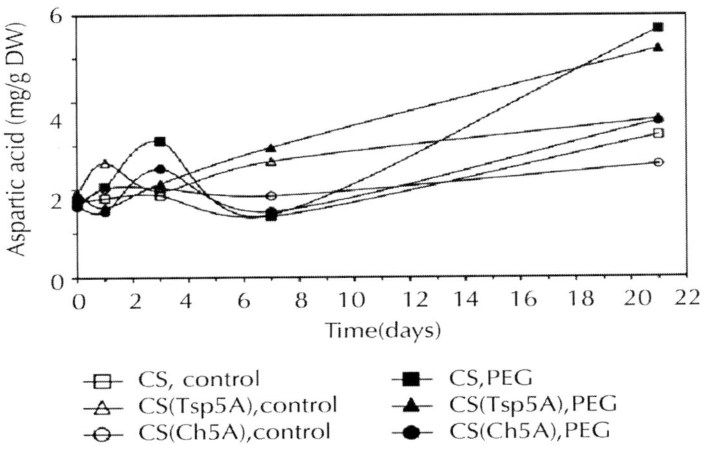

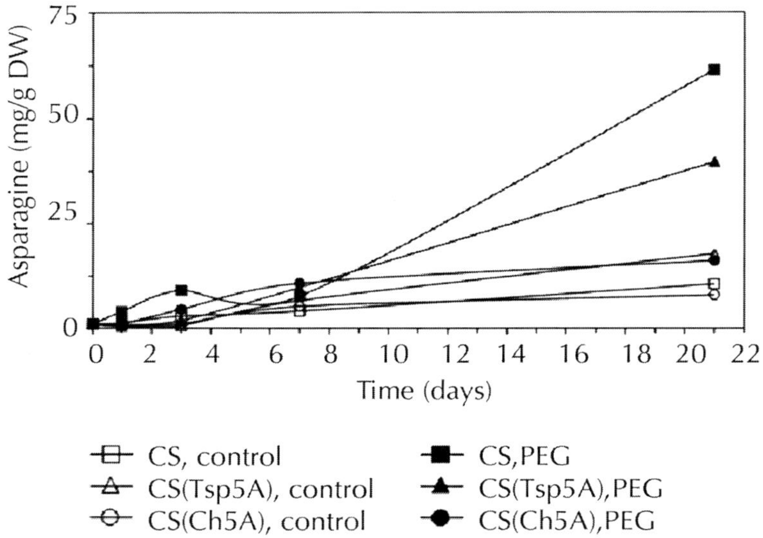
75
50
25
0
Asparagine (mg/g DW)
0 2 4 6 8 10 12 14 16 18 20 22
Time (days)
CS, control
CS(Tsp5A), control
CS(Ch5A), control
CS,PEG
CS(Tsp5A),PEG
CS(Ch5A),PEG

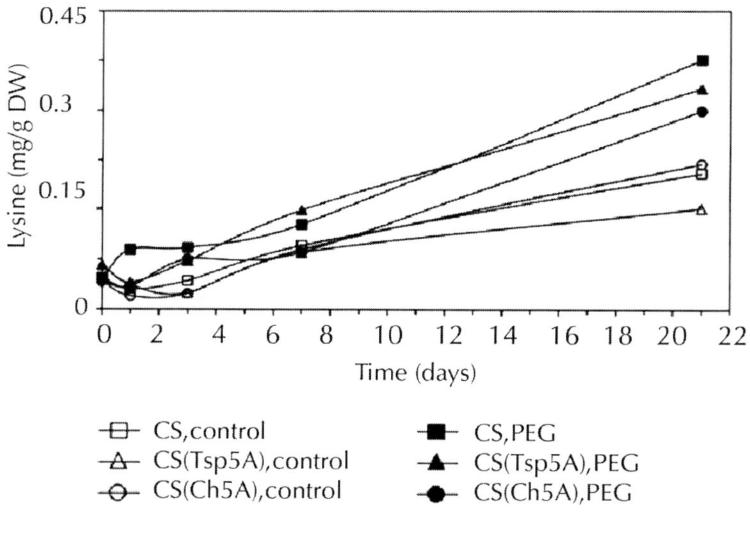
0.45
0.3
0.15
0
Lysine (mg/g DW)
0 2 4 6 8 10 12 14 16 18 20 22
Time (days)
CS,control
CS(Tsp5A),control
CS(Ch5A),control
CS,PEG
CS(Tsp5A),PEG
CS(Ch5A),PEG

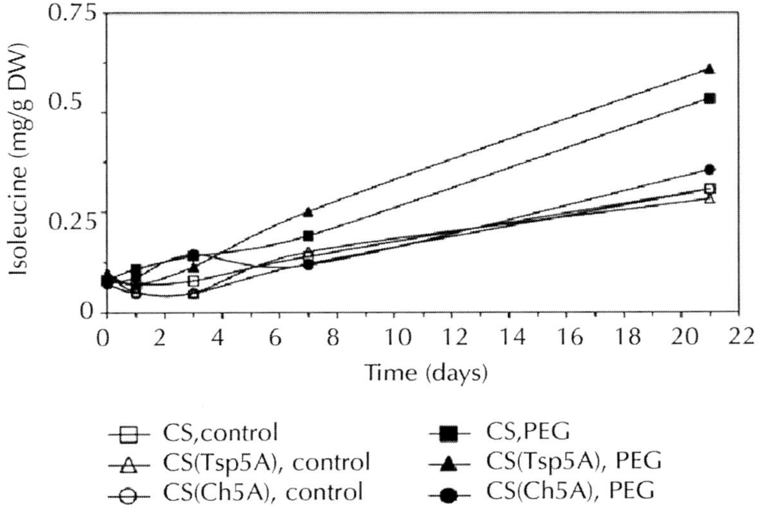

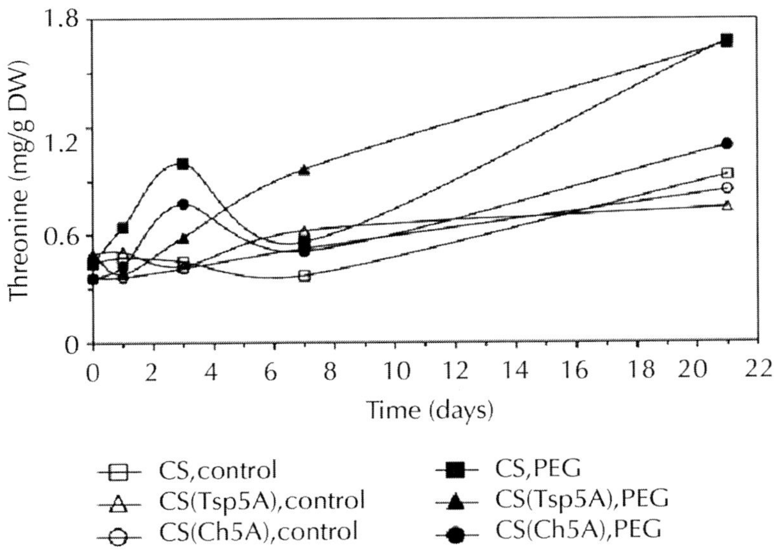

Figure 5: Effect of osmotic stress on the concentration of amino acids belonging to the aspartate family. The time course of changes was compared

under control and stress (15% PEG) conditions in the freezing-sensitive CS and CS (Tsp5A) and the freezing-tolerant CS (Ch5A) wheat genotypes. The standard deviations of the data were below 10% in each case (not shown on the graphs because of the better transparency). (a): Asp (SD: 0.505), (b): Asn (SD: 5.92), (c): Lys (SD: 0.037), (d): Ile (SD: 0.050), and (e): Thr (SD: 0.204).

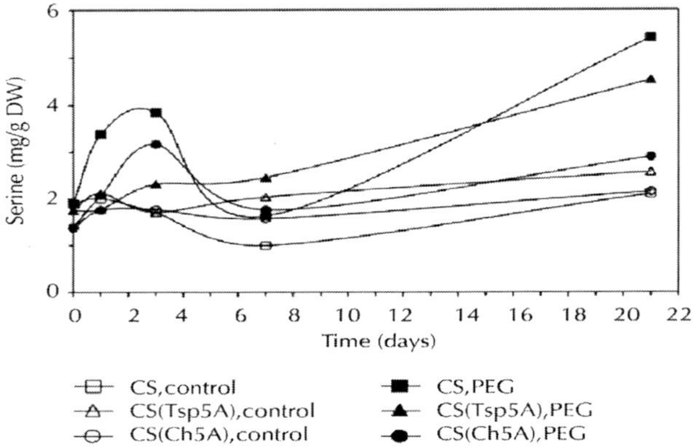

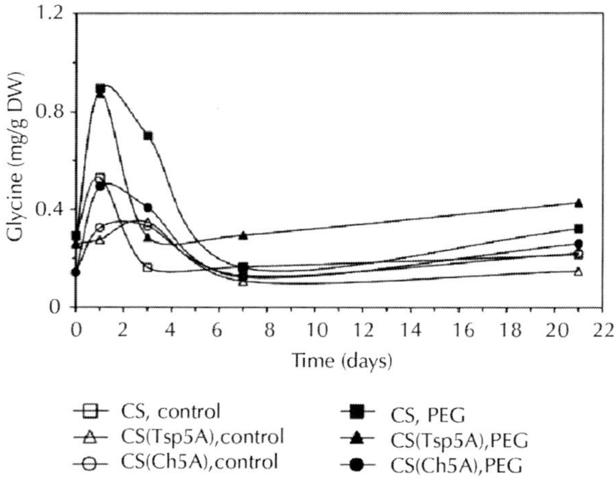

Figure 6: Effect of osmotic stress on the concentration of amino acids belonging to the serine family. The time course of changes was compared under control and stress (15% PEG) conditions in the freezing-sensitive CS and

CS (Tsp5A) and the freezing-tolerant CS (Ch5A) wheat genotypes. The standard deviations of the data were below 10% in each case (not shown on the graphs because of the better transparency). (A): Ser (SD: 0.644), and (b): Gly (SD: 0.159).

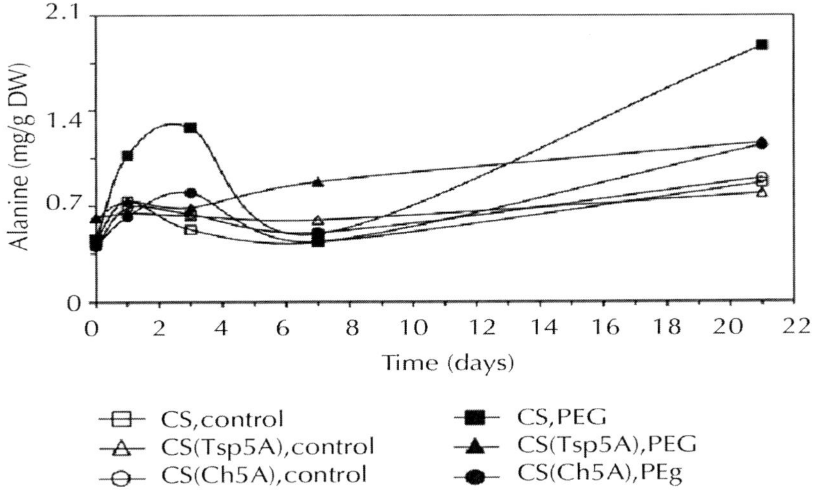

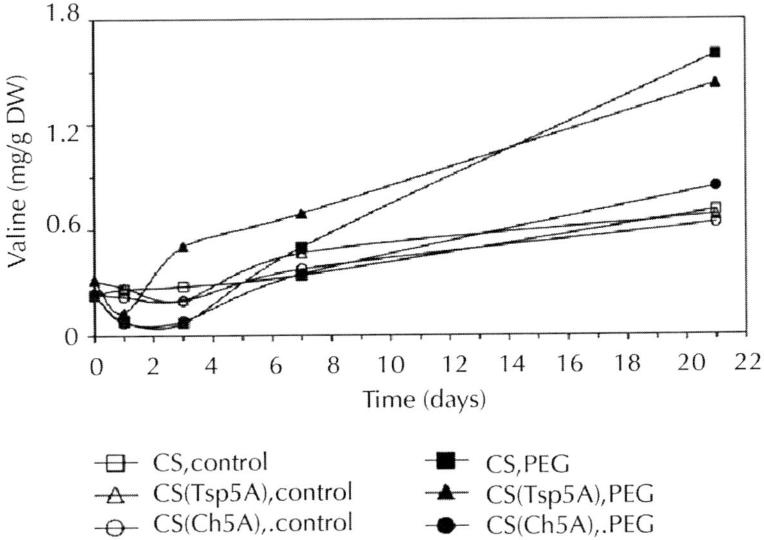

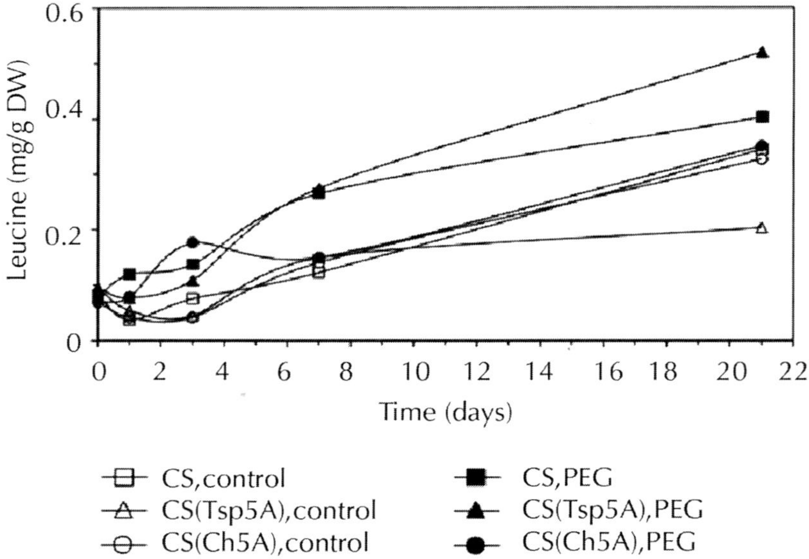

Figure 7: Effect of osmotic stress on the concentration of amino acids belonging to the pyruvate family. The time course of changes was compared under control and stress (15% PEG) conditions in the freezing-sensitive CS and CS (Tsp5A) and the freezing-tolerant CS(Ch5A) wheat genotypes. The standard deviations of the data were below 10% in each case (not shown on the graphs because of the better transparency). (A): Ala (SD: 0.261), (b): Val (SD: 0.146), and (c): Leu (SD: 0.058).

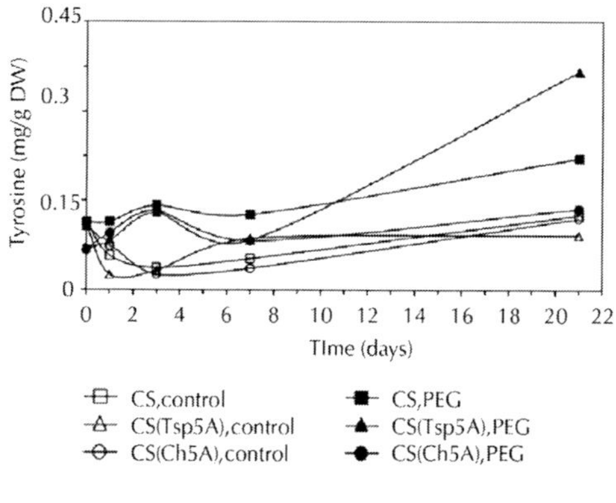

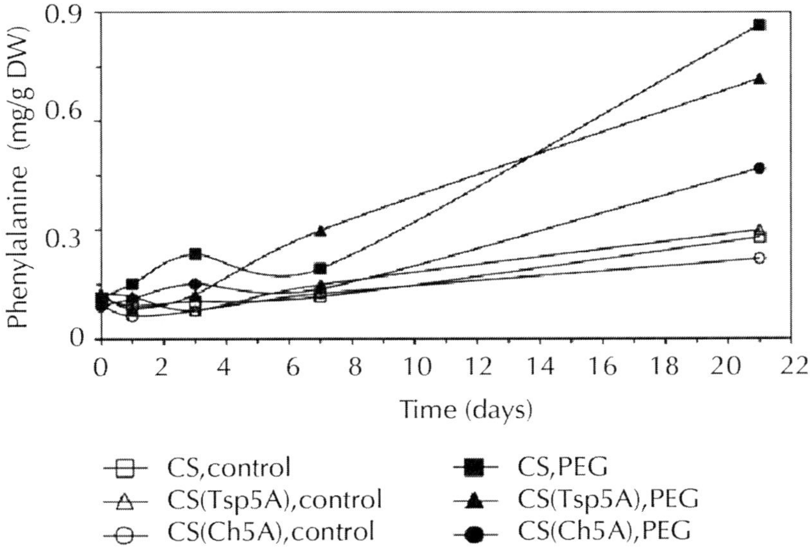
0.9
0.6
0.3
0
Phenylalanine (mg/g DW)
0
2
4
6
8
10
12
14
16
18
20
22
Time (days)
CS,control
CS(Tsp5A),control
CS(Ch5A),control
CS,PEG
CS(Tsp5A),PEG
CS(Ch5A),PEG

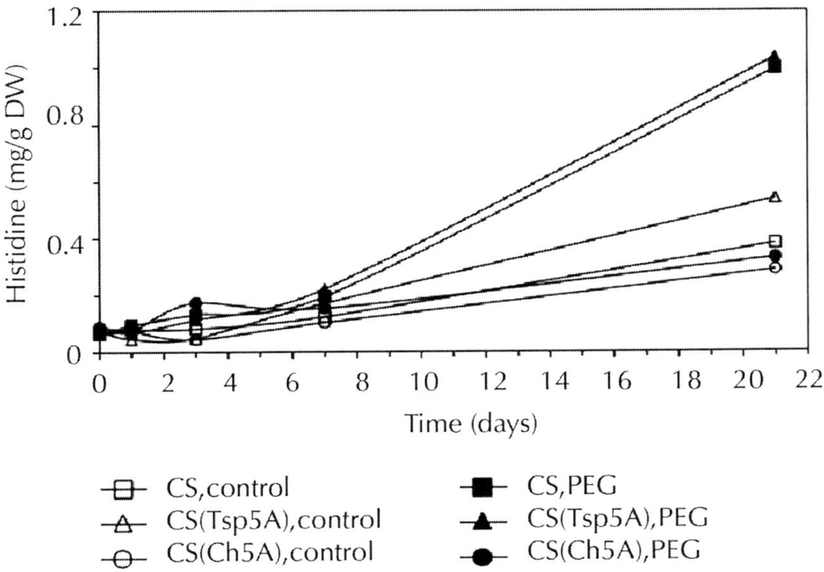
1.2
0.8
0.4
0
Histidine (mg/g DW)
0
2
4
6
8
10
12
14
16
18
20
22
Time (days)
CS,control
CS(Tsp5A),control
CS(Ch5A),control
CS,PEG
CS(Tsp5A),PEG
CS(Ch5A),PEG

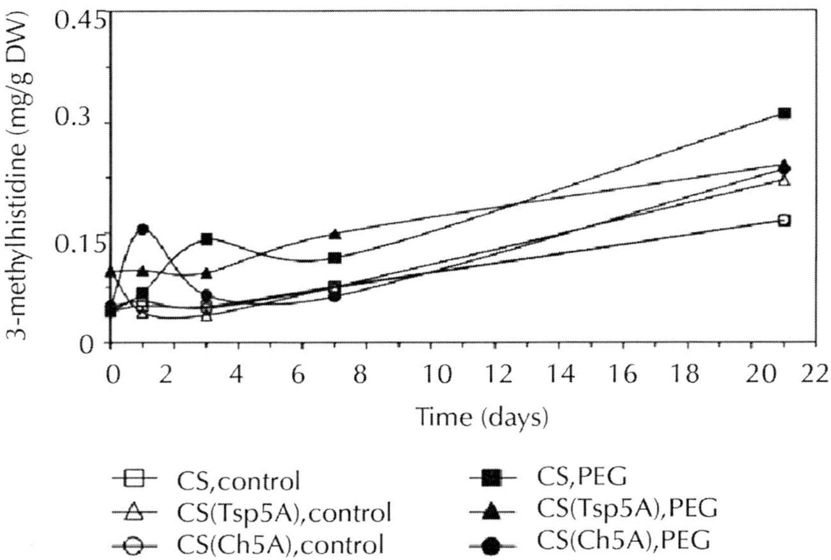

Figure 8: Effect of osmotic stress on the concentration of amino acids belonging to the aromatic amino acid family. The time course of changes was compared under control and stress (15% PEG) conditions in the freezing-sensitive CS and CS (Tsp5A) and the freezing-tolerant CS (Ch5A) wheat genotypes. The standard deviations of the data were below 10% in each case (not shown on the graphs because of the better transparency). (A): Tyr (SD: 0.046), (b): Phe (SD: 0.091), (c): His (SD: 0.122), and (d): 3-metil-His (SD: 0.026).

Changes of Glutamate Family during Osmotic Stress

The glutamate content exhibited a very rapid PEG-induced increase during the first three days of treatment, but later its concentration transiently decreased, then increased again in CS and CS(Ch5A) (Figure 4(a)). The changes in the glutamine content were very similar (Figure 4(b)). There was a peak in its concentration after 3 days and after a transient reduction its level increased again during the second half of the treatment. Although the Pro concentration exhibited a great increase (more than 10-fold) in all genotypes, its maximum level was

3.3-fold greater in CS (Tsp5A) than in CS (Ch5A) (Figure 4(c)). Slight changes occurred only in the Arg content during the first week of the treatment in all 3 genotypes, but later on it increased in CS and CS (Tsp5A) to high levels (Figure 4(d)). The gamma-aminobutyrate content exhibited first a transient decrease in CS (Tsp5A), and a transient increase in CS and CS (Ch5A) which was followed by a similar increase in all 3 genotypes (Figure 4(e)). The amounts of Glu and Gln were greater after 3 d in CS(Ch5A) and CS than in CS(Tsp5A), while after 21 d osmotic treatment higher levels of these amino acids were detected in CS(Tsp5A) and CS than in CS(Ch5A).

Changes of Aspartate Family during Osmotic Stress

The aspartate concentration exhibited a transient increase during the first days of the experiment with a local maximum after 3 d of treatment in CS (Ch5A) and CS (Figure 5(a)). Later a second increase was observed in these two genotypes, and the level gradually increased in CS (Tsp5A). Similarly to Asp, the greatest asparagine concentrations were detected after 3 weeks of osmotic stress in CS (Tsp5A) and CS (Figure 5(b)). A transient increase was found in the amount of this amino acid during the first week of the treatment in CS, while its concentration gradually increased in CS (Tsp5A). The Lys content increased gradually to similar level in all 3 genotypes after PEG treatment (Figure 5(c)). The Ile content reached higher concentrations after 3 weeks of osmotic stress in CS and CS (Tsp5A) than in CS (Ch5A) (Figure 5(d)). The Thr concentration exhibited a local maximum after 3 d in CS and CS(Ch5A), while it gradually increased throughout the experiment in CS(Tsp5A) (Figure 5(e)). The concentration of Asp and Thr was greater in CS (Ch5A) and CS than in CS (Tsp5A) after 3 d osmotic stress, and the level of Asp, Asn, Ile, and Thr was greater in CS (Tsp5A) and CS than in CS (Ch5A) after 21 d.

Changes of Other Amino Acid Families during Osmotic Stress

Two members of serine amino acid family could be detected. The Ser concentration exhibited a large transient increase during the first week of the treatment in CS and CS (Ch5A), while it gradually increased in CS (Tsp5A) during the 3-week osmotic stress (Figure 6(a)). The Ser content increased during the second half of the treatment again in CS. The Gly content reached a maximum after 1 d osmotic stress, then had a rapid decrease, and it was low after 1 week PEG-treatment. (Figure 6(b)).

All 3 members of the pyruvate family were studied. The Ala concentration showed greater changes only in CS, where it had a transient increase during the first week, and increased again during the second half of the experiment (Figure 7(a)). The Val concentration first decreased in all the genotypes, but subsequently it increased and this change was quicker in CS and CS(Tsp5A) than in CS(Ch5A) (Figure 7(b)). The Leu content increased in all three genotypes during the whole stress period (Figure 7(c)).

Four aromatic amino acids could be measured. The Tyr concentration, after an initial transient increase, exhibited a rapid increase in CS (Tsp5A) and a slow one in the other two genotypes (Figure 8(a)). The Phe content, except for a small transient decrease in CS and CS (Ch5A) after 3 days of PEG treatment, was increased by osmotic stress, and this change was much greater in CS and CS (Tsp5A) than in the other genotypes (Figure 8(b)). The time course of the alterations in His content was similar to that of Phe content (Figure 8(c)). The concentration of 3-methyl-His increased quickly during the first days of PEG treatment in CS and CS (Ch5A), which was followed by a decrease and subsequently by a slow increase (Figure 8(d)).

DISCUSSION

Interestingly, the fresh and dry weight data of control and treated plants did not differ at the end of the 3-week treatment with 15% PEG in the present experiment, however on the basis of shoot damage monitored after two weeks recovery phase following the 3-week osmotic stress in a previous study, the 3 genotypes had different level of tolerance [15, 16]. These controversial results can be observed by the phenomenon that the stress injuries appear very often not during the stress, but during the subsequent recovery phase. While during longer osmotic stress (7 and 21 d) the wheat variety Ch and the derived CS(Ch5A) chromosome substitution line seemed to be more tolerant to osmotic stress than the genotypes CS, Tsp and CS(Tsp5A) [15, 16], after short, 1-day exposure to osmotic stress only Ch proved to be more tolerant than the other genotypes [18]. Since dry weight data exhibited a smaller (5-fold) change than fresh weight data (10-fold) during the 3-week osmotic stress, the former one is more appropriate for usage as basis of the amino acid concentrations. In addition, the difference between control and treated plants was smaller in dry weight than in fresh weight at the end of the experiment.

Despite the great changes induced by osmotic stress in the amounts of most amino acids, the ratio of the amino acid families was similar after 3 weeks in the control and PEG-treated plants except for CS(Tsp5A), indicating that the alterations observed were coordinated. A similar coordination was found in the level of minor amino acids in the leaves of wheat and potato grown under various photosynthetic conditions [19]. In addition, a coordinated increase in the concentration of amino acids derived from pyruvate and oxalacetate was observed during cold shock in Arabidopsis [20]. The genetic modification of the Pro level affected the concentrations of several other amino acids in soybean [21]. These observations may be explained by the cross-pathway regulation of key enzymes of biosynthesis, but the involvement of protein degradation and storage cannot be excluded.

In the case of many amino acids (Glu, Gln, Asp, Thr, Ser, Gly, Ala, Tyr) a rapid transient increase was first observed with a local maximum after 3 days of PEG treatment, after which their levels increased again from the 7th day of osmotic stress. These results suggest that an elevation in the amino acid content is involved in both the short- and long-term response to osmotic stress. The initial increase was 1.5-2-fold greater in the case of several amino acids in the freezing-tolerant CS(Ch5A) than in the freezing-sensitive CS(Tsp5A) after osmotic stress (present study) and cold [12], indicating that the early accumulation of free amino acids contributes to a greater level of stress tolerance. After 3 weeks of stress other protective mechanisms may become more important, since the amino acid content was greater in the sensitive CS (Tsp5A). Thus, the level of the antioxidants ascorbate and glutathione (GSH) was greater after 3 weeks of cold in CS (Ch5A) than in CS (Tsp5A) [22]. The increase in GSH concentration may be due to the early stress-induced accumulation of Glu, which is a precursor of GSH.

The special role of Pro in the stress response, described in several publications [5, 23], was corroborated in the present study. The increase in the Pro content may originate from the induction of pyrroline 5-carboxylate synthetase, described at both the gene expression and activity levels in rice subjected to drought [6]. Besides its osmoprotective role, Pro may also function as an antioxidant. The stress-induced increase in its synthesis could be important for the maintenance of a low NADPH: $NADP^+$ ratio and could prevent the accumulation of NADPH, which occurs due to the inhibition of the Calvin cycle under stress conditions [23]. The recycling of $NADP^+$ is important as it provides on electron acceptor for photosynthesis, thus preventing the accumulation of reactive oxygen species.

In contrast to several other amino acids, there was no further great increase in the Glu content after the initial transient rise. This can be explained by its intensive use for the synthesis of Pro, polyamines, and GSH during the long-term (3-week) stress response. This assumption is confirmed by the observation that the spermine content increased in wheat after 3 weeks of osmotic stress [16]. As in the case of the Glu content in wheat, a relationship was found between the activities of enzymes involved in Glu synthesis and stress tolerance in the leaves of the resurrection plant Sporobolus stapfianus [3]. The high demand for Glu under stress conditions due to its use for the synthesis of Pro, polyamines, and GSH can be met by the conversion of Gln to Glu, since Gln accumulation was also induced by osmotic stress.

The protective role of Arg during osmotic stress, indicated by the fact that its level changed to different extents in wheat genotypes with different tolerance, may be due to its osmoprotective function, observed in yeast [8], or to its use for polyamine synthesis. Positively charged polyamines are involved in the stress response through their interaction with negatively charged macromolecules, such as DNA, RNA, proteins, and phospholipids, resulting in changes in the physical and chemical properties of the membranes, in the structure of nucleic acids and in enzyme activities [24].

The comparison of two chromosome 5A substitution lines with different levels of stress tolerance revealed that this chromosome affects both the short- and long-term changes in the free amino acid concentration. This effect is probably based on the regulatory role of the Cbf/Dreb (C-repeat binding/drought-responsive element binding) transcription factors mapped at the Fr2 (frost-resistance 2) locus on the long arm of chromosome 5A in the response to drought and frost [25]. Cbfs localised on chromosome 5A affect the expression of genes involved in the amino acid metabolism, as shown by a comparison of cold-induced changes in the transcriptome of the Cs (Tsp5A) and Cs (Ch5A) substitution lines [13]. Like those of amino acids, the levels of polyamines were also affected by chromosome 5A during osmotic stress in wheat [26]. Although in the present study we focused on chromosome 5A because of its importance in the stress response [13], the influence of other chromosomes on osmotic stress-induced amino acid accumulation cannot be excluded, since several chromosomes were shown to affect osmoregulation in wheat [27]. Thus, chromosome 5D had greater effect on PEG-induced carbohydrate accumulation than chromosome 5A [28].

In summary, coordinated changes in the levels of individual amino acids indicate that they are involved in both the short- and long-term response to osmotic stress. The alterations differed for the two chromosomes 5A substitution lines, depending on the stress tolerance of the chromosome donor genotype.

ACKNOWLEDGMENTS

The authors wish to thank A. Horváth and N. Csabai for their help in plant cultivation and treatment and K. Hetes-L rincz for technical work in the analysis of free amino acids. This work was supported by the European Union (AGRISAFE 203288-EU-FP7-REGPOT2007-1), by the Hungarian Scientific Research Fund (OTKA K83642), and by the Hungarian National Development Agency (TÁMOP-4.2.2.B-10/1-2010-0025).

REFERENCES

1. V. K. Rai, "Role of amino acids in plant responses to stresses," Biologia Plantarum, vol. 45, no. 4, pp. 481–487, 2002
2. A. Ogawa and A. Yamauchi, "Root osmotic adjustment under osmotic stress in maize seedlings 2. Mode of accumulation of several solutes for osmotic adjustment in the root," Plant Production Science, vol. 9, no. 1, pp. 39–46, 2006

3. T. Martinelli, A. Whittaker, A. Bochicchio, C. Vazzana, A. Suzuki, and C. Masclaux-Daubresse, "Amino acid pattern and glutamate metabolism during dehydration stress in the "resurrection" plantSporobolus stapfianus: a comparison between desiccation-sensitive and desiccation-tolerant leaves,"Journal of Experimental Botany, vol. 58, no. 11, pp. 3037–3046, 2007.
4. N. Verbruggen and C. Hermans, "Proline accumulation in plants: a review," Amino Acids, vol. 35, no. 4, pp. 753–759, 2008.
5. Y. Yoshiba, T. Kiyosue, K. Nakashima, K. Yamaguchi-Shinozaki, and K. Shinozaki, "Regulation of levels of proline as an osmolyte in plants under water stress," Plant and Cell Physiology, vol. 38, no. 10, pp. 1095–1102, 1997.
6. N. L. Choudhary, R. K. Sairam, and A. Tyagi, "Expression of 1-pyrroline-5-carboxylate synthetase gene during drought in rice (Oryza sativa L.)," Indian Journal of Biochemistry and Biophysics, vol. 42, no. 6, pp. 366–370, 2005.
7. D. Verdoy, T. Coba de la Peña, F. J. Redondo, M. M. Lucas, and J. J. Pueyo, "Transgenic Medicago truncatula plants that accumulate proline display nitrogen-fixing activity with enhanced tolerance to osmotic stress," Plant, Cell and Environment, vol. 29, no. 10, pp. 1913–1923, 2006
8. S. Xu, J. Zhou, L. Liu, and J. Chen, "Arginine: a novel compatible solute to protect Candida glabrataagainst hyperosmotic stress," Process Biochemistry, vol. 46, no. 6, pp. 1230–1235, 2011
9. M. B. Herrera-Rodríguez, R. Pérez-Vicente, and J. M. Maldonado, "Expression of asparagine synthetase genes in sunflower (Helianthus annuus) under various environmental stresses," Plant Physiology and Biochemistry, vol. 45, no. 1, pp. 33–38, 2007
10. H. Wang, D. Liu, J. Sun, and A. Zhang, "Asparagine synthetase gene TaASN1 from wheat is up-regulated by salt stress, osmotic stress and ABA," Journal of Plant Physiology, vol. 162, no. 1, pp. 81–89, 2005.
11. H. M. Cai, Y. Zhou, J. H. Xiao, X. H. Li, Q. F. Zhang, and X. M. Lian, "Overexpressed glutamine synthetase gene modifies nitrogen metabolism and abiotic stress responses in rice," Plant Cell Reports, vol. 28, no. 3, pp. 527–537, 2009.
12. Z. Kovács, L. Simon-Sarkadi, C. Sovány, K. Kirsch, G. Galiba, and

G. Kocsy, "Differential effects of cold acclimation and abscisic acid on free amino acid composition in wheat," Plant Science, vol. 180, no. 1, pp. 61–68, 2011.

13. G. Kocsy, B. Athmer, D. Perovic et al., "Regulation of gene expression by chromosome 5A during cold hardening in wheat," Molecular Genetics and Genomics, vol. 283, no. 4, pp. 351–363, 2010.
14. A. Vágújfalvi, I. Kerepesi, G. Galiba, T. Tischner, and J. Sutka, "Frost hardiness depending on carbohydrate changes during cold acclimation in wheat," Plant Science, vol. 144, no. 2, pp. 85–92, 1999.
15. G. Kocsy, G. Szalai, and G. Galiba, "Effect of osmotic stress on glutathione and hydroxymethylglutathione accumulation in wheat," Journal of Plant Physiology, vol. 161, no. 7, pp. 785–794, 2004.
16. Z. Kovács, L. Simon-Sarkadi, A. Sz cs, and G. Kocsy, "Differential effects of cold, osmotic stress and abscisic acid on polyamine accumulation in wheat," Amino Acids, vol. 38, no. 2, pp. 623–631, 2010.
17. D. R. Hoagland and D. I. Arnon, "The water-culture method for growing plants without soil,"California Agricultural Experiment Station Circular, vol. 347, pp. 1–39, 1950.
18. E. Horváth, M. Pál, G. Szalai, E. Páldi, and T. Janda, "Exogenous 4-hydroxybenzoic acid and salicylic acid modulate the effect of short-term drought and freezing stress on wheat plants," Biologia Plantarum, vol. 51, no. 3, pp. 480–487, 2007.
19. G. Noctor, L. Novitskaya, P. J. Lea, and C. H. Foyer, "Co-ordination of leaf minor amino acid contents in crop species: significance and interpretation," Journal of Experimental Botany, vol. 53, no. 370, pp. 939–945, 2002.
20. F. Kaplan, J. Kopka, D. W. Haskell et al., "Exploring the temperature-stress metabolome ofArabidopsis," Plant Physiology, vol. 136, no. 4, pp. 4159–4168, 2004.
21. G. Kocsy, R. Laurie, G. Szalai et al., "Genetic manipulation of proline levels affects antioxidants in soybean subjected to simultaneous drought and heat stresses," Physiologia Plantarum, vol. 124, no. 2, pp. 227–235, 2005

22. A. Soltész, I. Tímár, I. Vashegyi et al., "Redox changes during cold acclimation affect freezing tolerance but not the vegetative/reproductive transition of the shoot apex in wheat," Plant Biology, vol. 13, no. 5, pp. 757–766, 2011
23. L. Szabados and A. Savouré, "Proline: a multifunctional amino acid," Trends in Plant Science, vol. 15, no. 2, pp. 89–97, 2010
24. R. Alcázar, F. Marco, J. C. Cuevas et al., "Involvement of polyamines in plant response to abiotic stress," Biotechnology Letters, vol. 28, no. 23, pp. 1867–1876, 2006
25. G. Galiba, A. Vágújfalvi, C. Li, A. Soltész, and J. Dubcovsky, "Regulatory genes involved in the determination of frost tolerance in temperate cereals," Plant Science, vol. 176, no. 1, pp. 12–19, 2009
26. G. Galiba, G. Kocsy, R. Kaur-Sawhney, J. Sutka, and A. W. Galston, "Chromosomal localization of osmotic and salt stress-induced differential alterations in polyamine content in wheat," Plant Science, vol. 92, no. 2, pp. 203–211, 1993
27. G. Galiba, L. Simon-Sarkadi, G. Kocsy, A. Salgó, and J. Sutka, "Possible chromosomal location of genes determining the osmoregulation of wheat," Theoretical and Applied Genetics, vol. 85, no. 4, pp. 415–418, 1992
28. G. Galiba, I. Kerepesi, J. W. Snape, and A. Vágújfalvi, "Deletion analysis of genes regulating cold and peg-induced carbohydrate accumulation in hydroponically raised wheat seedlings," Acta Agronomica Hungarica, vol. 53, no. 4, pp. 359–370, 2005.

CHAPTER 8

Mechanisms of Salinity Tolerance in Plants: Physiological, Biochemical and Molecular Characterization

Bhaskar Gupta[1] and Bingru Huang [2]

[1]Department of Biological Sciences (Section Biotechnology), Presidency University, 86/1 College Street, Kolkata 700073, India

[2]Department of Plant Biology and Pathology, Rutgers University, New Brunswick, NJ 08901, USA

ABSTRACT

Salinity is a major abiotic stress limiting growth and productivity of plants in many areas of the world due to increasing use of poor quality of water for irrigation and soil salinization. Plant adaptation or tolerance to salinity stress involves complex physiological traits, metabolic pathways, and molecular or gene networks. A comprehensive understanding on how plants respond to salinity stress at different levels and an integrated approach of combining molecular tools with physiological and biochemical techniques are imperative for the development of salt-tolerant varieties of plants in salt-affected areas. Recent research has identified various adaptive responses to salinity stress at molecular, cellular, metabolic, and physiological levels, although mechanisms underlying salinity tolerance are far from being completely understood. This paper provides a comprehensive review of major research advances on biochemical, physiological, and molecular mechanisms regulating plant adaptation and tolerance to salinity stress.

INTRODUCTION

A major challenge towards world agriculture involves production of 70% more food crop for an additional 2.3 billion people by 2050 worldwide [1]. Salinity is a major stress limiting the increase in the demand for food crops. More than 20% of cultivated land worldwide (~ about 45 hectares) is affected by salt stress and the amount is increasing day by day. Plants on the basis of adaptive evolution can be classified roughly into two major types: the halophytes (that can withstand salinity) and the glycophytes (that cannot withstand salinity and eventually die). Majority of major crop species belong to this second category. Thus salinity is one of the most brutal environmental stresses that hamper crop productivity worldwide [2, 3].

Salinity stress involves changes in various physiological and metabolic processes, depending on severity and duration of the stress, and ultimately inhibits crop production [4–7]. Initially soil salinity is known to represses plant growth in the form of osmotic stress which is then followed by ion toxicity [4, 5]. During the initial phases of salinity stress, water absorption capacity of root systems decreases and water loss from leaves is accelerated due to osmotic stress of high salt accumulation in soil and plants, and therefore salinity stress is also considered as hyperosmotic stress [6]. Osmotic stress in the initial stage of salinity stress causes various physiological changes, such as interruption of membranes, nutrient imbalance, impairs the ability to detoxify reactive oxygen species (ROS), differences in the antioxidant enzymes and decreased photosynthetic activity, and decrease in stomatal aperture [3, 5]. Salinity stress is also considered as a hyperionic stress. One of the most detrimental effects of salinity stress is the accumulation of Na^+ and Cl^- ions in tissues of plants exposed to soils with high NaCl concentrations. Entry of both Na^+ and Cl^- into the cells causes severe ion imbalance and excess uptake might cause significant physiological disorder(s). High Na^+ concentration inhibits uptake of K^+ ions which is an essential element for growth and development that results into lower productivity and may even lead to death [4]. In response to salinity stress, the production of ROS, such as singlet oxygen, superoxide, hydroxyl radical, and hydrogen peroxide, is enhanced [8–12]. Salinity-induced ROS formation can lead to oxidative damages in various cellular components such as proteins, lipids, and DNA, interrupting vital cellular functions of plants.

Genetic variations in salt tolerance exist, and the degree of salt tolerance varies with plant species and varieties within a species. Among major crops, barley (Hordeum vulgare) shows a greater degree of salt tolerance than rice (Oryza sativa) and wheat (Triticum aestivum). The degree of variation is even more pronounced in the case of dicotyledons ranging from Arabidopsis thaliana, which is very sensitive towards salinity, to halophytes such as Mesembryanthemum crystallinum, Atriplex sp., Thellungiella salsuginea (previously known as T. halophila) [3, 13, 14]. In the last two decades sumptuous amount of research has been done in order to understand the mechanism of salt tolerance in model plant Arabidopsis [15]. Genetic variations and differential responses to salinity stress in plants differing in stress tolerance enable plant biologists to identify physiological mechanisms, sets of genes, and gene products that are involved in increasing stress tolerance and to incorporate them in suitable species to yield salt tolerant varieties.

The main aim of this review is to discuss research advances on the complex physiological and molecular mechanisms that are involved in plant salinity tolerance.

PHYSIOLOGICAL AND BIOCHEMICAL MECHANISMS OF SALT TOLERANCE

Plants develop various physiological and biochemical mechanisms in order to survive in soils with high salt concentration. Principle mechanisms include, but are not limited to (1), ion homeostasis and compartmentalization(2), ion transport and uptake,(3) biosynthesis of osmoprotectants and compatible solutes,(4)activation of antioxidant enzyme and (5)synthesis of antioxidant compounds, synthesis of polyamines,(6)generation of nitric oxide (NO), and,(7)hormone modulation. Research advances elucidating these mechanisms are discussed below.

Ion Homeostasis and Salt Tolerance

Maintaining ion homeostasis by ion uptake and compartmentalization is not only crucial for normal plant growth but is also an essential process for growth during salt stress [16–18]. Irrespective of their nature, both glycophytes and halophytes cannot tolerate high salt concentration in their cytoplasm. Hence, the excess salt is either transported to the vacuole or sequestered in older tissues which eventually are sacrificed, thereby protecting the plant from salinity stress [19, 20].

Major form of salt present in the soil is NaCl, so the main focus of research is the study about the transport mechanism of Na^+ ion and its compartmentalization. The Na^+ ion that enters the cytoplasm is then transported to the vacuole via Na^+/H^+ antiporter. Two types of H^+ pumps are present in the vacuolar membrane: vacuolar type H^+-ATPase (V-ATPase) and the vacuolar pyrophosphatase (V-PPase) [21–23]. Of these, V-ATPase is the most dominant H^+ pump present within the plant cell. During nonstress conditions it plays an important role in maintaining solute homeostasis, energizing secondary transport and facilitating vesicle fusion. Under stressed condition the survivability of the plant depends upon the activity of V-ATPase [21]. In a study performed by De Lourdes Oliveira Otoch et al. [22] in hypocotyls of Vigna unguiculataseedlings, it was observed that the activity of V-ATPase pump increased when exposed to salinity stress but under similar conditions, activity of V-PPase was inhibited, whereas in the case of halophyte Suaeda salsa, V-ATPase activity was upregulated and V-PPase played a minor role [23].

Increasing evidence demonstrates the roles of a Salt Overly Sensitive (SOS) stress signalling pathway in ion homeostasis and salt tolerance [24, 25]. The SOS signalling pathway (Figure 1) consists of three major proteins, SOS1, SOS2, and SOS3. SOS1, which encodes a plasma membrane Na^+/H^+ antiporter, is essential in regulating Na^+ efflux at cellular level. It also facilitates long distance transport of Na^+ from root to shoot. Overexpression of this protein confers salt tolerance in plants [26, 27]. SOS2 gene, which encodes a serine/threonine kinase, is activated by salt stress elicited Ca^+ signals. This protein consists of a well-developed N-terminal catalytic domain and a C-terminal regulatory domain [28]. The third type of protein involved in the SOS stress signalling pathway is the SOS3 protein which is a myristoylated Ca^+ binding protein and contains a myristoylation site at its N-terminus. This site plays an essential role in conferring salt tolerance [29]. C-terminal regulatory domain of SOS2 protein contains a FISL motif (also known as NAF domain), which is about 21 amino acid long sequence, and serves as a site of interaction for Ca^{2+} binding SOS3 protein (Figure1). This interaction between SOS2 and SOS3 protein results in the activation of the kinase [30]. The activated kinase then phosphorylates SOS1 protein thereby increasing its transport activity which was initially identified in yeast [31]. SOS1 protein is characterised by a long cytosolic C-terminal tail, about 700 amino acids long, comprising a putative nucleotide binding motif and an autoinhibitory domain. This autoinhibitory domain is the target site for SOS2 phosphorylation (Figure 1). Besides conferring salt tolerance it also regulates pH homeostasis, membrane vesicle trafficking, and vacuole functions [32, 33]. Thus with the increase in the concentration of Na^+ there is a sharp increase in the intracellular Ca^{2+} level which in turn facilitates its binding with SOS3 protein. Ca^{2+} modulates intracellular Na^+ homeostasis along with SOS proteins. The SOS3 protein then interacts and activates SOS2 protein by releasing its self-inhibition. The SOS3-SOS2 complex is then loaded onto plasma membrane where it phosphorylates SOS1 (Figure 1). The phosphorylated SOS1 results in the increased Na^+ efflux, reducing Na^+ toxicity [34].

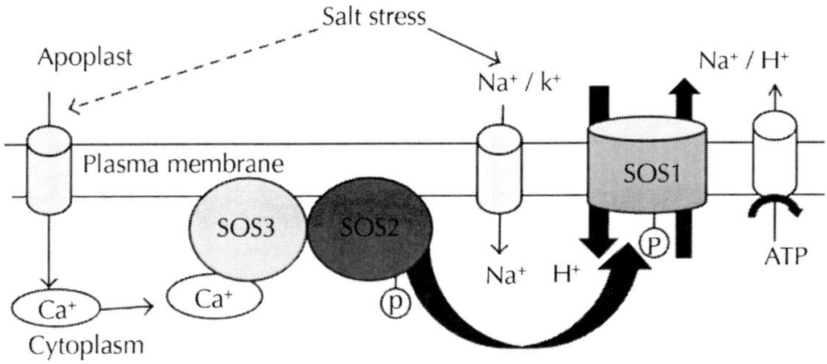

Figure 1: Model of SOS pathway for salinity stress responses.

Many plants have developed an efficient method to keep the ion concentration in the cytoplasm in a low level. Membranes along with their associated components play an integral role in maintaining ion concentration within the cytosol during the period of stress by regulating ion uptake and transport [35]. The transport phenomenon is carried out by different carrier proteins, channel proteins, antiporters and symporters. Maintaining cellular Na^+/K^+ homeostasis is pivotal for plant survival in saline environments. Ma et al. [36] have reported that Arabidopsis NADPH oxidases AtrbohD and AtrbohF function in ROS-dependent regulation of Na^+/K^+ homeostasis in Arabidopsis under salt stress. Plants maintain a high level of K^+ within the cytosol of about 100 mM ideal for cytoplasmic enzyme activities. Within the vacuole K^+ concentration ranges between 10 mM and 200 mM. The vacuole serves as the largest pool of K^+ within the plant cell. K^+ plays a major role in maintaining the turgor within the cell. It is transported into the plant cell against the concentration gradient via K^+ transporter and membrane channels. High affinity K^+ uptake mechanisms are mediated by K^+ transporters when the extracellular K^+ concentration is low, whereas low affinity uptake is carried out by K^+ channels when the extracellular K^+ concentration is high. Thus uptake mechanism is primarily determined by the concentration of K^+ available in the soil. On the other hand a very low concentration of Na^+ ion (about 1 mM or less) is maintained in the cytosol. During salinity stress, due to increased concentration of Na^+ in the soil, Na^+ ion competes with K^+ for the transporter as they both share the same transport mechanism, thereby decreasing the uptake of K^+ [3, 35].

A large number of genes and proteins, such as HKT and NHX, encoding K^+ transporters and channels have been identified and cloned in various plant species. During salt stress expression of some low abundance transcripts is enhanced which are found to be involved in K^+ uptake. This was observed in the halophyteMesembryanthemum crystallinum [37]. Transporters located on the plasma membrane, belonging to the HKT (histidine kinase transporter) family, also play an essential role in salt tolerance by regulating transportation of Na^+ and K^+. Class 1 HKT transporters that have been identified in Arabidopsis protect the plant from the adverse effects of salinity by preventing excess accumulation Na^+ in leaves. Similar results were observed in the experiment which was carried out with rice where class 1 HKT transporter removes excess Na^+ from xylem, thus protecting the photosynthetic leaf tissues from the toxic effect of Na^+ [38]. Intracellular NHX proteins are Na^+, K^+/H^+ antiporters involved in K^+ homeostasis, endosomal pH regulation, and salt tolerance. Barragán et al. [39] showed that tonoplast-localized NHX proteins (NHX1 and NHX2: the two major tonoplast-localized NHX isoforms) are essential for active K^+ uptake at the tonoplast, for turgor regulation, and for stomatal function. In fact more such NHX isoforms have been identified and their roles in ion (Na^+, K^+, H^+) homeostasis established from different plant species (e.g., LeNHX3 and LeNHX4 from tomato) [40].

Compatible Solute Accumulation and Osmotic Protection

Compatible solutes, also known as compatible osmolytes, are a group of chemically diverse organic compounds that are uncharged, polar, and soluble in nature and do not interfere with the cellular metabolism even at high concentration. They mainly include proline [41–45], glycine betaine [46, 47], sugar [48, 49], and polyols [50–53]. Organic osmolytes are synthesised and accumulated in varying amounts amongst different plant species. For example, quaternary ammonium compound beta alanine betaine's accumulation is restricted among few members of Plumbaginaceae [54], whereas accumulation of amino acid proline occurs in taxonomically diverse sets of plants [53]. The concentration of compatible solutes within the cell is maintained either by irreversible synthesis of the compounds or by a combination of synthesis and degradation. The biochemical pathways and genes involved in these processes have been thoroughly studied. As their accumulation is proportional to the external osmolarity, the major functions of these osmolytes are to protect the structure and to maintain osmotic balance within the cell via continuous water influx [24].

Amino acids such as cysteine, arginine, and methionine, which constitute about 55% of total free amino acids, decrease when exposed to salinity stress, whereas proline concentration rises in response to salinity stress [55]. Proline accumulation is a well-known measure adopted for alleviation of salinity stress [53, 56, and 57]. Intracellular proline which is accumulated during salinity stress not only provides tolerance towards stress but also serves as an organic nitrogen reserve during stress recovery. Proline is synthesised either from glutamate or ornithine. In osmotically stressed cell glutamate functions as the primary precursor. The biosynthetic pathway comprises two major enzymes, pyrroline carboxylic acid synthetase and pyrroline carboxylic acid reductase. Both these regulatory steps are used to overproduce proline in plants [35]. It functions as an O_2quencher thereby revealing its antioxidant capability. This was observed in a study carried out by Matysik et al. [56]. Ben Ahmed et al. [57] observed that proline supplements enhanced salt tolerance in olive (Olea europaea) by amelioration of some antioxidative enzyme activities, photosynthetic activity, and plant growth and the preservation of a suitable plant water status under salinity conditions. It has been reported that proline improves salt tolerance in Nicotiana tabacum by increasing the activity of enzymes involved in antioxidant defence system [58]. Deivanai et al. [59] also demonstrated that rice seedlings from seeds pretreated with 1 mM proline exhibited improvement in growth during salt stress.

Glycine betaine is an amphoteric quaternary ammonium compound ubiquitously found in microorganisms, higher plants and animals, and is electrically neutral over a wide range of pH. It is highly soluble in water but also contains nonpolar moiety constituting 3-methyl groups. Because of its unique structural features it interacts both with hydrophobic and hydrophilic domains of the macromolecules, such as enzymes and protein complexes. Glycine betaine is a nontoxic cellular osmolyte that raises the osmolarity of the cell during stress period; thus it plays an important function in stress mitigation. Glycine betaine also protects the cell by osmotic adjustment [60], stabilizes proteins [61], and protects the photosynthetic apparatus from stress damages [62] and reduction of ROS [52, 53]. Accumulation of glycine betaine is found in a wide variety of plants belonging to different taxonomical background. Glycine betaine is synthesised within the cell from either choline or glycine. Synthesis of glycine betaine from choline is a 2-step reaction involving two or more enzymes. In the first step choline is oxidised to betaine aldehyde which is then again oxidised in the next step to form glycine betaine. In higher plants the first conversion is carried out by the enzyme choline monooxygenase (CMO), whereas the next step is catalysed by betaine aldehyde dehydrogenase (BADH) [63]. Another pathway which is observed in some plants, mainly halophytic, demonstrated the synthesis of glycine betaine from glycine. Here glycine betaine is synthesized by three successive N-methylation and the reactions are catalysed by two S-adenosyl methionine dependent methyl transferases, glycine sarcosine N-methyl transferase (GSMT), and sarcosine dimethylglycine N-methyl transferase (SDMT). These two enzymes have overlapping functions as GSMT catalyses the first and the second step while SDMT catalyses the second and third step [63]. Rahman et al. [64] reported the positive effect of glycine betaine on the ultrastructure of Oryza sativa seedlings when exposed to salt stress. Under stressed condition (150 mM NaCl) the ultrastructure of the seedling shows several damages such as swelling of thylakoids, disintegration of grana and intergranal lamellae, and disruption of mitochondria. However, these damages were largely prevented when seedlings were pretreated with glycine betaine. When glycine betaine is applied as a foliar spray in a plant subjected to stress, it led to pigment stabilization and increase in photosynthetic rate and growth [62, 63].

Polyols are compounds with multiple hydroxyl functional groups available for organic reactions. Sugar alcohols are a class of polyols functioning as compatible solutes, as low molecular weight chaperones, and as ROS scavenging compounds [52]. They can be classified into two major types, cyclic (e.g., pinitol) and acyclic (e.g., mannitol). Mannitol synthesis is induced in plants during stressed period via action of NADPH dependent mannose-6-phosphate reductase. These compatible solutes function as a protector or stabilizer of enzymes or membrane structures that are sensitive to dehydration or ionically induced damage. It was found that the transformation with bacterial mltd gene that encodes for mannitol-1-phosphate dehydrogenase in both Arabidopsis and tobacco (Nicotiana tabacum) plants confer salt tolerance, thereby maintaining normal growth and development when subjected to high level of salt stress [65, 66]. Pinitol is accumulated within the plant cell when the plant is subjected to salinity stress. The biosynthetic pathway consists of two major steps, methylation of myo-inositol which results in formation of an intermediate compound, ononitol, which undergoes epimerization to form pinitol. Inositol methyl transferase enzyme encoded by imt gene plays major role in the synthesis of pinitol. Transformation of imt gene in plants shows a result similar to that observed in the case of mltd gene. Thus it can be said that pinitol also plays a significant role in stress alleviation. Accumulation of polyols, either straight-chain metabolites such as mannitol and sorbitol or cyclic polyols such as myo-inositol and its methylated derivatives, is correlated with tolerance to drought and/or salinity, based on polyol distribution in many species, including microbes, plants, and animals [49].

Accumulations of carbohydrates such as sugars (e.g., glucose, fructose, fructans, and trehalose) and starch occur under salt stress [67]. The major role played by these carbohydrates in stress mitigation involves osmoprotection, carbon storage, and scavenging of reactive oxygen species. It was observed that salt stress increases the level of reducing sugars (sucrose and fructans) within the cell in a number of plants belonging to different species [48]. Besides being a carbohydrate reserve, trehalose accumulation protects organisms against several physical and chemical stresses including salinity stress. They play an osmoprotective role in physiological responses [63]. Sucrose content was found to increase in tomato (Solanum lycopersicum) under salinity due to increased activity of sucrose phosphate synthase [68]. Sugar content, during salinity stress, has been reported to both increase and decrease in various rice genotype [69]. In rice roots it has been observed that starch content decreased in response to salinity while it remained fairly unchanged in the shoot. Decrease in starch content and increase in reducing and nonreducing sugar content were noted in leaves of Bruguiera parviflora [67].

Antioxidant Regulation of Salinity Tolerance

Abiotic and biotic stress in living organisms, including plants, can cause overflow, deregulation, or even disruption of electron transport chains (ETC) in chloroplasts and mitochondria. Under these conditions molecular oxygen (O_2) acts as an electron acceptor, giving rise to the accumulation of ROS. Singlet oxygen ($^1O_2$), the hydroxyl radical (OH^-), the superoxide radical (O_2^-), and hydrogen peroxide (H_2O_2) are all strongly oxidizing compounds and therefore potentially harmful for cell integrity [70]. Antioxidant metabolism, including antioxidant enzymes and nonenzymatic compounds, play critical parts in detoxifying ROS induced by salinity stress. Salinity tolerance is positively correlated with the activity of antioxidant enzymes, such as superoxide dismutase (SOD), catalase (CAT), glutathione peroxidise (GPX), ascorbate peroxidase (APX), and glutathione reductase (GR) and with the accumulation of nonenzymatic antioxidant compounds [71, 72]. Gill et al. [73] and Tuteja et al. [74] have recently reported a couple of helicase proteins (e.g., DESD-box helicase and OsSUV3 dual helicase) functioning in plant salinity tolerance by improving/maintaining photosynthesis and antioxidant machinery. Kim et al. [75] showed that silicon (Si) application to rice root zone influenced the hormonal and antioxidant responses under salinity stress. The results showed that Si treatments significantly increased rice plant growth compared to controls under salinity stress. Si treatments reduced the sodium accumulation resulting in low electrolytic leakage and lipid peroxidation compared to control plants under salinity stress. Enzymatic antioxidant (catalase, peroxidase, and polyphenol oxidase) responses were more pronounced in control plants than in Si-treated plants under salinity stress.

Anthocyanin is a flavonoid whose accumulation in plant exposed to salt stress has been largely documented. Van Oosten et al. [76] isolated the anthocyanin-impaired-response-1 (air1) mutant that is unable to accumulate anthocyanins under salt stress. The air1 mutant showed a defect in anthocyanin production in response to salt stress but not to other stresses such as high light, low phosphorous, high temperature, or drought stress. This specificity indicated that air1 mutation did not affect anthocyanin biosynthesis but rather its regulation in response to salt stress. The discovery and characterization of AIR1 opens avenues to dissect the connections between abiotic stress and accumulation of antioxidants in the form of flavonoids and anthocyanins.

Ascorbate is one of the major antioxidants present within the cell. Pea plants grown under saline (150 mM NaCl) stress showed an enhancement of both APX activity and S-nitrosylated APX, as well as an increase of H_2O_2, NO, and S-nitrosothiol (SNO) content that can justify the induction of the APX activity. Proteomic data have shown that APX is one of the potential targets of PTMs mediated by NO-derived molecules [77]. Using recombinant pea cytosolic APX, the impact of peroxynitrite ($ONOO^-$) and S-nitrosoglutathione (GSNO), which are known to mediate protein nitration and S-nitrosylation processes, respectively, was analysed. While peroxynitrite inhibits APX activity, GSNO enhances its enzymatic activity. The results provide new insight into the molecular mechanism of the regulation of APX, which can be both inactivated by irreversible nitration and activated by reversible S-nitrosylation [77]. Exogenous application of ascorbate mitigates the adverse effects of salinity stress in various plant species and promotes plant recovery from the stress [78, 79]. Another antioxidant in stress mitigation is glutathione, which can react with superoxide radical, hydroxyl radical, and hydrogen peroxide, thereby functioning as a free radical scavenger. It can also participate in the regeneration of ascorbate via ascorbate-glutathione cycle [80]. When applied exogenously glutathione helped to maintain plasma membrane permeability and cell viability during salinity stress in Allium cepa [81]. Application of glutathione and ascorbate was found to be effective in increasing the height of the plant, branch number, fresh and dry weight of herbs and flowers, and the content of carbohydrates, phenols, xanthophylls pigment, and mineral ion content when subjected to saline condition [82]. Many studies have found differences in levels of expression or activity of antioxidant enzymes; these differences are sometimes associated with the more tolerant genotype and sometimes with the more sensitive genotype. Munns and Tester [3] suggested that differences in antioxidant activity between genotypes may be due to genotypic differences in degrees of stomatal closure or in other responses that alter the rate of CO_2 fixation and differences that bring into play the processes that avoid photoinhibition and for which the plant has abundant capacity [3]. Roy et al. [83] in their recent review have argued that there are three main traits in plants, which help them in their adaptation to salinity stress: ion exclusion, tissue tolerance, and salinity tolerance. It seems that antioxidants have some role in tissue and salinity tolerance mechanism.

Roles of Polyamines in Salinity Tolerance

Polyamines (PA) are small, low molecular weight, ubiquitous, polycationic aliphatic molecules widely distributed throughout the plant kingdom. Polyamines play a variety of roles in normal growth and development such as regulation of cell proliferation, somatic embryogenesis, differentiation and morphogenesis, dormancy breaking of tubers and seed germination, development of flowers and fruit, and senescence [84–87]. It also plays a crucial role in abiotic stress tolerance including salinity and increases in the level of polyamines are correlated with stress tolerance in plants [88–91].

The most common polyamines that are found within the plant system are diamine putrescine (PUT), triamine spermidine (SPD), and tetra-amine spermine (SPM) [92–96]. The PA biosynthetic pathway has been thoroughly investigated in many organisms including plants and has been reviewed in details [97–104]. PUT is the smallest polyamine and is synthesised from either ornithine or arginine by the action of enzyme ornithine decarboxylase (ODC) and arginine decarboxylase (ADC), respectively [85, 105]. N-carbamoyl-putrescine is converted to PUT by the enzyme N-carbamoyl-putrescine aminohydrolase [106, 107]. The PUT thus formed functions as a primary substrate for higher polyamines such as SPD and SPM biosynthesis. The triamine SPD and tetramine SPM are synthesized by successive addition of aminopropyl group to PUT and SPD, respectively, by the enzymes spermidine synthase (SPDS) and spermine synthase (SPMS) [108, 109]. ODC pathway is the most common pathway for synthesis of polyamine found in plants. Most of the genes involved in the ODC pathway have been identified and cloned. However there are some plants where ODC pathway is absent; for instance in Arabidopsis polyamines are synthesized via ADC pathway [110–112]. All the genes involved in polyamine biosynthesis pathways have been identified from different plant species includingArabidopsis [113–115]. Polyamine biosynthesis pathway in Arabidopsis involves six major enzymes: ADC encoding genes (ADC1 and ADC2); SPDS (SPDS1 and SPDS2) and SAMDC (SAMDC1, SAMDC2, SAMDC3, SAMDC4) [115–118]. On the contrary, SPM synthase, thermospermine synthase, agmatine iminohydrolase and N-carbamoylputrescine amidohydrolase are represented by single genes only [119, 120].

Increase in endogenous polyamine level has been reported when the plant is exposed to salinity stress. Intracellular polyamine level is regulated by polyamine catabolism. Polyamines are oxidatively catabolised by amine oxidases which include copper binding diamine oxidases and FAD binding polyamine oxidases. These enzymes play a significant role in stress tolerance [121, 122]. The changes in cellular polyamine level due to stress provide possible implications in stress but do not provide evidence of their role in counteracting stress. Hence, to understand whether polyamines actually protect cells from stress-induced damages, exogenous application of polyamines, which is expected to increase endogenous polyamine, has been investigated before or during stress [123, 124]. Application of exogenous polyamine has been found to increases the level of endogenous polyamine during stress; the positive effects of polyamines have been associated with the maintenance of membrane integrity, regulation of gene expression for the synthesis of osmotically-active solutes, reduction in ROS production, and controlling accumulation of Na^+ and Cl^- ion in different organs [123–130]. It was observed that plant deficient in ADC1 and ADC2 is hypersensitive to stress [131]. InArabidopsis, expression of ADC and SPMS increases when exposed to salinity stress. Whereas mutants of polyamine biosynthetic genes show sensitivity to salinity [132]. Overproduction of PUT, SPD, and SPM in rice, tobacco, and Arabidopsis enhances salt tolerance [133]. Salt stress regulates polyamine biosynthesis and catabolism by acting as a cellular signal in hormonal pathways thereby regulating abscisic acid (ABA) in response to stress [134]. Additionally, SPM and SPD are regarded as potent inducers of NO which is another important signalling molecule [135] and its involvement in salinity tolerance is discussed below. It has been reported that exogenous application of polyamines could alleviate salt-induced reduction in photosynthetic efficiency, but this effect depends on polyamine concentration and types and level of stress [136]. When the seedling of Sorghum bicolor treated with 0.25 mM SPM is subjected to salt stress it shows improvement in growth and partial increase in the activity of peroxidase and glutathione reductase enzyme with a concomitant decrease in the level of membrane lipid peroxidation [137]. Li et al. [138] performed 2-DE gel electrophoresis and MALDI-TOF/TOF MS with cytosolic proteins to understand the effect of exogenous SPD on proteomic changes under normal and NaCl stress of 3 days old cucumber seedling leaves. Many changes were observed in the levels of proteins involved in energy and metabolic pathways, protein metabolic, stress defense, and other functional proteins. They observed that increased salt tolerance by exogenous SPD would contribute to higher expressions of proteins involved in the SAMs metabolism, protein biosynthesis, and defense mechanisms on antioxidant and detoxification. Li et al. [138] also argued that the regulation of Calvin cycle, protein folding assembly, and the inhibition of protein proteolysis by SPD might play important roles in salt tolerance.

Roles of Nitric Oxide in Salinity Tolerance

Nitric oxide (NO) is a small volatile gaseous molecule, which is involved in the regulation of various plant growth and developmental processes, such as root growth, respiration, stomata closure, flowering, cell death, seed germination and stress responses, as well as a stress signalling molecule [139–143]. NO directly or indirectly triggers expression of many redox-regulated genes. NO reacts with lipid radicals thus preventing lipid oxidation, exerting a protective effect by scavenging superoxide radical and formation of peroxynitrite that can be neutralised by other cellular processes. It also helps in the activation of antioxidant enzymes (SOD, CAT, GPX, APX, and GR) [144].

Exogenous NO application has been found to play roles in stress mitigation [145–147], but the effects depend on NO concentration. Exogenous application of sodium nitroprusside (SNP), a NO donor, on Lupinus luteusseedlings subjected to salt stress enhanced seed germination and root growth [148]. Seed germination was promoted at concentrations between 0.1 and 800 μM SNP in a dose-dependent manner. The stimulation was most pronounced after 18 and 24 h and ceased after 48 h of imbibition. The promoting effect of NO on seed germination persisted even in the presence of heavy metals (Pb and Cd) and NaCl. Kopyra and Gwó d [148] further showed that the pretreatment of L. luteus seedlings for 24 h with 10 μM SNP resulted in efficient reduction of the detrimental effect of the abiotic stressors on root growth and morphology. Pretreatment of maize seedlings with 100 μM SNP increases dry matter of roots and shoots under salinity stress; however, when the concentration of SNP was increased to 1000 μM shoot and root dry weight decreased [149]. Thus, this experiment highlighted both the protective effects of low NO concentration and the toxic effect of high NO concentration on plants.

The positive effects of NO on salinity tolerance or stress mitigation have been attributed to antioxidant activities and modulation of ROS detoxification system [150]. Improved plant growth under salinity stress by exogenous application of NO was associated with increases in antioxidant enzymes such as SOD, CAT, GPX, APX, and GR [151], and suppression of malondialdehyde (MDA) production or lipid peroxidation [152]. Effects of NO on salinity tolerance are also related to its regulation of plasma membrane H^+-ATPase and Na^+/K^+ ratio [143]. NO stimulates H^+-ATPase (H^+-PPase), thereby producing a H^+ gradient and offering the force for Na^+/H^+ exchange. Such an increase of Na^+/H^+ exchange may contribute to K^+ and Na^+ homeostasis [149]. Although NO acts as a signal molecule under salt stress and induces salt resistance by increasing PM H^+-ATPase activity, research results from Zhang et al. [153] with calluses from Populus euphratica also indicated NO cannot activate purified PM H^+-ATPase activity, at least in vitro. They initially hypothesized ABA or H_2O_2 might be downstream signal molecules to regulate the activity of PM H^+-ATPase. Further results indicated H_2O_2 content increased greatly under salt stress. Since H_2O_2 might be the candidate downstream signal molecule, Zhang et al. [153] tested PM H^+-ATPase activity and K to Na ratio in calluses by adding H_2O_2. The results suggested that H_2O_2 inducing an increased PM H^+-ATPase activity resulted in an increased K to Na ratio leading to NaCl stress adaptation.

Hormone Regulation of Salinity Tolerance

ABA is an important phytohormone whose application to plant ameliorates the effect of stress condition(s). It has long been recognized as a hormone which is upregulated due to soil water deficit around the root. Salinity stress causes osmotic stress and water deficit, increasing the production of ABA in shoots and roots [154–158]. The accumulation of ABA can mitigate the inhibitory effect of salinity on photosynthesis, growth, and translocation of assimilates [158, 159]. The positive relationship between ABA accumulation and salinity tolerance has been at least partially attributed to the accumulation of K^+, Ca^{2+} and compatible solutes, such as proline and sugars, in vacuoles of roots, which counteract with the uptake of Na^+ and Cl^- [160, 161]. ABA is a vital cellular signal that modulates the expression of a number of salt and water deficit-responsive genes. Fukuda and Tanaka [162] demonstrated the effects of ABA on the expression of two genes, HVP1 and HVP10, for vacuolar H^+-inorganic pyrophosphatase, and of HvVHA-A, for the catalytic subunit (subunit A) of vacuolar H^+-

ATPase in Hordeum vulgare under salinity stress. ABA treatment in wheat induced the expression of MAPK4-like, TIP 1, and GLP 1 genes under salinity stress [163].

Some other compounds having hormonal properties, such as salicylic acid (SA) and brassinosteroids (BR), also participate in plant abiotic stress responses [164, 165]. Under salinity stress endogenous level of SA increased along with the increase in the activity of salicylic acid biosynthetic enzyme in rice seedling [166]. Jayakannan et al. [167] have recently shown that SA improves salinity tolerance in Arabidopsis by restoring membrane potential and preventing salt-induced K^+ loss via a guard cell outward rectifying K(+) (GORK) channel.Arabidopsis seedling pretreated with SA showed upregulation of H^+-ATPase activity, thereby improving K^+retention during salt stress; SA pretreatment did not prevent accumulation of Na^+ in roots but somehow helped to reduce the concentration of accumulated Na^+ in the shoot [167]. The application of SA also promoted salinity tolerance in barley, as manifested by increases in the content of chlorophyll and carotenoid and maintaining membrane integrity, which was associated with more K^+ and soluble sugar accumulation in the root under saline condition [168]. Nazar et al. [169] have argued that SA alleviates decreases in photosynthesis under salt stress by enhancing nitrogen and sulfur assimilation and antioxidant metabolism differentially in mung bean cultivars. The negative effects of salinity may also be mitigated by BR [170–173]. Application of BR enhanced the activity of antioxidant enzymes (SOD, POX, APX, and GPX) and the accumulation of nonenzymatic antioxidant compounds (tocopherol, ascorbate, and reduced glutathione) [170]. Both BRs and SA are ubiquitous in the plant kingdom, affecting plant growth and development in many different ways, and are known to improve plant stress tolerance. Ashraf et al. [173] have reviewed and discussed the current knowledge and possible applications of BRs and SA that could be used to mitigate the harmful effects of salt stress in plants. They have also discussed the roles of exogenous applications of BRs and SA in the regulation of various biochemical and physiological processes leading to improved salt tolerance in plants.

TRANSCRIPTIONAL REGULATION AND GENE EXPRESSION OF SALINITY TOLERANCE

Regulation of gene expression in salinity stress includes a wide array of mechanisms that are used by plants to upregulate or downregulate (increase or decrease) the production of specific gene products (protein or RNA). Various mechanisms of gene regulation have been identified during the central dogma, from transcriptional initiation, to RNA processing, and to the posttranslational modification of a protein.

Transcriptomic analysis provides detailed knowledge about the gene expression at mRNA level, which is widely used to screen candidate genes involved in stress responses. Genomic approaches play a significant role in encoding, cloning, and characterization of important genes. A huge number of salt-responsive transcription factors and genes which are either upregulated or downregulated in response to salinity stress have been identified and characterized using transcriptomic and genomic approaches.

Transcription factors are considered as most important regulators that control gene expressions. Among them, bZIP, WRKY, AP2, NAC, C2H2 zinc finger gene, and DREB families comprise a large number of stress-responsive members. These transcription factor genes are capable of controlling the expression of a broad range of target genes by binding to the specific cis-acting element in the promoters of these genes. Johnson et al. [174] observed that the expression of bZIP genes were upregulated in salt-sensitive wheat cultivar, when exposed to long-term salinity, but decreased in salt-tolerant variety. Overexpression of a NAC transcription factor in both rice and wheat confers salt tolerance, thereby predicting their role in stress mitigation [175]. In rice transcriptional regulators that have been demonstrated to play a significant role in abiotic stress responses involve DREB1/CBF, DREB2, and AREB/ABF [176–180]. Transcriptions factors such as OsNAC5 and ZFP179 show an upregulation under salinity stress, which may regulate the synthesis and accumulation of proline, sugar, and LEA proteins that in turn play an integral role in stress tolerance [181]. In Arabidopsis, salt stress results in the upregulation of AtWRKY8 which directly binds with the promoter of RD29A, suggesting it to be as one of the target genes of AtWRKY8 [182].

A large number of genes and transcription factors are upregulated in response to salinity in different plant species, which serve diverse functions [183–192]. Examples of salt-responsive genes are listed in the Table 1, and these genes are mainly classified into the following functional categories: ion transport or homeostasis (e.g., SOS genes, AtNHX1, and H^+-ATPase), senescence-associated genes (e.g., SAG), molecular chaperones (e.g., HSP genes), and dehydration-related transcription factors (e.g., DREB). Among stress-responsive genes, the SOS gene family, which we have already discussed in Section 2.1, is believed to play a very intriguing role in ion homeostasis, thereby conferring salt tolerance [24–37, 190, 191]. Some ROS-scavenging and osmotic-regulating genes are also upregulated by salinity in some plant species. For example, a continuous exposure of rice plants to salinity for about 24 hours resulted in upregulation of glutathione-S-transferase and ascorbate peroxidase, both of which were known to play an active role in ROS scavenging, and with the increase in duration of exposure to salinity stress, upregulation of metallothionein and water channel proteins was also observed [192]. Halophyte plant species Spartina alterniflora when subjected to salt stress exhibits upregulation of 10 genes associated with osmotic regulation [193].

Table 1: Examples of upregulated genes in response to salinity stress

Species	NaCl concentration (mM)	Gene name	Gene functions	References
Brassica juncea and Brassica campestris	25 and 50	SOS1 SOS2 SOS3 AtNHX1	(i) Plasma membrane Na^+/K^+ antiporter. (ii) Protein kinase. (iii) Calcium-binding protein. (iv) Vacuolar Na^+/K^+ antiporter.	[183]

Oryza sativa	50	PRP SAG HSPC025	(i) Proline-rich proteins and cell-wall protection. (ii) Senescence associated genes, regulatory processes, and cellular signal transduction. (iii) Heat-shock proteins, protein stabilizing.	[184]
Oryza sativa	100	OsHSP23.7 OsHSP71.1, OsHSP80.2	Heat-shock proteins, molecular chaperones, folding, assembling, and transporting proteins.	[185]
Arabidopsis thaliana	150	AtSKIP	Transcription factor, transcriptional pre-initiation, splicing, and polyadenylation.	[186]
Oryza sativa	200	OsHsp17.0, OsHsp23.7	Heat-shock proteins, molecular chaperones, and folding, assembling, and transporting proteins.	[187]
Carrot	300	DcHsp17.7	Cell viability and membrane stability under heat stress.	[188]
Arabidopsis thaliana	300	JcDREB	Transcription factor	[189]

Recently, Schmidt et al. [194] identified SALT-RESPONSIVE ERF1 (SERF1), a rice (Oryza sativa) transcription factor gene that showed a root-specific induction upon salt and H_2O_2 treatment. Loss of SERF1 impaired the salt-inducible expression of genes encoding members of a mitogen-activated protein kinase (MAPK) cascade and salt tolerance-mediating TFs. Furthermore, they showed that SERF1-dependent genes are H_2O_2responsive and demonstrated that SERF1 binds to the promoters of MAPK KINASE KINASE6 (MAP3K6), MAPK5, DEHYDRATION-RESPONSIVE ELEMENT BINDING2A (DREB2A), and ZINC FINGER PROTEIN179 (ZFP179) in vitro and in vivo. SERF1 also directly induces its own gene expression. In addition, it was observed that SERF1 is a phosphorylation target of MAPK5, resulting in enhanced transcriptional activity of SERF1 toward its direct target genes. Finally, they demonstrated that the plants deficient for SERF1 are more sensitive to salt stress compared with the wild type, while constitutive overexpression of SERF1 improves salinity tolerance.

There are some transcription factors which are regulated by different kinases and have been found to be significant players of plant adaptation to salinity stress. Serra et al. [195] showed that OsRMC encodes a receptor-like kinase described as a negative regulator of salt stress responses in rice. Two transcription factors, OsEREBP1 and OsEREBP2, belonging to the AP2/ERF family were shown to bind to the same GCC-like DNA motif in OsRMC promoter and to negatively regulate its gene expression. Serra et al. [195] further revealed thatOsEREBP1 transcript level is not significantly affected by salt, ABA, or severe cold (5°C) and is only slightly regulated by drought and moderate cold. On the other hand, the OsEREBP2 transcript level increased after cold, ABA, drought, and high salinity treatments, indicating that OsEREBP2 may play a central role mediating the response to different abiotic stresses. Gene expression analysis in rice varieties with contrasting salt tolerance further suggests that OsEREBP2 is involved in salt stress response in rice. A bZIP class of ABRE binding transcription factor, known as OSBZ8, has also been identified from rice and has been shown to be highly expressed in salt tolerant cultivars than in salt sensitive one [196]. Moreover, OSBZ8 has been shown to be activated/phosphorylated by a SNF-1 group of serine/threonine kinase in the presence of Spd during salinity stress [197].

High-throughput sequencing for transcript profiling in plants has revealed that alternative splicing affects a much higher proportion of the transcriptome than was previously assumed. Alternative splicing is involved in most plant processes and is particularly prevalent in plants exposed to environmental stress. The identification of mutations in predicted splicing factors and spliceosomal proteins that affect cell fate, the circadian clock, plant defense, and tolerance/sensitivity to abiotic stress all points to a fundamental role of splicing/alternative splicing in plant growth, development, and responses to external cues [198]. A suite of Ser/Arg-rich proteins that are key regulators of alternative splicing undergoes alternative themselves in response to various abiotic stresses, such as salt stress [198–200]. PRMT5, a type II protein Arg methyltransferase that symmetrically dimethylates Arg side chains, also impacts splicing/alternative in Arabidopsis. The prmt5 mutant, also known as shk1 kinase binding protein1 (skb1), is sensitive to salt [201]. It was proposed that PRMT5/SKB1 affects plant development and the salt response by altering the methylation status of H4R3sme2 (for symmetric dimethylation of histone H4 arginine 3) and LSm4 and thus linking transcription to pre-mRNA splicing [201]. A nuclear coactivator, At-SKIP (Ski-interacting protein), expression was found to increase in response to salt, mannitol, and ABA treatment, and At-SKIP overexpression or antisense lines show altered tolerance to a plethora of abiotic stress factors [186], and it is likely that a role in alternative splicing contributes to these phenotypes [198].

The small ubiquitin-like modifier (SUMO) is a crucial regulator of signaling proteins in eukaryotes. Attachment of SUMO onto substrates is reversible, and SUMO proteases, which specifically cleave the SUMO-substrate linkages, play a vital regulatory role during SUMOylation. Conti et al. [202] have identified two SUMO proteases, Overly Tolerant to Salt1 (OTS1) and OTS2, which are localized in the nucleus and act redundantly to regulate salt stress responses in Arabidopsis thaliana. Cui et al. [203] have identified anArabidopsis endoplasmic reticulum (ER-) associated protein degradation (ERAD) component called Ubiquitin conjugase UBC32 that functions in BR-mediated salt stress tolerance. More and more such reports of sumoylation and other ubiquitin like posttranslational modifications during plant salinity stress are coming up.

Downregulated genes are emerging now as essential components of the response to salinity. For example downregulation of -carotene hydroxylase increases -carotene and total carotenoids enhancing salt stress tolerance in transgenic cultured cells of sweet potato [204]. It seems that mutual regulation mechanism exists between different genes and proteins and signals underlying different processes of plant adaptation to abiotic stress.

In addition to protein coding genes, recently discovered microRNAs (miRNAs) and endogenous small interfering RNAs (siRNAs) have emerged as important players in plant stress responses. Initial clues suggesting that small RNAs are involved in plant stress responses stem from studies showing stress regulation of miRNAs and endogenous siRNAs, as well as from target predictions for some miRNAs [205, 206].

PROTEOMIC AND METABOLIC RESPONSES TO SALINITY STRESS

Genomics technologies have helped to address the mutagenicity of the plant abiotic stress responses. Analysis of genome sequences, and specific transcript collections and their dynamic changes, has provided a more global picture of stress-dependent responses at the cell, tissue, and whole plant level and moved the field from a single-gene approach toward an understanding of interactions between multiple components in cells, facilitating the dissection of abiotic stress circuits and coexpression hubs [5–8, 207–209]. However, directly focusing on genes may not accurately portray conditions in the cell at a particular state and time during stress due to regulation at the RNA and protein level, including posttranslational regulation. Proteomics, and in particular quantitative proteomics, is emerging as a powerful technique to be applied to the field of crop abiotic stress tolerance research; it has the potential to allow rapid identification and quantification of novel stress- and tolerance-related proteins. Understanding the dynamics of expression and posttranslational modifications of these proteins, and gaining direct insight into their function and interactions, can provide essential information that can be applied to engineer stress-tolerant crops with novel traits through biomarker selection and transgenic strategies [209, 210]. Available data suggest that several common stress responsive proteins are expressed in response to various abiotic stresses in different plant species. About 2171 proteins from 34 plant species have been identified and characterized as salt-responsive proteins, which are either upregulated or downregulated by salinity stress [211]. Based on gene ontology, BLAST alignment and literature information, salt-responsive proteins can be grouped into 14 functional categories. Specifically, there appears to be a general regulation of proteins involved in carbohydrate, nitrogen, and energy metabolisms, with particular emphasis on glycolytic and Krebs cycle enzymes. Moreover, as discussed earlier, salinity and other abiotic stresses lead to metabolic imbalances that lead to ROS generation. Therefore, it is not surprising that plant root or shoot proteomics show the expression of ROS scavenging proteins like SOD, CAT, GPX, APX, and GR [209, 212]. Other proteins that are identified in multiple studies are those involved in protein synthesis, processing, turnover, and degradation, as well as cytoskeleton stability. For photosynthetic processes, there appears to be a general decrease in levels of chlorophyll biosynthesis related proteins but an increase in proteins involved in the light-dependent reactions. Some of the proteins identified are indicative of a general stress-responsive pathway in plants. Less common are proteins identified in the categories of signaling, trafficking, transport, and cell structure [209]. Plant lamina or root membrane proteomics, including that of plasma membrane, mitochondrial, and thylakoid membrane, have revealed the up-/downregulation of a plethora of proteins. These include receptor proteins that perceive the stress, membrane bound signaling, and regulatory proteins that function to relay the stress, vesicle trafficking proteins, and transport proteins that function to maintain ion and water homeostasis, and drive sequestration and/or removal of toxic compounds from the cell, membrane bound kinases, and intrinsic proteins [213–219].

Another significant research approach in plant system biology is the metabolomics which involves the study of metabolome. Higher plants have a remarkable ability to synthesize a vast array of metabolites that differ in chemical complexity and biological functions playing an indispensible role in stress alleviation [229, 230]. Examples of plant metabolites that are involved in salinity tolerance include polyols such as mannitol and sorbitol, dimethylsulfonium compounds, glycine betaine, sugars such as sucrose, trehalose and fructans, or amino acids such as proline that serve as an osmolyte or osmoprotectant [231]. Plants when subjected to salinity stress show an increase in the concentration of these osmolytes thus playing significant role in stress mitigation. We have already discussed their role in Section 2.2. Advances in analytical chemistry, such as MS based methods and NMR, and sophisticated "data processing and mining" techniques, have allowed the plant biologists to venture into hitherto unexplored domains and generate extensive metabolic profiles due to various environmental stimuli including salinity. Results indicate that the metabolic processes are highly specific for given tissues, species, and plant-environment interactions. The clusters of identified compounds not only serve as base in the quest of novel defense compounds but also as markers for the characterization of the plants' defensive state. The latter is especially useful in agronomic applications where meaningful markers are essential for crop protection [232].

BIOENGINEERING FOR IMPROVING SALINITY TOLERANCE

Genetic transformation technology enables scientists to achieve gene transfer in precise and predictable manner. Hence genetic engineering approaches would be useful to manipulate the osmoprotectants biosynthetic pathways for accumulating such molecules that act by scavenging ROS, reducing lipid peroxidation, maintaining protein structure and functions [233–235]. Many works on the transformation of plants for improving salinity tolerance focus on genes controlling ion transport, as regulation of Na^+ uptake and compartmentalization is a critically important mechanism for plant survival under salinity stress, and many candidate genes controlling this mechanism have been identified. Engineering plants for overexpression of genes encoding for antiporters is identified as an effective method for generating salt-tolerant plants (Table2 and relative references in Table 2). Gene expression studies using constitutive promoters provide limited biological information compared with the use of inducible promoters or cell type-specific promoters. The choice of promoters can significantly affect the results from a transgenic manipulation. Thus salt tolerant crops could be engineered by (1) successful fine-tuning of the stress response by engineering novel regulatory targets;(2) proper understanding of posttranslational modifications which regulate plant growth performance under stress; (3) overexpression of miRNAs or their targets; (4) maintaining hormone homeostasis to avoid pleiotropic effects under stress; and (5) applying plant synthetic biology approaches to improve genetic engineering strategies [236].

Table 2: Improving plant salt tolerance through engineering genes for various membrane antiporters

Transgenic host	Gene engineered	Source	Improved functions under salinity stress	References
Arabidopsis	Vacuolar Na^+/H^+ antiporter Ms NHX1	Alfalfa (Medicago sativa)	Increased osmotic balance. MDA content rises.	[220]
Rice	Vacuolar Na^+/H^+ Antiporter Pg NHX1	Pennisetum glaucum	Elaborate root system.	[221]
Wheat	Vacuolar Na^+/H^+ Antiporter At NHX1	Arabidopsis thaliana L.	Increase in grain yield and biomass production. Accumulation of K^+ in leaf. Reduced aggregation of Na^+.	[222]
Tobacco	Vacuolar Na^+/H^+ antiporter GhNHX1	Gossypium hirsutum	Na^+ compartmentalization.	[223]
Tomato	Vacuolar Na^+/H^+ antiporter AtNHX1	Arabidopsis thaliana L.	Over production of vacuolar Na^+/H^+ antiporter.	[224]
Tobacco	Vacuolar Na^+/H^+ antiporter AlNHXI	Aeluropus littoralis	Compartmentalization of Na in roots. Maintenance of K^+/Na^+ ratio in the leaf.	[225]

Brassica	Vacuolar Na^+/H^+ antiporter AtNHX1	Arabidopsis thaliana L.	Increased proline content. Improved growth rate. Mitigate the toxic effect of Na^+.	[226]
Arabidopsis	Plasma membrane Na^+/H^+ antiporter SOS1	Arabidopsis thaliana L. (wild type)	Improved germination rate, root growth, and chlorophyll content. Reduced accumulation of Na^+.	[227]
Maize	Vacuolar Na^+/H^+ antiporter AtNHX1	Arabidopsis thaliana L.	Increased rate of germination.	[228]

CONCLUSIONS AND FUTURE RESEARCH PERSPECTIVES

Salinity tolerance involves a complex of responses at molecular, cellular, metabolic, physiological, and whole-plant levels. Extensive research through cellular, metabolic, and physiological analysis has elucidated that among various salinity responses, mechanisms or strategies controlling ion uptake, transport and balance, osmotic regulation, hormone metabolism, antioxidant metabolism, and stress signalling play critical roles in plant adaptation to salinity stress. Taking advantage of the latest advancements in the field of genomic, transcriptomic, proteomic, and metabolomic techniques, plant biologists are focusing on the development of a complete profile of genes, proteins, and metabolites responsible for different mechanisms of salinity tolerance in different plant species. However, there is lack of the integration of information from genomic, transcriptomic, proteomic, and metabolomics studies, and the combined approach is essential for the determination of the key pathways or processes controlling salinity tolerance. In addition, in spite of the significant progress in the understanding of plant stress responses, there is still a large gap in our knowledge of transmembrane ion transport, sensor and receptor in the signalling transduction, molecules in long distance signalling, and metabolites in energy supply. The future focus should be on the study of intercellular and intracellular molecular interaction involved in salinity stress response. Genetic engineering has been proved to be an efficient approach to the development of salinity-tolerant plants, and this approach will become more powerful as more candidate genes associated with salinity tolerance are identified and widely utilized.

ACKNOWLEDGMENTS

Bhaskar Gupta acknowledges the financial help from DBT (Govt. of India) for the DBT-RGYI grant and Government of West Bengal for the WB-DST grant.

REFERENCES

1. FAO, High Level Expert Forum—How to Feed the World in 2050, Economic and Social Development, Food and Agricultural Organization of the United Nations, Rome, Italy, 2009.
2. T. J. Flowers, "Improving crop salt tolerance," Journal of Experimental Botany, vol. 55, no. 396, pp. 307–319, 2004.
3. R. Munns and M. Tester, "Mechanisms of salinity tolerance," Annual Review of Plant Biology, vol. 59, pp. 651–681, 2008.
4. R. A. James, C. Blake, C. S. Byrt, and R. Munns, "Major genes for Na^+ exclusion, Nax1 and Nax2 (wheat HKT1;4 and HKT1;5), decrease Na^+ accumulation in bread wheat leaves under saline and waterlogged conditions," Journal of Experimental Botany, vol. 62, no. 8, pp. 2939–2947, 2011
5. A. Rahnama, R. A. James, K. Poustini, and R. Munns, "Stomatal conductance as a screen for osmotic stress tolerance in durum wheat growing in saline soil," Functional Plant Biology, vol. 37, no. 3, pp. 255–263, 2010.
6. R. Munns, "Genes and salt tolerance: bringing them together," New Phytologist, vol. 167, no. 3, pp. 645–663, 2005.

7. J. Rozema and T. Flowers, "Ecology: crops for a salinized world," Science, vol. 322, no. 5907, pp. 1478–1480, 2008.
8. K. Apel and H. Hirt, "Reactive oxygen species: metabolism, oxidative stress, and signal transduction,"Annual Review of Plant Biology, vol. 55, pp. 373–399, 2004.
9. S. Mahajan and N. Tuteja, "Cold, salinity and drought stresses: an overview," Archives of Biochemistry and Biophysics, vol. 444, no. 2, pp. 139–158, 2005.
10. P. Ahmad, "Growth and antioxidant responses in mustard (Brassica juncea L.) plants subjected to combined effect of gibberellic acid and salinity," Archives of Agronomy and Soil Science, vol. 56, no. 5, pp. 575–588, 2010.
11. P. Ahmad and M. N. V. Prasad, Abiotic Stress Responses in Plants: Metabolism, Productivity and Sustainability, Springer, New York, NY, USA, 2012.
12. P. Ahmad and S. Umar, Oxidative Stress: Role of Antioxidants in Plants, Studium Press, New Delhi, India, 2011.
13. Q. Pang, S. Chen, S. Dai, Y. Chen, Y. Wang, and X. Yan, "Comparative proteomics of salt tolerance inArabidopsis thaliana and Thellungiella halophila," Journal of Proteome Research, vol. 9, no. 5, pp. 2584–2599, 2010.
14. E. Ábrahám, I. P. Salamó, C. Koncz, and L. Szabados, "Identification of Arabidopsis and Thellungiellagenes involved in salt tolerance by novel genetic system," Acta Biologica Szegediensis, vol. 55, no. 1, pp. 53–57, 2011.
15. J. L. Zhang and H. Shi, "Physiological and molecular mechanisms of plant salt tolerance," Photosynthesis Research, vol. 115, pp. 1–22, 2013.
16. N. X. Niu Xiaomu, R. A. Bressan, P. M. Hasegawa, and J. M. Pardo, "Ion homeostasis in NaCl stress environments," Plant Physiology, vol. 109, no. 3, pp. 735–742, 1995.
17. R. Serrano, J. M. Mulet, G. Rios et al., "A glimpse of the mechanisms of ion homeostasis during salt stress," Journal of Experimental Botany, vol. 50, pp. 1023–1036, 1999.
18. P. M. Hasegawa, "Sodium (Na^+) homeostasis and salt tolerance of plants," Environmental and Experimental Botany, vol. 92, pp. 19–31, 2013.

19. M. P. Reddy, S. Sanish, and E. R. R. Iyengar, "Photosynthetic studies and compartmentation of ions in different tissues of Salicornia brachiata Roxb. under saline conditions," Photosynthetica, vol. 26, pp. 173–179, 1992.
20. J.-K. Zhu, "Regulation of ion homeostasis under salt stress," Current Opinion in Plant Biology, vol. 6, no. 5, pp. 441–445, 2003.
21. K. J. Dietz, N. Tavakoli, C. Kluge et al., "Significance of the V-type ATPase for the adaptation to stressful growth conditions and its regulation on the molecular and biochemical level," Journal of Experimental Botany, vol. 52, no. 363, pp. 1969–1980, 2001.
22. M. De Lourdes Oliveira Otoch, A. C. Menezes Sobreira, M. E. Farias De Aragão, E. G. Orellano, M. Da Guia Silva Lima, and D. Fernandes De Melo, "Salt modulation of vacuolar H^+-ATPase and H^+-Pyrophosphatase activities in Vigna unguiculata," Journal of Plant Physiology, vol. 158, no. 5, pp. 545–551, 2001.
23. B. Wang, U. Lüttge, and R. Ratajczak, "Effects of salt treatment and osmotic stress on V-ATPase and V-PPase in leaves of the halophyte Suaeda salsa," Journal of Experimental Botany, vol. 52, no. 365, pp. 2355–2365, 2001.
24. P. M. Hasegawa, R. A. Bressan, J.-K. Zhu, and H. J. Bohnert, "Plant cellular and molecular responses to high salinity," Annual Review of Plant Biology, vol. 51, pp. 463–499, 2000.
25. D. Sanders, "Plant biology: the salty tale of Arabidopsis," Current Biology, vol. 10, no. 13, pp. R486–R488, 2000.
26. H. Shi, M. Ishitani, C. Kim, and J.-K. Zhu, "The Arabidopsis thaliana salt tolerance gene SOS1 encodes a putative Na^+/H^+ antiporter," Proceedings of the National Academy of Sciences of the United States of America, vol. 97, no. 12, pp. 6896–6901, 2000.
27. H. Shi, F. J. Quintero, J. M. Pardo, and J.-K. Zhu, "The putative plasma membrane Na^+/H^+ antiporter SOS1 controls long-distance Na^+ transport in plants," Plant Cell, vol. 14, no. 2, pp. 465–477, 2002.
28. J. Liu, M. Ishitani, U. Halfter, C.-S. Kim, and J.-K. Zhu, "The Arabidopsis thaliana SOS2 gene encodes a protein kinase that is required for salt tolerance," Proceedings of the National Academy

of Sciences of the United States of America, vol. 97, no. 7, pp. 3730–3734, 2000. · View at Google Scholar·

29. M. Ishitani, J. Liu, U. Halfter, C.-S. Kim, W. Shi, and J.-K. Zhu, "SOS3 function in plant salt tolerance requires N-myristoylation and calcium binding," Plant Cell, vol. 12, no. 9, pp. 1667–1677, 2000.
30. Y. Guo, Q.-S. Qiu, F. J. Quintero et al., "Transgenic Evaluation of Activated Mutant Alleles of SOS2 Reveals a Critical Requirement for Its Kinase Activity and C-Terminal Regulatory Domain for Salt Tolerance in Arabidopsis thaliana," Plant Cell, vol. 16, no. 2, pp. 435–449, 2004.
31. F. J. Quintero, M. Ohta, H. Shi, J.-K. Zhu, and J. M. Pardo, "Reconstitution in yeast of the ArabidopsisSOS signaling pathway for Na^+ homeostasis," Proceedings of the National Academy of Sciences of the United States of America, vol. 99, no. 13, pp. 9061–9066, 2002.
32. F. J. Quintero, J. Martinez-Atienza, I. Villalta et al., "Activation of the plasma membrane Na/H antiporter salt-overly-sensitive 1 (SOS1) by phosphorylation of an auto-inhibitory C-terminal domain,"Proceedings of the National Academy of Sciences of the United States of America, vol. 108, no. 6, pp. 2611–2616, 2011.
33. D.-H. Oh, S. Y. Lee, R. A. Bressan, D.-J. Yun, and H. J. Bohnert, "Intracellular consequences of SOS1 deficiency during salt stress," Journal of Experimental Botany, vol. 61, no. 4, pp. 1205–1213, 2010.
34. J. Martínez-Atienza, X. Jiang, B. Garciadeblas et al., "Conservation of the salt overly sensitive pathway in rice," Plant Physiology, vol. 143, no. 2, pp. 1001–1012, 2007.
35. R. K. Sairam and A. Tyagi, "Physiology and molecular biology of salinity stress tolerance in plants,"Current Science, vol. 86, no. 3, pp. 407–421, 2004.
36. L. Ma, H. Zhang, L. Sun et al., "NADPH oxidase AtrbohD and AtrbohF function in ROS-dependent regulation of Na^+/K^+ homeostasis in Arabidopsis under salt stress," Journal of Experimental Botany, vol. 63, no. 1, pp. 305–317, 2012.
37. H. E. Yen, S.-M. Wu, Y.-H. Hung, and S.-K. Yen, "Isolation of 3 salt-induced low-abundance cDNAs from light-grown callus of

Mesembryanthemum crystallinum by suppression subtractive hybridization,"Physiologia Plantarum, vol. 110, no. 3, pp. 402–409, 2000.

38. J. I. Schroeder, E. Delhaize, W. B. Frommer et al., "Using membrane transporters to improve crops for sustainable food production," Nature, vol. 497, pp. 60–66, 2013.
39. V. Barragán, E. O. Leidi, Z. Andrés et al., "Ion exchangers NHX1 and NHX2 mediate active potassium uptake into vacuoles to regulate cell turgor and stomatal function in Arabidopsis," Plant Cell, vol. 24, no. 3, pp. 1127–1142, 2012.
40. F. J. Gálvez, M. Baghour, G. Hao, O. Cagnac, M. P. Rodríguez-Rosales, and K. Venema, "Expression of LeNHX isoforms in response to salt stress in salt sensitive and salt tolerant tomato species," Plant Physiology and Biochemistry, vol. 51, pp. 109–115, 2012.
41. M. A. Hoque, M. N. A. Banu, E. Okuma et al., "Exogenous proline and glycinebetaine increase NaCl-induced ascorbate-glutathione cycle enzyme activities, and proline improves salt tolerance more than glycinebetaine in tobacco Bright Yellow-2 suspension-cultured cells," Journal of Plant Physiology, vol. 164, no. 11, pp. 1457–1468, 2007.
42. P. Ahmad, C. A. Jaleel, M. A. Salem, G. Nabi, and S. Sharma, "Roles of enzymatic and nonenzymatic antioxidants in plants during abiotic stress," Critical Reviews in Biotechnology, vol. 30, no. 3, pp. 161–175, 2010.
43. M. A. Hossain, S. Munemasa, M. Uraji, Y. Nakamura, I. C. Mori, and Y. Murata, "Involvement of endogenous abscisic acid in methyl jasmonate-induced stomatal closure in Arabidopsis," Plant Physiology, vol. 156, no. 1, pp. 430–438, 2011.
44. N. Nounjan, P. T. Nghia, and P. Theerakulpisut, "Exogenous proline and trehalose promote recovery of rice seedlings from salt-stress and differentially modulate antioxidant enzymes and expression of related genes," Journal of Plant Physiology, vol. 169, no. 6, pp. 596–604, 2012.
45. M. A. Tahir, T. Aziz, M. Farooq, and G. Sarwar, "Silicon-induced changes in growth, ionic composition, water relations, chlorophyll contents and membrane permeability in two salt-stressed wheat

genotypes,"Archives of Agronomy and Soil Science, vol. 58, no. 3, pp. 247–256, 2012.

46. M. A. Khan, I. A. Ungar, and A. M. Showalter, "Effects of sodium chloride treatments on growth and ion accumulation of the halophyte haloxylon recurvum," Communications in Soil Science and Plant Analysis, vol. 31, no. 17-18, pp. 2763–2774, 2000.
47. Y. Wang and N. Nii, "Changes in chlorophyll, ribulose bisphosphate carboxylase-oxygenase, glycine betaine content, photosynthesis and transpiration in Amaranthus tricolor leaves during salt stress,"Journal of Horticultural Science and Biotechnology, vol. 75, no. 6, pp. 623–627, 2000.
48. I. Kerepesi and G. Galiba, "Osmotic and salt stress-induced alteration in soluble carbohydrate content in wheat seedlings," Crop Science, vol. 40, no. 2, pp. 482–487, 2000.
49. H. J. Bohnert, D. E. Nelson, and R. G. Jensen, "Adaptations to environmental stresses," Plant Cell, vol. 7, no. 7, pp. 1099–1111, 1995.
50. C. W. Ford, "Accumulation of low molecular weight solutes in water-stressed tropical legumes,"Phytochemistry, vol. 23, no. 5, pp. 1007–1015, 1984.
51. M. Dopp, F. Larher, and P. Weigel, "Osmotic adaption in Australian mangroves," Vegetatio, vol. 61, no. 1–3, pp. 247–253, 1985.
52. M. Ashraf and M. R. Foolad, "Roles of glycine betaine and proline in improving plant abiotic stress resistance," Environmental and Experimental Botany, vol. 59, no. 2, pp. 206–216, 2007.
53. S. C. Saxena, H. Kaur, P. Verma, et al., "Osmoprotectants: potential for crop improvement under adverse conditions," in Plant Acclimation to Environmental Stress, pp. 197–232, Springer, New York, NY, USA, 2013.
54. A. D. Hanson, B. Rathinasabapathi, J. Rivoal, M. Burnet, M. O. Dillon, and D. A. Gage, "Osmoprotective compounds in the Plumbaginaceae: a natural experiment in metabolic engineering of stress tolerance,"Proceedings of the National Academy of Sciences of the United States of America, vol. 91, no. 1, pp. 306–310, 1994.

55. F. El-Shintinawy and M. N. El-Shourbagy, "Alleviation of changes in protein metabolism in NaCl-stressed wheat seedlings by thiamine," Biologia Plantarum, vol. 44, no. 4, pp. 541–545, 2001.
56. J. Matysik, A. Alia, B. Bhalu, and P. Mohanty, "Molecular mechanisms of quenching of reactive oxygen species by proline under stress in plants," Current Science, vol. 82, no. 5, pp. 525–532, 2002.
57. C. Ben Ahmed, B. Ben Rouina, S. Sensoy, M. Boukhriss, and F. Ben Abdullah, "Exogenous proline effects on photosynthetic performance and antioxidant defense system of young olive tree," Journal of Agricultural and Food Chemistry, vol. 58, no. 7, pp. 4216–4222, 2010.
58. M. A. Hoque, M. N. A. Banu, Y. Nakamura, Y. Shimoishi, and Y. Murata, "Proline and glycinebetaine enhance antioxidant defense and methylglyoxal detoxification systems and reduce NaCl-induced damage in cultured tobacco cells," Journal of Plant Physiology, vol. 165, no. 8, pp. 813–824, 2008.
59. S. Deivanai, R. Xavier, V. Vinod, K. Timalata, and O. F. Lim, "Role of exogenous proline in ameliorating salt stress at early stage in two rice cultivars," Journal of Stress Physiology & Biochemistry, vol. 7, pp. 157–174, 2011.
60. M. A. A. Gadallah, "Effects of proline and glycinebetaine on Vicia faba responses to salt stress," Biologia Plantarum, vol. 42, no. 2, pp. 249–257, 1999.
61. P. Makela, J. Karkkainen, and S. Somersalo, "Effect of glycinebetaine on chloroplast ultrastructure, chlorophyll and protein content, and RuBPCO activities in tomato grown under drought or salinity,"Biologia Plantarum, vol. 43, no. 3, pp. 471–475, 2000.
62. S. Cha-Um and C. Kirdmanee, "Effect of glycinebetaine on proline, water use, and photosynthetic efficiencies, and growth of rice seedlings under salt stress," Turkish Journal of Agriculture and Forestry, vol. 34, no. 6, pp. 517–527, 2010.
63. R. Ahmad, C. J. Lim, and S.-Y. Kwon, "Glycine betaine: a versatile compound with great potential for gene pyramiding to improve crop plant performance against environmental stresses," Plant Biotechnology Reports, vol. 7, pp. 49–57, 2013.

64. S. Rahman, H. Miyake, and Y. Takeoka, "Effects of exogenous glycinebetaine on growth and ultrastructure of salt-stressed rice seedlings (Oryza sativa L.)," Plant Production Science, vol. 5, no. 1, pp. 33–44, 2002.
65. M. L. Binzel, F. D. Hess, R. A. Bressan, and P. M. Hasegawa, "Intracellular compartmentation of ions in salt adapted tobacco cells," Plant Physiology, vol. 86, pp. 607–614, 1988.
66. J. C. Thomas, M. Sepahi, B. Arendall, and H. J. Bohnert, "Enhancement of seed germination in high salinity by engineering mannitol expression in Arabidopsis thaliana," Plant, Cell and Environment, vol. 18, no. 7, pp. 801–806, 1995.
67. A. K. Parida, A. B. Das, and P. Mohanty, "Investigations on the antioxidative defence responses to NaCl stress in a mangrove, Bruguiera parviflora: differential regulations of isoforms of some antioxidative enzymes," Plant Growth Regulation, vol. 42, no. 3, pp. 213–226, 2004.
68. Z. Gao, M. Sagi, and S. H. Lips, "Carbohydrate metabolism in leaves and assimilate partitioning in fruits of tomato (Lycopersicon esculentum L.) as affected by salinity," Plant Science, vol. 135, no. 2, pp. 149–159, 1998.
69. A. N. M. Alamgir and M. Yousuf Ali, "Effect of salinity on leaf pigments, sugar and protein concentrations and chloroplast ATPase activity of rice (Oryza sativa L.)," Bangladesh Journal of Botany, vol. 28, no. 2, pp. 145–149, 1999.
70. F. Groß, J. Durner, and F. Gaupels, "Nitric oxide, antioxidants and prooxidants in plant defence responses," Frontiers in Plant Science, vol. 4, article 419, 2013.
71. K. Asada, "The water-water cycle in chloroplasts: scavenging of active oxygens and dissipation of excess photons," Annual Review of Plant Biology, vol. 50, pp. 601–639, 1999.
72. K. J. Gupta, M. Stoimenova, and W. M. Kaiser, "In higher plants, only root mitochondria, but not leaf mitochondria reduce nitrite to NO, in vitro and in situ," Journal of Experimental Botany, vol. 56, no. 420, pp. 2601–2609, 2005.
73. S. S. Gill, M. Tajrishi, M. Madan, and N. Tuteja, "A DESD-box helicase functions in salinity stress tolerance by improving photosynthesis and antioxidant machinery in rice (Oryza sativa

L. cv. PB1),"Plant Molecular Biology, vol. 82, no. 1-2, pp. 1–22, 2013.

74. N. Tuteja, R. K. Sahoo, B. Garg, and R. Tuteja, "OsSUV3 dual helicase functions in salinity stress tolerance by maintaining photosynthesis and antioxidant machinery in rice (Oryza sativa L. cv. IR64),"The Plant Journal, vol. 76, no. 1, pp. 115–127, 2013.
75. Y. H. Kim, A. Latif Khan, M. Waqas, et al., "Silicon application to rice root zone influenced the phytohormonal and antioxidant responses under salinity stress," Journal of Plant Growth Regulation, 2013.
76. M. J. Van Oosten, A. Sharkhuu, G. Batelli, R. A. Bressan, and A. Maggio, "The Arabidopsis thalianamutant air1 implicates SOS3 in the regulation of anthocyanins under salt stress," Plant Molecular Biology, vol. 83, pp. 405–415, 2013.
77. J. C. Begara-Morales, B. Sánchez-Calvo, M. Chaki, et al., "Dual regulation of cytosolic ascorbate peroxidase (APX) by tyrosine nitration and S-nitrosylation," Journal of Experimental Botany, vol. 65, no. 2, pp. 527–538, 2014.
78. S. Agarwal and R. Shaheen, "Stimulation of antioxidant system and lipid peroxidation by abiotic stresses in leaves of Momordica charantia," Brazilian Journal of Plant Physiology, vol. 19, no. 2, pp. 149–161, 2007.
79. N. Munir and F. Aftab, "Enhancement of salt tolerance in sugarcane by ascorbic acid pretreatment,"African Journal of Biotechnology, vol. 10, no. 80, pp. 18362–18370, 2011.
80. C. H. Foyer, H. Lopez-Delgado, J. F. Dat, and I. M. Scott, "Hydrogen peroxide- and glutathione-associated mechanisms of acclimatory stress tolerance and signalling," Physiologia Plantarum, vol. 100, no. 2, pp. 241–254, 1997.
81. K. H. Aly-Salama and M. M. Al-Mutawa, "Glutathione-triggered mitigation in salt-induced alterations in plasmalemma of onion epidermal cells," International Journal of Agriculture and Biology, vol. 11, no. 5, pp. 639–642, 2009.
82. A. Rawia Eid, L. S. Taha, and S. M. M. Ibrahiem, "Alleviation of adverse effects of salinity on growth, and chemical constituents of marigold plants by using glutathione and ascorbate," Journal of Applied Sciences Research, vol. 7, pp. 714–721, 2011.

83. S. J. Roy, S. Negrão, and M. Tester, "Salt resistant crop plants," Current Opinion in Biotechnology, vol. 26, pp. 115–124, 2014.
84. J. M. Knott, P. Römer, and M. Sumper, "Putative spermine synthases from Thalassiosira pseudonanaand Arabidopsis thaliana synthesize thermospermine rather than spermine," FEBS Letters, vol. 581, no. 16, pp. 3081–3086, 2007.
85. K. Gupta, A. Dey, and B. Gupta, "Plant polyamines in abiotic stress responses," Acta Physiologiae Plantarum, vol. 35, no. 7, pp. 2015–2036, 2013.
86. M. Panicot, E. G. Minguet, A. Ferrando et al., "A polyamine metabolon involving aminopropyl transferase complexes in Arabidopsis," Plant Cell, vol. 14, no. 10, pp. 2539–2551, 2002. View at Publisher
87. A. W. Galston, R. Kaur-Sawhney, T. Altabella, and A. F. Tiburcio, "Plant polyamines in reproductive activity and response to abiotic stress," Botanica Acta, vol. 110, no. 3, pp. 197–207, 1997.
88. K. Gupta, A. Dey, and B. Gupta, "Polyamines and their role in plant osmotic stress tolerance," inClimate Change and Plant Abiotic Stress Tolerance, N. Tuteja and S. S. Gill, Eds., pp. 1053–1072, Wiley-VCH, Weinheim, Germany, 2013.
89. J. Yang, J. Zhang, K. Liu, Z. Wang, and L. Liu, "Involvement of polyamines in the drought resistance of rice," Journal of Experimental Botany, vol. 58, no. 6, pp. 1545–1555, 2007.
90. M. D. Groppa and M. P. Benavides, "Polyamines and abiotic stress: recent advances," Amino Acids, vol. 34, no. 1, pp. 35–45, 2008.
91. Z. Kovács, L. Simon-Sarkadi, A. Szucs, and G. Kocsy, "Differential effects of cold, osmotic stress and abscisic acid on polyamine accumulation in wheat," Amino Acids, vol. 38, no. 2, pp. 623–631, 2010.
92. R. Alcázar, J. C. Cuevas, J. Planas et al., "Integration of polyamines in the cold acclimation response,"Plant Science, vol. 180, no. 1, pp. 31–38, 2011.
93. J. Martin-Tanguy, "Metabolism and function of polyamines in plants: recent development (new approaches)," Plant Growth Regulation, vol. 34, no. 1, pp. 135–148, 2001. · ·

94. V. V. Kuznetsov and N. I. Shevyakova, "Polyamines and stress tolerance of plants," Plant Stress, vol. 1, pp. 50–71, 2007.
95. S. S. Hussain, M. Ali, M. Ahmad, and K. H. M. Siddique, "Polyamines: natural and engineered abiotic and biotic stress tolerance in plants," Biotechnology Advances, vol. 29, no. 3, pp. 300–311, 2011.
96. S. Shu, S. R. Guo, and L. Y. Yuan, "A review: polyamines and photosynthesis," in Advances in Photosynthesis—Fundamental Aspects, M. M. Najafpour, Ed., pp. 439–464, InTech, Rijeka, Croatia, 2012.
97. A. I. Alet, D. H. Sánchez, J. C. Cuevas et al., "New insights into the role of spermine in Arabidopsis thaliana under long-term salt stress," Plant Science, vol. 182, no. 1, pp. 94–100, 2012. · ·
98. R. Alcázar, J. L. García-Martínez, J. C. Cuevas, A. F. Tiburcio, and T. Altabella, "Overexpression of ADC2 in Arabidopsis induces dwarfism and late-flowering through GA deficiency," Plant Journal, vol. 43, no. 3, pp. 425–436, 2005.
99. C. Bortolotti, A. Cordeiro, R. Alcázar et al., "Localization of arginine decarboxylase in tobacco plants,"Physiologia Plantarum, vol. 120, no. 1, pp. 84–92, 2004.
100. C. Illingworth, M. J. Mayer, K. Elliott, C. Hanfrey, N. J. Walton, and A. J. Michael, "The diverse bacterial origins of the Arabidopsis polyamine biosynthetic pathway," FEBS Letters, vol. 549, no. 1–3, pp. 26–30, 2003.
101. J. M. Knott, P. Römer, and M. Sumper, "Putative spermine synthases from Thalassiosira pseudonanaand Arabidopsis thaliana synthesize thermospermine rather than spermine," FEBS Letters, vol. 581, no. 16, pp. 3081–3086, 2007.
102. T. Kusano, K. Yamaguchi, T. Berberich, and Y. Takahashi, "The polyamine spermine rescuesArabidopsis from salinity and drought stresses (Plant Signaling and Behavior)," Plant Signaling and Behavior, vol. 2, no. 4, pp. 251–252, 2007.
103. J. Martin-Tanguy, "Metabolism and function of polyamines in plants: recent development (new approaches)," Plant Growth Regulation, vol. 34, no. 1, pp. 135–148, 2001.
104. J. L. Rambla, F. Vera-Sirera, M. A. Blázquez, J. Carbonell, and A. Granell, "Quantitation of biogenic tetraamines in Arabidopsis

thaliana," Analytical Biochemistry, vol. 397, no. 2, pp. 208–211, 2010.

105. M. Hasanuzzaman, K. Nahar, and M. Fujita, "Regulatory role of polyamines in growth, development and abiotic stress tolerance in plants," in Plant Adaptation to Environmental Change: Significance of Amino Acids and Their Derivatives, pp. 157–193, 2014.

106. R. Alcázar, J. Planas, T. Saxena et al., "Putrescine accumulation confers drought tolerance in transgenicArabidopsis plants over-expressing the homologous Arginine decarboxylase 2 gene," Plant Physiology and Biochemistry, vol. 48, no. 7, pp. 547–552, 2010.

107. A. Bouchereau, A. Aziz, F. Larher, and J. Martin-Tanguy, "Polyamines and environmental challenges: recent development," Plant Science, vol. 140, no. 2, pp. 103–125, 1999.

108. R. Alcázar, F. Marco, J. C. Cuevas et al., "Involvement of polyamines in plant response to abiotic stress,"Biotechnology Letters, vol. 28, no. 23, pp. 1867–1876, 2006.

109. R. Fluhr and A. K. Mattoo, "Ethylene-biosynthesis and perception," Critical Reviews in Plant Sciences, vol. 15, pp. 479–523, 1996.

110. T. Kusano, K. Yamaguchi, T. Berberich, and Y. Takahashi, "Advances in polyamine research in 2007,"Journal of Plant Research, vol. 120, no. 3, pp. 345–350, 2007.

111. K. D. Allen, "Assaying gene content in Arabidopsis," Proceedings of the National Academy of Sciences of the United States of America, vol. 99, pp. 9568–9572, 2002.

112. C. Hanfrey, S. Sommer, M. J. Mayer, D. Burtin, and A. J. Michael, "Arabidopsis polyamine biosynthesis: absence of ornithine decarboxylase and the mechanism of arginine decarboxylase activity," Plant Journal, vol. 27, no. 6, pp. 551–560, 2001.

113. K. Urano, Y. Yoshiba, T. Nanjo et al., "Characterization of Arabidopsis genes involved in biosynthesis of polyamines in abiotic stress responses and developmental stages," Plant, Cell and Environment, vol. 26, no. 11, pp. 1917–1926, 2003. · ·

114. C. Ge, X. Cui, Y. Wang et al., "BUD2, encoding an S-adenosylmethionine decarboxylase, is required forArabidopsis

growth and development," Cell Research, vol. 16, no. 5, pp. 446–456, 2006.

115. T. Janowitz, H. Kneifel, and M. Piotrowski, "Identification and characterization of plant agmatine iminohydrolase, the last missing link in polyamine biosynthesis of plants," FEBS Letters, vol. 544, no. 1–3, pp. 258–261, 2003.
116. M. B. Watson and R. L. Malmberg, "Regulation of Arabidopsis thaliana (L.) Heynh arginine decarboxylase by potassium deficiency stress," Plant Physiology, vol. 111, no. 4, pp. 1077–1083, 1996.
117. Y. Hanzawa, A. Imai, A. J. Michael, Y. Komeda, and T. Takahashi, "Characterization of the spermidine synthase-related gene family in Arabidopsis thaliana," FEBS Letters, vol. 527, no. 1–3, pp. 176–180, 2002.
118. T. Hashimoto, K. Tamaki, K.-I. Suzuki, and Y. Yamada, "Molecular cloning of plant spermidine synthases," Plant and Cell Physiology, vol. 39, no. 1, pp. 73–79, 1998.
119. K. Urano, Y. Yoshiba, T. Nanjo, T. Ito, K. Yamaguchi-Shinozaki, and K. Shinozaki, "Arabidopsis stress-inducible gene for arginine decarboxylase AtADC2 is required for accumulation of putrescine in salt tolerance," Biochemical and Biophysical Research Communications, vol. 313, no. 2, pp. 369–375, 2004.
120. Y. Hanzawa, A. Imai, A. J. Michael, Y. Komeda, and T. Takahashi, "Characterization of the spermidine synthase-related gene family in Arabidopsis thaliana," FEBS Letters, vol. 527, no. 1–3, pp. 176–180, 2002
121. T. Takahashi and J.-I. Kakehi, "Polyamines: ubiquitous polycations with unique roles in growth and stress responses," Annals of Botany, vol. 105, no. 1, pp. 1–6, 2010.
122. A. Cona, G. Rea, R. Angelini, R. Federico, and P. Tavladoraki, "Functions of amine oxidases in plant development and defence," Trends in Plant Science, vol. 11, no. 2, pp. 80–88, 2006.
123. A. Tisi, R. Angelini, and A. Cona, "Wound healing in plants: cooperation of copper amine oxidase and flavin-containing polyamine oxidase," Plant Signaling and Behavior, vol. 3, no. 3, pp. 204–206, 2008.

124. E. Navakoudis, C. Lütz, C. Langebartels, U. Lütz-Meindl, and K. Kotzabasis, "Ozone impact on the photosynthetic apparatus and the protective role of polyamines," Biochimica et Biophysica Acta—General Subjects, vol. 1621, no. 2, pp. 160–169, 2003.
125. P. Roy, K. Niyogi, D. N. SenGupta, and B. Ghosh, "Spermidine treatment to rice seedlings recovers salinity stress-induced damage of plasma membrane and PM-bound H^+-ATPase in salt-tolerant and salt-sensitive rice cultivars," Plant Science, vol. 168, no. 3, pp. 583–591, 2005.
126. A. Roychoudhury, S. Basu, and D. N. Sengupta, "Amelioration of salinity stress by exogenously applied spermidine or spermine in three varieties of indica rice differing in their level of salt tolerance," Journal of Plant Physiology, vol. 168, no. 4, pp. 317–328, 2011.
127. X. Wang, G. Shi, Q. Xu, and J. Hu, "Exogenous polyamines enhance copper tolerance of Nymphoides peltatum," Journal of Plant Physiology, vol. 164, no. 8, pp. 1062–1070, 2007.
128. R. M. Ali, "Role of putrescine in salt tolerance of Atropa belladonna plant," Plant Science, vol. 152, no. 2, pp. 173–179, 2000.
129. I. Afzal, F. Munir, C. M. Ayub, S. M. A. Basra, A. Hameed, and A. Nawaz, "Changes in antioxidant enzymes, germination capacity and vigour of tomato seeds in response of priming with polyamines,"Seed Science and Technology, vol. 37, no. 3, pp. 765–770, 2009.
130. J.-C. Yiu, L.-D. Juang, D. Y.-T. Fang, C.-W. Liu, and S.-J. Wu, "Exogenous putrescine reduces flooding-induced oxidative damage by increasing the antioxidant properties of Welsh onion," Scientia Horticulturae, vol. 120, no. 3, pp. 306–314, 2009.
131. K. Hussain, K. Nawaz, A. Majeed et al., "Role of exogenous salicylic acid applications for salt tolerance in violet (Viola Odorata L.)," Sarhad Journal of Agriculture, vol. 27, pp. 171–175, 2011.
132. K. Yamaguchi, Y. Takahashi, T. Berberich et al., "The polyamine spermine protects against high salt stress in Arabidopsis thaliana," FEBS Letters, vol. 580, no. 30, pp. 6783–6788, 2006.
133. M. Roy and R. Wu, "Overexpression of S-adenosylmethionine decarboxylase gene in rice increases polyamine level and

enhances sodium chloride-stress tolerance," Plant Science, vol. 163, no. 5, pp. 987–992, 2002.

134. N. I. Shevyakova, L. I. Musatenko, L. A. Stetsenko, et al., "Effects of abscisic acid on the contents of polyamines and proline in common bean plants under salt stress," Russian Journal of Plant Physiology, vol. 60, pp. 200–211, 2013.

135. P. N. Moschou, K. A. Paschalidis, and K. A. Roubelakis-Angelakis, "Plant polyamine catabolism: the state of the art," Plant Signaling and Behavior, vol. 3, no. 12, pp. 1061–1066, 2008.

136. J. Duan, J. Li, S. Guo, and Y. Kang, "Exogenous spermidine affects polyamine metabolism in salinity-stressed Cucumis sativus roots and enhances short-term salinity tolerance," Journal of Plant Physiology, vol. 165, no. 15, pp. 1620–1635, 2008.

137. Y. Y. Chai, C. D. Jiang, L. Shi, T. S. Shi, and W. B. Gu, "Effects of exogenous spermine on sweet sorghum during germination under salinity," Biologia Plantarum, vol. 54, no. 1, pp. 145–148, 2010.

138. B. Li, L. He, S. Guo, et al., "Proteomics reveal cucumber Spd-responses under normal condition and salt stress," Plant Physiology and Biochemistry, vol. 67, pp. 7–14, 2013.

139. M. Delledonne, Y. Xia, R. A. Dixon, and C. Lamb, "Nitric oxide functions as a signal in plant disease resistance," Nature, vol. 394, no. 6693, pp. 585–588, 1998.

140. L. Lamattina, C. García-Mata, M. Graziano, and G. Pagnussat, "Nitric oxide: the versatility of an extensive signal molecule," Annual Review of Plant Biology, vol. 54, pp. 109–136, 2003. View at Publisher·

141. A. Besson-Bard, A. Pugin, and D. Wendehenne, "New insights into nitric oxide signaling in plants,"Annual Review of Plant Biology, vol. 59, pp. 21–39, 2008.

142. M.-G. Zhao, L. Chen, L.-L. Zhang, and W.-H. Zhang, "Nitric reductase-dependent nitric oxide production is involved in cold acclimation and freezing tolerance in Arabidopsis," Plant Physiology, vol. 151, no. 2, pp. 755–767, 2009.

143. N. M. Crawford, "Mechanisms for nitric oxide synthesis in plants," Journal of Experimental Botany, vol. 57, no. 3, pp. 471–478, 2006. · ·

144. A. Bajgu, "Nitric oxide: role in plants under abiotic stress," in Physiological Mechanisms and Adaptation Strategies in Plants Under Changing Environment, pp. 137–159, Springer, 2014.

145. C. H. Sung and J. K. Hong, "Sodium nitroprusside mediates seedling development and attenuation of oxidative stresses in Chinese cabbage," Plant Biotechnology Reports, vol. 4, no. 4, pp. 243–251, 2010.

146. K. K. Hossain, R. D. Itoh, G. Yoshimura et al., "Effects of nitric oxide scavengers on thermoinhibition of seed germination in Arabidopsis thaliana," Russian Journal of Plant Physiology, vol. 57, no. 2, pp. 222–232, 2010.

147. J. Xiong, G. Fu, L. Tao, and C. Zhu, "Roles of nitric oxide in alleviating heavy metal toxicity in plants,"Archives of Biochemistry and Biophysics, vol. 497, no. 1-2, pp. 13–20, 2010.

148. M. Kopyra and E. A. Gwó d , "Nitric oxide stimulates seed germination and counteracts the inhibitory effect of heavy metals and salinity on root growth of Lupinus luteus," Plant Physiology and Biochemistry, vol. 41, no. 11-12, pp. 1011–1017, 2003.

149. Y. Zhang, L. Wang, Y. Liu, Q. Zhang, Q. Wei, and W. Zhang, "Nitric oxide enhances salt tolerance in maize seedlings through increasing activities of proton-pump and Na^+/H^+ antiport in the tonoplast,"Planta, vol. 224, no. 3, pp. 545–555, 2006.

150. S. Mishra, A. B. Jha, and R. S. Dubey, "Arsenite treatment induces oxidative stress, upregulates antioxidant system, and causes phytochelatin synthesis in rice seedlings," Protoplasma, vol. 248, no. 3, pp. 565–577, 2011.

151. L. Zhao, F. Zhang, J. Guo, Y. Yang, B. Li, and L. Zhang, "Nitric oxide functions as a signal in salt resistance in the calluses from two ecotypes of reed," Plant Physiology, vol. 134, no. 2, pp. 849–857, 2004.

152. A. M. Nalousi, S. Ahmadiyan, A. Hatamzadeh, and M. Ghasemnezhad, "Protective role of exogenous nitric oxide against oxidative stress induced by salt stress in bell-pepper (Capsicum annum L.),"American-Eurasian Journal of Agricultural & Environmental Science, vol. 12, no. 8, pp. 1085–1090, 2012.

153. F. Zhang, Y. Wang, Y. Yang, H. Wu, D. Wang, and J. Liu, "Involvement of hydrogen peroxide and nitric oxide in salt

resistance in the calluses from Populus euphratica," Plant, Cell and Environment, vol. 30, no. 7, pp. 775–785, 2007.
154. T. He and G. R. Cramer, "Abscisic acid concentrations are correlated with leaf area reductions in two salt-stressed rapid-cycling Brassica species," Plant and Soil, vol. 179, no. 1, pp. 25–33, 1996.
155. C. Cabot, J. V. Sibole, J. Barceló, and C. Poschenrieder, "Abscisic acid decreases leaf Na^+ exclusion in salt-treated Phaseolus vulgaris L," Journal of Plant Growth Regulation, vol. 28, no. 2, pp. 187–192, 2009.
156. G. R. Cramer and S. A. Quarrie, "Abscisic acid is correlated with the leaf growth inhibition of four genotypes of maize differing in their response to salinity," Functional Plant Biology, vol. 29, no. 1, pp. 111–115, 2002.
157. D.-J. Kang, Y.-J. Seo, J.-D. Lee et al., "Jasmonic acid differentially affects growth, ion uptake and abscisic acid concentration in salt-tolerant and salt-sensitive rice cultivars," Journal of Agronomy and Crop Science, vol. 191, no. 4, pp. 273–282, 2005.
158. L. P. Popova, Z. G. Stoinova, and L. T. Maslenkova, "Involvement of abscisic acid in photosynthetic process in Hordeum vulgare L. during salinity stress," Journal of Plant Growth Regulation, vol. 14, no. 4, pp. 211–218, 1995.
159. W. D. Jeschke, A. D. Peuke, J. S. Pate, and W. Hartung, "Transport, synthesis and catabolism of abscisic acid (ABA) in intact plants of castor bean (Ricinus communis L.) under phosphate deficiency and moderate salinity," Journal of Experimental Botany, vol. 48, no. 314, pp. 1737–1747, 1997.
160. S. Chen, J. Li, S. Wang, A. Hüttermann, and A. Altman, "Salt, nutrient uptake and transport, and ABA of Populus euphratica; a hybrid in response to increasing soil NaCl," Trees—Structure and Function, vol. 15, no. 3, pp. 186–194, 2001.
161. A. R. Gurmani, A. Bano, S. U. Khan, J. Din, and J. L. Zhang, "Alleviation of salt stress by seed treatment with abscisic acid (ABA), 6-benzylaminopurine (BA) and chlormequat chloride (CCC) optimizes ion and organic matter accumulation and increases yield of rice (Oryza sativa L.)," Australian Journal of Crop Science, vol. 5, no. 10, pp. 1278–1285, 2011.

162. A. Fukuda and Y. Tanaka, "Effects of ABA, auxin, and gibberellin on the expression of genes for vacuolar H^+-inorganic pyrophosphatase, H^+-ATPase subunit A, and Na^+/H^+ antiporter in barley,"Plant Physiology and Biochemistry, vol. 44, no. 5-6, pp. 351–358, 2006.
163. B. C. Keskin, A. T. Sarikaya, B. Yüksel, and A. R. Memon, "Abscisic acid regulated gene expression in bread wheat (Triticum aestivum L.)," Australian Journal of Crop Science, vol. 4, no. 8, pp. 617–625, 2010.
164. C. Fragnire, M. Serrano, E. Abou-Mansour, J.-P. Métraux, and F. L'Haridon, "Salicylic acid and its location in response to biotic and abiotic stress," FEBS Letters, vol. 585, no. 12, pp. 1847–1852, 2011.
165. S. D. Clause and J. M. Sasse, "Brassinosteroids: essential regulators of plant growth and development,"Annual Review of Plant Biology, vol. 49, pp. 427–451, 1998.
166. H. Sawada, I.-S. Shim, and K. Usui, "Induction of benzoic acid 2-hydroxylase and salicylic acid biosynthesis-Modulation by salt stress in rice seedlings," Plant Science, vol. 171, no. 2, pp. 263–270, 2006.
167. M. Jayakannan, J. Bose, O. Babourina, et al., "Salicylic acid improves salinity tolerance in Arabidopsis by restoring membrane potential and preventing salt-induced K^+ loss via a GORK channel," Journal of Experimental Botany, vol. 64, no. 8, pp. 2255–2268, 2013.
168. M. A. El-Tayeb, "Response of barley grains to the interactive effect of salinity and salicylic acid," Plant Growth Regulation, vol. 45, no. 3, pp. 215–224, 2005.
169. R. Nazar, N. Iqbal, S. Syeed, and N. A. Khan, "Salicylic acid alleviates decreases in photosynthesis under salt stress by enhancing nitrogen and sulfur assimilation and antioxidant metabolism differentially in two mungbean cultivars," Journal of Plant Physiology, vol. 168, no. 8, pp. 807–815, 2011.
170. A. A. A. El-Mashad and H. I. Mohamed, "Brassinolide alleviates salt stress and increases antioxidant activity of cowpea plants (Vigna sinensis)," Protoplasma, vol. 249, no. 3, pp. 625–635, 2012.

171. S. Anuradha and S. Seeta Ram Rao, "Effect of brassinosteroids on salinity stress induced inhibition of seed germination and seedling growth of rice (Oryza sativa L.)," Plant Growth Regulation, vol. 33, no. 2, pp. 151–153, 2001.
172. P. Krishna, "Brassinosteroid-Mediated Stress Responses," Journal of Plant Growth Regulation, vol. 22, no. 4, pp. 289–297, 2003
173. M. Ashraf, N. A. Akram, R. N. Arteca, and M. R. Foolad, "The physiological, biochemical and molecular roles of brassinosteroids and salicylic acid in plant processes and salt tolerance," Critical Reviews in Plant Sciences, vol. 29, no. 3, pp. 162–190, 2010.
174. R. R. Johnson, R. L. Wagner, S. D. Verhey, and M. K. Walker-Simmons, "The abscisic acid-responsive kinase PKABA1 interacts with a seed-specific abscisic acid response element-binding factor, TaABF, and phosphorylates TaABF peptide sequences," Plant Physiology, vol. 130, no. 2, pp. 837–846, 2002.
175. K. Nakashima, L.-S. P. Tran, D. Van Nguyen et al., "Functional analysis of a NAC-type transcription factor OsNAC6 involved in abiotic and biotic stress-responsive gene expression in rice," Plant Journal, vol. 51, no. 4, pp. 617–630, 2007.
176. J. Mizoi, K. Shinozaki, and K. Yamaguchi-Shinozaki, "AP2/ERF family transcription factors in plant abiotic stress responses," Biochimica et Biophysica Acta—Gene Regulatory Mechanisms, vol. 1819, no. 2, pp. 86–96, 2012.
177. Y. Fujita, M. Fujita, K. Shinozaki, and K. Yamaguchi-Shinozaki, "ABA-mediated transcriptional regulation in response to osmotic stress in plants," Journal of Plant Research, vol. 124, no. 4, pp. 509–525, 2011.
178. A. RoyChoudhury, B. Gupta, and D. N. Sengupta, "Trans-acting factor designated OSBZ8 interacts with both typical abscisic acid responsive elements as well as abscisic acid responsive element-like sequences in the vegetative tissues of indica rice cultivars," Plant Cell Reports, vol. 27, no. 4, pp. 779–794, 2008.
179. Y. Ito, K. Katsura, K. Maruyama et al., "Functional analysis of rice DREB1/CBF-type transcription factors involved in cold-responsive gene expression in transgenic rice," Plant and Cell Physiology, vol. 47, no. 1, pp. 141–153, 2006.

180. Y. Fujita, T. Yoshida, and K. Yamaguchi-Shinozaki, "Pivotal role of the AREB/ABF-SnRK2 pathway in ABRE-mediated transcription in response to osmotic stress in plants," Physiologia Plantarum, vol. 147, no. 1, pp. 15–27, 2013.
181. S.-Y. Song, Y. Chen, J. Chen, X.-Y. Dai, and W.-H. Zhang, "Physiological mechanisms underlying OsNAC5-dependent tolerance of rice plants to abiotic stress," Planta, vol. 234, no. 2, pp. 331–345, 2011.
182. Y. Hu, L. Chen, H. Wang, L. Zhang, F. Wang, and D. Yu, "Arabidopsis transcription factor WRKY8 functions antagonistically with its interacting partner VQ9 to modulate salinity stress tolerance," The Plant Journal, vol. 74, pp. 730–745, 2013.
183. K. Chakraborty, R. K. Sairam, and R. C. Bhattacharya, "Differential expression of salt overly sensitive pathway genes determines salinity stress tolerance in Brassica genotypes," Plant Physiology and Biochemistry, vol. 51, pp. 90–101, 2012.
184. P. Roshandel and T. Flowers, "The ionic effects of NaCl on physiology and gene expression in rice genotypes differing in salt tolerance," Plant and Soil, vol. 315, no. 1-2, pp. 135–147, 2009.
185. J. Zou, A. Liu, X. Chen et al., "Expression analysis of nine rice heat shock protein genes under abiotic stresses and ABA treatment," Journal of Plant Physiology, vol. 166, no. 8, pp. 851–861, 2009.
186. G.-H. Lim, X. Zhang, M.-S. Chung et al., "A putative novel transcription factor, AtSKIP, is involved in abscisic acid signalling and confers salt and osmotic tolerance in Arabidopsis," New Phytologist, vol. 185, no. 1, pp. 103–113, 2010. · ·
187. J. Zou, C. Liu, A. Liu, D. Zou, and X. Chen, "Overexpression of OsHsp17.0 and OsHsp23.7 enhances drought and salt tolerance in rice," Journal of Plant Physiology, vol. 169, no. 6, pp. 628–635, 2012.
188. N.-H. Song and Y.-J. Ahn, "DcHsp17.7, a small heat shock protein in carrot, is tissue-specifically expressed under salt stress and confers tolerance to salinity," New Biotechnology, vol. 28, no. 6, pp. 698–704, 2011.
189. M. Tang, X. Liu, H. Deng, and S. Shen, "Over-expression of JcDREB, a putative AP2/EREBP domain-containing transcription factor gene in woody biodiesel plant Jatropha curcas, enhances

salt and freezing tolerance in transgenic Arabidopsis thaliana," Plant Science, vol. 181, no. 6, pp. 623–631, 2011.

190. J. Liu, M. Ishitani, U. Halfter, C.-S. Kim, and J.-K. Zhu, "The Arabidopsis thaliana SOS2 gene encodes a protein kinase that is required for salt tolerance," Proceedings of the National Academy of Sciences of the United States of America, vol. 97, no. 7, pp. 3730–3734, 2000.

191. H. Shi, M. Ishitani, C. Kim, and J.-K. Zhu, "The Arabidopsis thaliana salt tolerance gene SOS1 encodes a putative Na^+/H^+ antiporter," Proceedings of the National Academy of Sciences of the United States of America, vol. 97, no. 12, pp. 6896–6901, 2000.

192. S. Kawasaki, C. Borchert, M. Deyholos et al., "Gene expression profiles during the initial phase of salt stress in rice," Plant Cell, vol. 13, no. 4, pp. 889–905, 2001.

193. N. Baisakh, P. K. Subudhi, and N. P. Parami, "cDNA-AFLP analysis reveals differential gene expression in response to salt stress in a halophyte Spartina alterniflora Loisel," Plant Science, vol. 170, no. 6, pp. 1141–1149, 2006.

194. R. Schmidt, D. Mieulet, H.-M. Hubberten, et al., "SALT-RESPONSIVE ERF1 regulates reactive oxygen species-dependent signaling during the initial response to salt stress in rice," The Plant Cell, vol. 25, no. 6, pp. 2115–2131, 2013.

195. T. Serra, D. D. Figueiredo, A. M. Cordeiro, et al., "OsRMC, a negative regulator of salt stress response in rice, is regulated by two AP2/ERF transcription factors," Plant Molecular Biology, vol. 82, pp. 439–455, 2013.

196. K. Mukherjee, A. R. Choudhury, B. Gupta, et al., "An ABRE-binding factor, OSBZ8, is highly expressed in salt tolerant cultivars than in salt sensitive cultivars of indica rice," BMC Plant Biology, vol. 6, article 18, no. 1, 2006.

197. B. Gupta, K. Gupta, and D. N. Sengupta, "Spermidine-mediated in vitro phosphorylation of transcriptional regulator OSBZ8 by SNF1-type serine/threonine protein kinase SAPK4 homolog in indica rice," Acta Physiologiae Plantarum, vol. 34, no. 4, pp. 1321–1336, 2012.

198. D. Staiger and J. W. Brown, "Alternative splicing at the intersection of biological timing, development, and stress responses," The Plant Cell, vol. 25, no. 10, pp. 3640–3656, 2013.
199. S. G. Palusa, G. S. Ali, and A. S. N. Reddy, "Alternative splicing of pre-mRNAs of Arabidopsisserine/arginine-rich proteins: regulation by hormones and stresses," Plant Journal, vol. 49, no. 6, pp. 1091–1107, 2007.
200. N. Tanabe, K. Yoshimura, A. Kimura, Y. Yabuta, and S. Shigeoka, "Differential expression of alternatively spliced mRNAs of Arabidopsis SR protein homologs, atSR30 and atSR45a, in response to environmental stress," Plant and Cell Physiology, vol. 48, no. 7, pp. 1036–1049, 2007.
201. Z. Zhang, S. Zhang, Y. Zhang et al., "Arabidopsis floral initiator SKB1 confers high salt tolerance by regulating transcription and pre-mRNA splicing through altering histone H4R3 and small nuclear ribonucleoprotein LSM4 methylation," Plant Cell, vol. 23, no. 1, pp. 396–411, 2011.
202. L. Conti, G. Price, E. O'Donnell, B. Schwessinger, P. Dominy, and A. Sadanandom, "Small ubiquitin-like modifier proteases overly tolerant to Salt1 and -2 regulate salt stress responses in Arabidopsis," Plant Cell, vol. 20, no. 10, pp. 2894–2908, 2008.
203. F. Cui, L. Liu, Q. Zhao et al., "Arabidopsis ubiquitin conjugase UBC32 is an ERAD component that functions in brassinosteroid-mediated salt stress tolerance," Plant Cell, vol. 24, no. 1, pp. 233–244, 2012.
204. S. H. Kim, Y. O. Ahn, M.-J. Ahn, H.-S. Lee, and S.-S. Kwak, "Down-regulation of -carotene hydroxylase increases -carotene and total carotenoids enhancing salt stress tolerance in transgenic cultured cells of sweetpotato," Phytochemistry, vol. 74, pp. 69–78, 2012.
205. R. Sunkar, V. Chinnusamy, J. Zhu, and J.-K. Zhu, "Small RNAs as big players in plant abiotic stress responses and nutrient deprivation," Trends in Plant Science, vol. 12, no. 7, pp. 301–309, 2007.
206. F. Xie, C. N. Stewart Jr., F. A. Taki, et al., "High-throughput deep sequencing shows that microRNAs play important roles in switchgrass responses to drought and salinity stress," Plant Biotechnology Journal, 2013.

207. H. J. Bohnert, Q. Gong, P. Li, and S. Ma, "Unraveling abiotic stress tolerance mechanisms—getting genomics going," Current Opinion in Plant Biology, vol. 9, no. 2, pp. 180–188, 2006.

208. D. J. Weston, L. E. Gunter, A. Rogers, and S. D. Wullschleger, "Connecting genes, coexpression modules, and molecular signatures to environmental stress phenotypes in plants," BMC Systems Biology, vol. 2, article 16, 2008. · ·

209. B. Barkla, T. Castellanos-Cervantes, J. L. de León, et al., "Elucidation of salt stress defense and tolerance mechanisms of crop plants using proteomics—current achievements and perspectives," Proteomics, vol. 13, no. 12-13, pp. 1885–1900, 2013. · View at Google Scholar

210. B. Gupta, A. Sengupta, J. Saha, and K. Gupta, "Plant abiotic stress: "Omics" approach," Journal of Plant Biochemistry & Physiology, vol. 1, article e108, 2013. · View at Google Scholar

211. H. Zhang, B. Han, T. Wang et al., "Mechanisms of plant salt response: insights from proteomics,"Journal of Proteome Research, vol. 11, no. 1, pp. 49–67, 2012. · View at Google Scholar·

212. S. S. Gill and N. Tuteja, "Reactive oxygen species and antioxidant machinery in abiotic stress tolerance in crop plants," Plant Physiology and Biochemistry, vol. 48, no. 12, pp. 909–930, 2010.

213. B. J. Barkla, R. Vera-Estrella, and O. Pantoja, "Progress and challenges for abiotic stress proteomics of crop plants," Proteomics, vol. 13, no. 12-13, pp. 1801–1815, 2013. · View at Google Scholar

214. Y. Osakabe, K. Yamaguchi-Shinozaki, K. Shinozaki, and L. S. Tran, "Sensing the environment: key roles of membrane-localized kinases in plant perception and response to abiotic stress," Journal of Experimental Botany, vol. 64, no. 2, pp. 445–458, 2013.

215. G. Friso, L. Giacomelli, A. J. Ytterberg et al., "In-depth analysis of the thylakoid membrane proteome ofArabidopsis thaliana chloroplasts: new proteins, new functions, and a plastid proteome database," Plant Cell, vol. 16, no. 2, pp. 478–499, 2004.

216. A. Marmagne, M.-A. Roue, M. Ferro et al., "Identification of new intrinsic proteins in Arabidopsisplasma membrane proteome,"

Molecular and Cellular Proteomics, vol. 3, no. 7, pp. 675–691, 2004.

217. D. Takahashi, B. Li, T. Nakayama, Y. Kawamura, and M. Uemura, "Plant plasma membrane proteomics for improving cold tolerance," Frontiers in Plant Science, vol. 4, article 90, 2013.

218. D. Ghosh and J. Xu, "Abiotic stress responses in plant roots: a proteomics perspective," Frontiers in Plant Science, vol. 5, article 6, 2014. · View at Google Scholar

219. S. T. Kim, S. G. Kim, G. K. Agrawal, S. Kikuchi, R. Rakwal, et al., "Rice proteomics: a model system for crop improvement and food security," Proteomics, vol. 14, no. 4-5, pp. 593–610, 2014. View at Publisher· View at Google Scholar

220. B.-Y. An, Y. Luo, J.-R. Li, W.-H. Qiao, X.-S. Zhang, and X.-Q. Gao, "Expression of a vacuolar Na^+/H^+antiporter gene of alfalfa enhances salinity tolerance in transgenic Arabidopsis," Acta Agronomica Sinica, vol. 34, no. 4, pp. 557–564, 2008.

221. D. Verma, S. L. Singla-Pareek, D. Rajagopal, M. K. Reddy, and S. K. Sopory, "Functional validation of a novel isoform of Na^+/H^+ antiporter from Pennisetum glaucum for enhancing salinity tolerance in rice,"Journal of Biosciences, vol. 32, no. 3, pp. 621–628, 2007.

222. Z.-Y. Xue, D.-Y. Zhi, G.-P. Xue, H. Zhang, Y.-X. Zhao, and G.-M. Xia, "Enhanced salt tolerance of transgenic wheat (Tritivum aestivum L.) expressing a vacuolar Na^+/H^+ antiporter gene with improved grain yields in saline soils in the field and a reduced level of leaf Na^+," Plant Science, vol. 167, no. 4, pp. 849–859, 2004.

223. C.-A. Wu, G.-D. Yang, Q.-W. Meng, and C.-C. Zheng, "The cotton GhNHX1 gene encoding a novel putative tonoplast Na^+/H^+ antiporter plays an important role in salt stress," Plant and Cell Physiology, vol. 45, no. 5, pp. 600–607, 2004.

224. H.-X. Zhang and E. Blumwald, "Transgenic salt-tolerant tomato plants accumulate salt in foliage but not in fruit," Nature Biotechnology, vol. 19, no. 8, pp. 765–768, 2001.

225. G.-H. Zhang, Q. Su, L.-J. An, and S. Wu, "Characterization and expression of a vacuolar Na^+/H^+antiporter gene from the

monocot halophyte Aeluropus littoralis," Plant Physiology and Biochemistry, vol. 46, no. 2, pp. 117–126, 2008.

226. H.-X. Zhang, J. N. Hodson, J. P. Williams, and E. Blumwald, "Engineering salt-tolerant Brassica plants: characterization of yield and seed oil quality in transgenic plants with increased vacuolar sodium accumulation," Proceedings of the National Academy of Sciences of the United States of America, vol. 98, no. 22, pp. 12832–12836, 2001.
227. H. Shi, B.-H. Lee, S.-J. Wu, and J.-K. Zhu, "Overexpression of a plasma membrane Na^+/H^+ antiporter gene improves salt tolerance in Arabidopsis thaliana," Nature Biotechnology, vol. 21, no. 1, pp. 81–85, 2003.
228. X.-Y. Yin, A.-F. Yang, K.-W. Zhang, and J.-R. Zhang, "Production and analysis of transgenic maize with improved salt tolerance by the introduction of AtNHX1 gene," Acta Botanica Sinica, vol. 46, no. 7, pp. 854–861, 2004.
229. K. Saito and F. Matsuda, "Metabolomics for functional genomics, systems biology, and biotechnology,"Annual Review of Plant Biology, vol. 61, pp. 463–489, 2010.
230. R. A. Dixon and D. Strack, "Phytochemistry meets genome analysis, and beyond," Phytochemistry, vol. 62, no. 6, pp. 815–816, 2003.
231. V. Shulaev, D. Cortes, G. Miller, and R. Mittler, "Metabolomics for plant stress response," Physiologia Plantarum, vol. 132, no. 2, pp. 199–208, 2008.
232. D. Balmer, V. Flors, G. Glauser, and B. Mauch-Mani, "Metabolomics of cereals under biotic stress: current knowledge and techniques," Frontiers in Plant Science, vol. 4, article 82, 2013. ·View at Google Scholar.
233. M. Ashraf, "Biotechnological approach of improving plant salt tolerance using antioxidants as markers,"Biotechnology Advances, vol. 27, no. 1, pp. 84–93, 2009.
234. G. M. Abogadallah, "Antioxidative defense under salt stress," Plant Signaling and Behavior, vol. 5, no. 4, pp. 369–374, 2010.
235. S. Saxena, H. Kaur, P. Verma, et al., "Osmoprotectants: potential for crop improvement under adverse conditions," in Plant

Acclimation to Environmental Stress, N. Tuteja and S. Singh Gill, Eds., pp. 197–232, Springer, New York, NY, USA, 2013.

236. J. V. Cabello, A. F. Lodeyro, and M. D. Zurbriggen, "Novel perspectives for the engineering of abiotic stress tolerance in plants," Current Opinion in Biotechnology, vol. 26, pp. 62–70, 2014.

The SbSOS1 Gene from the Extreme Halophyte *Salicornia brachiata* Enhances Na^+ Loading in Xylem and Confers Salt Tolerance in Transgenic Tobacco

Narendra Singh Yadav, Pushp Sheel Shukla, Anupama Jha, Pradeep K Agarwal, and Bhavanath Jha*

Discipline of Marine Biotechnology and Ecology, CSIR-Central Salt and Marine Chemicals Research Institute, (Council of Scientific and Industrial Research), G.B. Road, Bhavnagar, Gujarat, 364002, India

ABSTRACT

Background

Soil salinity adversely affects plant growth and development and disturbs intracellular ion homeostasis resulting cellular toxicity. The Salt Overly Sensitive 1 (SOS1) gene encodes a plasma membrane Na^+/H^+ antiporter that plays an important role in imparting salt stress tolerance to plants. Here, we report the cloning and characterization of the SbSOS1 gene from Salicornia brachiata, an extreme halophyte.

Results

The SbSOS1 gene is 3774 bp long and encodes a protein of 1159 amino acids. SbSOS1 exhibited a greater level of constitutive expression in roots than in shoots and was further increased by salt stress. Overexpressing the S. brachiata SbSOS1 gene in tobacco conferred high salt tolerance, promoted seed germination and increased root length, shoot length, leaf area, fresh weight, dry weight, relative water content (RWC), chlorophyll, K^+/Na^+ ratio, membrane stability index, soluble sugar, proline and amino acid content relative to wild type (WT) plants. Transgenic plants exhibited reductions in electrolyte leakage, reactive oxygen species (ROS) and MDA content in response to salt stress, which probably occurred because of reduced cytosolic Na^+ content and oxidative damage. At higher salt stress, transgenic tobacco plants exhibited reduced Na^+ content in root and leaf and higher concentrations in stem and xylem sap relative to WT, which suggests a role of SbSOS1 in Na^+ loading to xylem from root and leaf tissues. Transgenic lines also showed increased K^+ and Ca^{2+} content in root tissue compared to WT, which reflect that SbSOS1 indirectly affects the other transporters activity.

Conclusions

Overexpression of SbSOS1 in tobacco conferred a high degree of salt tolerance, enhanced plant growth and altered physiological and biochemical parameters in response to salt stress. In addition to Na^+ efflux outside the plasma membrane, SbSOS1 also helps to maintain variable Na^+ content in different organs and also affect the other transporters activity indirectly. These results broaden the role of SbSOS1 in planta and suggest that this gene could be used to develop salt-tolerant transgenic crops.

BACKGROUND

The productivity of over one-third of the arable land in the world is affected by the salinity of the soil [1]. More than 800 million ha of land worldwide are salt-affected [2]. High salinity adversely affects plant growth and development by disturbing intracellular ion homeostasis, which results in membrane dysfunction, attenuation of metabolic activity and secondary effects that inhibit growth and induce cell death [3]. Nearly all enzyme activity is reduced severely at NaCl concentrations above 0.3 M because of disruption of the electrostatic forces that maintain protein structure [4]. NaCl stress also significantly damages photosynthetic mechanisms through a combination of superoxide and H_2O_2-mediated oxidation [5]. Plants adapt to environmental stresses via a plethora of responses, including the activation of molecular networks that regulate stress perception, signal transduction and the expression of both stress-related genes and metabolites. Plants have stress-specific adaptive responses as well as responses which protect the plants from more than one environmental stress [6]. Numerous abiotic stress-related genes, as well as some transcription factors and regulatory sequences in plant promoters, have been characterised [7]. Plants employ three different strategies to prevent and adapt to high Na^+ concentrations: 1) active Na^+ efflux, 2) Na^+ compartmentalisation in vacuoles, and 3) Na^+ influx prevention [8, 9]. Antiporters are an important group of genes that play a pivotal role in ion homeostasis in plants. Na^+/H^+ antiporters (NHX1 and SOS1) maintain the appropriate concentration of ions in the cytosol, thereby minimising cytotoxicity. NHX1 are located in tonoplasts and reduce cytosolic Na^+concentration by pumping in the vacuole [10], whereas SOS1 is localised at the plasma membrane and extrudes Na^+ in apoplasts [11]. Both of these antiporters are driven by the proton motive force generated by the H^+-ATPase [12].

The discovery of, and pioneer studies on, sos mutants in Arabidopsis uncovered a new pathway for ion homeostasis that promotes tolerance to salt stress. The sos mutants were specifically hypersensitive to high external concentrations of Na^+ or Li^+ and were unable to grow at low external K^+ concentrations [13, 14]. The SOS pathway consists of three proteins: SOS3, a calcium sensor protein [15]; SOS2, a serine/threonine protein kinase [16]; and SOS1 [17]. During salt stress, cellular Ca^{2+} levels are altered and CIPK and CBL interacting proteins are activated. Calcineurin B-like proteins (CBL) participate in salt stress-mediated signal transduction to control the influx and efflux of Na^+ [18]. The calcineurin B-like (regulatory) Ca^{2+} sensor SOS3 (Salt overly sensitive 3) has been cloned from Arabidopsis [15]. SOS3 interacts with and activates the serine/threonine protein kinase SOS2 [16, 19]. This interaction has been reported to recruit SOS2 to the plasma membrane where it interacts with SOS1 [20]. The A. thaliana SOS1 gene was ectopically expressed for the first time in Arabidopsis and reduced the accumulation of Na^+ in the presence of salt stress [21]. Similar results were obtained when SOD2 and NhaA, which are plasma membrane Na^+/H^+ antiporters from Schizosaccharomyces pombe and Escherichia coli, respectively, were overexpressed in Arabidopsis [22] and rice [23]. Heterologous expression of different plant SOS1 genes suppressed the Na^+ sensitivity of the yeast mutant (AXT3K) (Arabidopsis thaliana, [11]; Arabidopsis thaliana, [24]; Cymodocea nodosa, [25]; Oryza sativa,[26]; and Solanum lycopersicum,[27]). Additionally, Wu et al. [28] and Garciadeblas et al. [25] showed that expression of PeSOS1 and CnSOS1 partially suppressed the salt-sensitive phenotypes of the EP432 bacterial strain (nhaA,nhaB), which lacks activity of the two Na^+/H^+ antiporters EcNhaA and EcNhaB.

Few studies have performed in planta overexpression of the SOS1 gene from Arabidopsis [21, 29], Thellungiella [30] and Puccinellia tenuiflora[31]. The study of the salt tolerance mechanisms of halophytic plants has emerged as an important area because these species are well-adapted to and can overcome soil salinity more efficiently than glycophytic plants [32]. Our lab studies an extreme halophyte, Salicornia brachiata Roxb. (Amaranthaceae), in an effort to identify and characterise novel genes that enable salt tolerance. S. brachiata, a leafless succulent annual halophyte, commonly grows in the Gujarat coastal marshes in India. Salicornia can grow in a wide range of salt concentrations (0.1–2.0 M) and can accumulate quantities of salt as high as 40% of its dry weight. This unique characteristic provides an advantage for the study of salt tolerance mechanisms. Salicornia accumulates salt in the pith region, which reflects the fact that antiporter genes are necessary to maintain homeostasis in extreme salinity. This report describes the isolation of the SbSOS1 gene from S. brachiata and overexpression in tobacco plant for functional validation.

METHODS

Plant Growth and Salt Treatments

Salicornia brachiata Roxb. Seeds were collected from dried plants obtained from the coastal area near Bhavnagar, Gujarat, India. The seeds were germinated in plastic pots containing garden soil and the plants were grown in natural conditions. At one month, the seedlings were carefully uprooted and transferred to hydroponic conditions (¼ major and minor MS stock) in a culture room with a dark/light cycle of 8/16 h at 25°C for one month. The plants were treated with various concentrations of NaCl (0.10, 0.25, 0.5, 1.0, 1.5, and 2.0 M) for 48 h. Upon completion of the treatments, shoot and root tissues were separately collected, frozen in liquid nitrogen and stored at −80°C.

Cloning Of Sbsos1 Gene

Degenerate primers were designed from conserved regions of the SOS1 gene. To identify conserved regions of the SOS1 gene for PCR amplification, SOS1 cDNA sequences fromArabidopsis (NM126259), Chenopodium quinoa (NM126259), Suaeda japonica (AJ717346), Mesembryanthemum crystallinum (EF207776) Solanum lycopersicum (AJ717346),Oryzasativa(AY785147),andTriticumaestivum(AY326952) wereretrievedfrom NCBI andaligned with the ClustalW program. On the basis of sequence conservation across species, five pairs corresponding degenerate primers were designed. Of these primers, DF2-DR2 (5′-GCTTGTCGTCACTTTCTTCG-3′and 5′-CM (A+C) CCAAATGCH (A+T+C) TCTAATGC-3′) and DF5-DR4 (5′-TGCTTACTGGGYGATGCT-3′and 5′-AGYCCCAAAGTACTTCCATG-3′) primer sets produced fragments of the desired size.

RNA was extracted from the shoots of S. brachiata using the GITC method. Total RNA (5 µg) was used in RT-PCR reactions and cDNA was synthesised with a Superscript RT II first strand cDNA synthesis kit (Invitrogen, San Diego, CA, USA). The SOS1 gene was amplified in the presence of 10 pmol of the degenerate primers, 2.5 U Taq and 1 µl cDNA according to the following sequence: initial denaturation at 95°C for 2 min, 35 cycles of 94°C for 1 min, 55°C for 1 min, 72°C for 2 min, and an additional 10 min polymerisation step at 72°C. The amplified fragments were purified from agarose gels and cloned into the pGEM-T Easy vector system II (Promega, Madison, Wisconsin) and transformed into DH5 Escherichia coli cells. After sequencing these RT-PCR fragments (1300 bp and 750 bp), four additional primers were designed to amplify the full-length cDNA via rapid amplification of cDNA ends (RACE).

The 5′-RACE reaction was performed according to the manufacturer's protocol (Invitrogen, San Diego, CA, USA). The first strand of cDNA was synthesised with a gene-specific primer GSP R1 (5′-AGAGTCAAGGGGGTTTCAATTC-3′) and Superscript RT II. The mRNA was removed with RNase H, and a homopolymeric tail was added to the 3′-end of the cDNA. PCR amplification was performed using a nested, gene-specific primer GSP R2 (5′-CTACAGCTACAGGATCAGTTGCAC-3′) and a dG-anchor primer AAP (5′-GGCCACGCGTCGACTAGTAC $(G)_{16}$ -3′).

To perform 3′-RACE reactions, the first strand of cDNA was synthesized at the poly (A) mRNA tail with PK1 oligo dT adapter primer (5′-CCAGTGAGCAGAGTGACGAGGACTCGAGCTCAAGC $(T)_{17}$-3′). Following the first strand of cDNA synthesis, PCR was performed with a gene-specific primer, GSP F1 (5′-AGTAGTAAAGACCAGGCAAGCAAC-3′), and an adaptor primer, PK2 (5′-CCAGTGAGCAGAGTGACG-3′). The 5′- and 3′-RACE products were cloned into a pGEM-T Easy vector. The cloned products of the original PCR and the 5′-and 3′-RACE reactions were sequenced, and contiguous sequences were assembled to obtain the full-length SbSOS1 gene. After determining the open reading frame, the full length SOS1 cDNA was amplified with AccuPrime Pfx DNA polymerase (Invitrogen), using SbSOS F (5′-ATCGG

GGTACC

ATATGGCAGCATCTCGAATTGA-3′) and SbSOS R (5′-ATTCC

CCCGGG

TCAAGGAGCTTGGCGGAA-3') primers. The amplification product was then cloned into a pJET1.2/blunt cloning vector (MBI Fermentas) and sequenced by primer walking (Macrogen Inc., Seoul, South Korea).

In silico Analysis

The NCBI database was searched for nucleotide and protein sequences. TMpred online software was used to predict transmembrane domains, and DNAMAN was used to align sequences. Conserved domains in SbSOS1 were identified via the BLASTp program (http://www.ncbi.nlm.nih.gov webcite). ExPASy tools (http://www.expasy.ch/tools/ webcite) was used to predict secondary structures. The relationship of SbSOS1 with SOS1 from other plant species was inferred by constructing phylogenetic tree using MEGA Ver. 4.0. The robustness in topology of the phylogenetic tree was assessed based on bootstrap value.

Quantitative Real Time PCR

Total RNA was isolated from control and NaCl-treated plant samples using GITC buffer and quantified with a Nanodrop spectrophotometer (USA). The cDNA was prepared with 5 μg of total RNA with a Superscript RT II first-strand cDNA synthesis kit (Invitrogen, San Diego, CA, USA). Real Time qPCR was performed in a Bio-Rad IQ5 detection system (Bio-Rad, U.S.A.) with 1x SYBR Green (Sigma, USA). The PCR reactions were carried out in 1x PCR buffer supplemented with 200 μM dNTPs, 1.25 U Taq DNA polymerase and 5 pmol of each gene-specific primer. The PCR reactions were performed under the following conditions: initial denaturation at 95°C for 5 min, 40 cycles at 95°C for 10 sec, 60°C for 30 sec and 72°C for 30 sec. At the end of the PCR cycles, the products were subjected to melt curve analysis to verify the specificity of PCR amplification. The amplified products were electrophoresed through 1.2% agarose gel to verify that they matched the predicted size. Three independent experiments were performed with three replicates each. Fold changes were calculated using the CT method [33], and CT values for individual variants were compared to those of a reference control (-tubulin). The SbSOS1-specific primer pair RT F (5'-GGAAGGTTTGGGGATGGTAT-3') and

RT R (5'-GTCCAGCAAGCAAAACCATT-3') were utilised for expression study of SbSOS1, whereas QBT F (5'-GGAGTCACCGAG GCAGAG-3') and QBT R (5'-ATCACATATCAGAAACCACAA-3') primers were used for an internal control.

Construction of Plant Transformation Vector and Tobacco Transformation

To perform plant transformation, SbSOS1 cDNA was PCR-amplified with AccuPrime Pfx DNA polymerase (Invitrogen) in conjunction with UA F (5'-TAACA

GGGCCC

ATGGCAGCATCTCGAATTGA-3') and UK R (5'-ATCGG

GGTACC

TCAAGGAGCTTGGCGGAA-3') primers, which contained ApaI and KpnI sites, respectively. The digested SbSOS1 gene was cloned as an ApaI/ KpnI fragment into the pRT 100 vector [34]. Thereafter, the entire cassette containing the CaMV 35S constitutive promoter and both theSbSOS1 gene and terminator were cloned into the pCAMBIA2301 vector at the PstI site. The resulting vector was mobilised into Agrobacterium tumefaciens (LBA 4404) and used to transform tobacco (Nicotiana tabacum cv. Xanthi) plants according to a standard protocol [35]. Putative transgenic plants regenerated directly from leaf edges in the presence of kanamycin (50 mg/l) were transferred in jam bottles that contained MS basal medium [36] supplemented with kanamycin (100 mg/l). The transgenic lines were screened via GUS assay, PCR amplification and semiquantitative PCR analysis. To obtain transgenic T_1 seedlings, seeds from the parental plants (T_0 seeds) were germinated on kanamycin-supplemented MS medium.

Semiquantitative PCR Analysis

Total RNA was isolated from WT and transgenic plant samples using GITC buffer and was quantified with a Nanodrop spectrophotometer (USA). The cDNA was prepared using 5 μg of total RNA with a Superscript RT III first-strand cDNA synthesis kit (Invitrogen, San Diego, CA). The synthesised cDNA (1 μl, diluted 1:5) was used as a template, and actin was used as an internal control for RT-PCR analysis. The SbSOS1-specific primer pair, RT F (5'-GGAAGGTTTGGGGATGGTAT-3') and RT R (5'-GTCCAGCAAGCAAACCATT-3'), was utilised for expression study of the SbSOS1, whereas QACT F (5'-CGTTTGGATCTTGCTGGTCGT-3') and QACT R (5'- CAGCAATGCCAGGGAACATAG −3') primers were used for actin. The PCR reactions were carried out in 1x PCR buffer supplemented with 200 μM dNTPs, 1.25 U Taq DNA polymerase and 5 pmol of each of the gene-specific primers according to the following conditions: an initial denaturation at 95°C for 5 min, 25 cycles at 94°C for 30 sec, 60°C for 30 sec and 72°C for 30 sec, followed by a final extension step at 72°C for 7 min. RT-PCR experiments were repeated three times, and the amplification products were analysed via agarose gel electrophoresis.

Histochemical GUS staining

GUS activity was visualised in leaf tissue with 1 mM 4-methyl-umbelliferyl-b-D-glucuronide as was described by Jefferson [37]. Leaves from control and transgenic plants were cut to 1 cm^2 in size and rinsed in 50 mM phosphate buffer (pH 7.0). Leaf sections were then incubated with 1 mM 4-methyl-umbelliferyl-b-D-glucuronide prepared in 50 mM phosphate buffer, vacuum infiltrated for 10 min and then incubated overnight at 37°C in the dark. The tissues were then rinsed with 80% ethanol for 4 h to remove chlorophyll.

Extraction of Genomic DNA and PCR Analysis

Genomic DNA was isolated from different T_0 lines via the CTAB (N-cetyl-N, N, N-trimethylammonium bromide) method. To verify the presence of the transgene, PCR was conducted with gene-specific primers for real-time PCR (RTF and RTR) and GUS-specific primers (gusA F 5'-GATCGCGAAAACTGTGGAAT-3' and gusA R 5'-TGAGCGTCGCAGAACATTAC-3').

Leaf Disc Assay and Estimation of Chlorophyll

Leaf discs from WT and two T_0 transgenic lines (L1 and L7) were used in the salt tolerance assay. The leaf discs were floated on ¼ MS (control) and different concentrations of NaCl for 15 days. The leaf discs were then homogenised thoroughly in 80% acetone and centrifuged at

3,000 ×g for 2–3 min. The O.D. of each supernatant was recorded at 663 and 645 nm, and the chlorophyll content was calculated per gram of fresh tissue weight [38]. The treatments were conducted under continuous exposure to white light at 25 ± 2°C.

Evaluation of Transgenic Plants Exposed To Salt Stress

To analyse the salt stress tolerance of SbSOS1-overexpressing tobacco plants, the seeds from T_0transgenic plants were germinated in MS medium supplemented with 0, 50, 100, 150, 200, or 300 mM NaCl in culture room conditions. The percentage of seed germination was scored 15 days after seed inoculation. T_1 seedlings were also analysed for different growth parameters under salt stress. At eight days, WT and kanamycin-positive T_1 seedlings (50 mg/l) were transferred to MS medium supplemented with 0, 100 or 200 mM NaCl in petridish. Shoot length, root length, leaf surface area, fresh weight, dry weight and relative water content (RWC) of the seedlings were measured after 30 days of growth. At one month, WT and kanamycin-positive T_1 seedlings were transferred into beakers containing 1/2 MS hydroponic culture and were maintained therein for 20 days prior to receiving stress treatments of 0, 100 or 200 mM NaCl. Leaf, shoot and root tissues were collected 45 days post-treatment and subjected to physiological and biochemical analyses. For ROS analysis 15-day-old WT and kanamycin-positive T_1 seedlings were transferred to MS medium supplemented with 0, 100, 200 and 300 mM NaCl in culture jars and maintained for 60 days.

Membrane Stability Index (MSI)

MSI was determined as described by Hayat et al. [39]. In brief, MSI was estimated by taking leaf samples in 10 ml of double-distilled water in two sets. One set was heated at 40°C for 30 min in a water bath and measured for electrical conductivity (C_1). The second set was boiled at 100°C for 10 min prior to having its conductivity (C_2) measured. MSI was calculated according to the formula [1] as described by Sairam [40]:

$$MSI = (1 - C_1/C_2) \qquad (1)$$

Electrolyte Leakage

Electrolyte leakage was measured as described by Lutts et al. [41]. Young leaves (3rd leaf from the top) of similar size were collected from four plants for each treatment and washed thoroughly with deionised water to remove surface-adhered electrolytes. The samples were placed in closed vials containing 10 ml of deionised water, incubated at 25°C on a rotary shaker for 24 h. Subsequently, the electrical conductivity of the solution (L_t) was determined using a SevenEasy conductivity meter (Mettler Toledo AG 8603, Switzerland). The samples were then autoclaved at 120°C for 20 min and cooled at 25°C before determining the final electrical conductivity (L_0). Electrolyte leakage was defined as follows:

$$\text{Electrolyte leakage (\%)} = (L_t/L_0) \times 100 \quad (2$$

Lipid Peroxidation

Lipid peroxidation was estimated by determining the concentration of malondialdehyde (MDA) produced by the thiobarbituric acid (TBA) reaction; this experiment was performed following the method described by Hodges et al. [42]. Leaf material (0.5 g) was homogenised in 15 ml of an 80% acetone solution. In one set, 1 ml of extract was added to 1 ml 0.5% (w/v) TBA in 20% (w/v) TCA. In another set, TBA was excluded. The mixture was incubated at 90°C for 30 min and then cooled at room temperature. The samples were centrifuged at 4,000 rpm for 3 min and the absorbance of the supernatants at 440 nm, 532 nm and 600 nm was determined. The MDA concentration was derived according to the following equations:

$$1)\ A = [(Abs_{532+TBA} - Abs_{600+TBA}) - (Abs_{532-TBA} - Abs_{600-TBA})] \quad (3)$$

$$2)\ B = [(Abs_{440+TBA} - Abs_{600-TBA})\ 0.0571] \quad (4)$$

$$3)\ \text{MDA equivalents (nmol gm}^{-1}) = (A - B/15{,}700)10^6 \quad (5)$$

In Vivo Localization and Quantification of O_2^- and H_2O_2 Content in the Leaves of T_1 Plants

In vivo detection of O_2^- and H_2O_2 was accomplished by histochemical staining with nitro blue tetrazolium (NBT) and 3, 3'- diaminobenzidine (DAB) as described by Shi et al. [43]. The presence of O_2^- and H_2O_2 in transgenic and WT leaves exposed to salt stress was detected by immersing the leaf samples in room temperature solutions of NBT (1 mg ml^{-1}) and DAB (1 mg ml^{-1}, pH 3.8) in 10 mM phosphate buffer (pH 7.8). To detect O_2^-, immersed leaves were illuminated for 12 h until blue spots appeared, which are indicative of formazan precipitates. To determine the localisation of H_2O_2, immersed leaves were incubated in the light at room temperature for 24 h until brown spots became visible; these spots occur due to the reaction of DAB with H_2O_2. Following incubation, the leaf chlorophyll was bleached in absolute ethanol to enable visualisation of the blue and brown spots.

O_2^- content was determined according to Liu et al. [44]. Leaf tissue was extracted in 10 ml of 65 mM potassium phosphate buffer (pH 7.8) and centrifuged at 5,000 ×g for 10 min. The reaction mixture containing 0.9 ml of 65 mM phosphate buffer (pH 7.8), 0.1 ml of 10 mM hydroxylamine hydrochloride, and 1 ml of the extract was incubated at 25°C for 20 min. Then 17 mM sulfanilamide and 7 mM -naphthylamine were added, further incubated at 25°C for 20 min and the absorbance was read at 530 nm. A standard curve (10–200 nmol) was prepared with $NaNO_2$, to calculate the production rate of O_2^-.

The H_2O_2 content in leaf samples was measured as described by Mukherjee and Choudhuri [45]. Leaf tissue was extracted with cold acetone to determine the H_2O_2 levels. 2 ml of the extract was mixed with 0.5 ml of 0.1% titanium dioxide in 20% (v:v) H_2SO_4 and the mixture was then centrifuged at 6,000 xg for 15 min. The intensity of yellow colour of the supernatant was measured at 415 nm and the concentration of H_2O_2 was calculated against the standard curve.

Measurement of Proline, Soluble Sugar and Amino Acid Content

Free proline content in the leaves was determined using acid ninhydrin as previously described [46] with minor modifications. Plant tissue (100 mg) was homogenised in 1.2 ml 3% aqueous sulphosalicylic acid and centrifuged at 13000 rpm for 10 min. After centrifugation, 500 µl of supernatant was diluted to 1 ml with distilled water, reacted with 1 ml of 2% ninhydrin in acetone and incubated at 90°C for 1 h. The samples were cooled on ice and 2 ml of toluene was added and vortexed for 2 min. The upper phase was aliquoted to determine absorbance at 520 nm in a T80+ UV–vis spectrophotometer (PG Instruments Ltd., U.K.). Proline content was calculated by comparing the value against a standard curve derived from known concentrations of L-proline (Sigma Aldrich, USA) and expressed as µg/mg of fresh weight. Total soluble sugars were analysed by treating 0.1 ml of the alcoholic extract with 3 ml of freshly prepared anthrone reagent (150 mg anthrone in 100 ml 72% (v/v) H_2SO_4). This mixture was placed in a boiling water bath for 10 min as described by Irigoyen et al. [47]. After the mixture was cooled, absorbance at 620 nm was measured with a T80+ UV–vis spectrophotometer. A calibration curve was constructed with glucose in the range 20–400 µg/ml (Sigma Aldrich, USA). Total amino acid content was determined as described by Shukla et al. [48]. Specifically, 1 ml of plant extract was treated with 1 ml of 0.2 M citrate buffer (pH 5.0), 1 ml of 80% ethanol and 1 ml ninhydrin (1%), followed by incubation at 95°C for 15 min. The samples were cooled and the absorbance at 570 nm was measured [49] with a T80+ UV–vis spectrophotometer.

Na^+, K^+ and Ca^{2+} Ion Content Analysis

Ion content was determined via the method described by Shukla et al. [48]. Plant tissue (0.2 g) was digested in 4 ml of a solution of perchloric and nitric acids (3:1). The solution was dried on a hot plate at 90°C and reconstituted to 25 ml with deionised water before filtration through a 0.2 µm microfiber filter. Xylem sap was collected according to the method described by Olias et al. [27]. The ion content of digested plant tissues (root, stem and leaf) and xylem sap was measured with an inductively coupled plasma optical emission spectrometer (Optima 2000DV, Perkin Elmer, and Germany).

Statistical Analyses

Each experiment was performed three times and data from 10 plants were recorded. One-way ANOVA with replicates was performed in Microsoft Excel. C.D. values were calculated at $P = 0.05$ to determine the significance of difference between the means of WT and transgenic plants of each treatment group. Mean values that were significantly different within treatment from each other are indicated by different letters. The SD was calculated to show the variation in the replicates.

RESULTS

Sbsos1 Encodes a Putative Na^+/H^+ Antiporter

The cDNA fragment spanning the entire open reading frame of SbSOS1 was cloned and sequenced (GenBank accession number: EU879059). The SbSOS1 cDNA was 3,774 bp long, contained a 90 bp 5′ UTR, a 3,480 bp open reading frame and a 204 bp 3′ UTR region. The cDNA encoded a polypeptide of 1,159 amino acid residues with a molecular mass of 128.4 kDa and an isoelectric point of 6.24. Hydrophobicity plot analysis predicted the presence of 11 strong transmembrane helices, beginning with the N-terminus outside and continuing till 447 residues. The remaining C-terminal amino acid residues included a hydrophilic region and were predicted to reside in the cytoplasm. The SbSOS1 protein sequences were compared with the conserved domain in NCBI database and two conserved regions were identified: Nhap, an Na^+/H^+ antiporter spanning the transmembrane domain (amino acids 42 to 447) and a cyclic nucleotide-binding domain (cNMP; amino acids 753 to 857) that is located at the C-terminal tail (Figure 1a). The PSIPRED protein structure prediction server was used to predict the secondary structure, which included 55 coils, 45 alpha-helices and 18 beta-strands. Amino acid alignments of the SbSOS1 sequence revealed a high degree of similarity to Suaeda japonica SjSOS1 (90%), Chenopodium quinoa CqSOS1 (83%), Mesembryanthemum crystallinum McSOS1

(75%), Populus euphratica PeSOS1 (68%), Solanum lycopersicum SlSOS1 (66%), Oryza sativa OsSOS1 (62%), Thellungiella halophila ThSOS1 (61%), Brassica napusBnSOS1 (61%), Triticum aestivum TaSOS1 (60%) and Arabidopsis thaliana AtSOS1 (57%). Phylogenetic relationships between these genes were examined by comparing the SbSOS1 protein to 21 other species (Figure 1b), which revealed that SbSOS1 was in close proximity to other halophytes, Suaeda japonica and Chenopodium quinoa at 100 bootstrap values resolving Mesembryanthemum crystallinum at another branch of 99 bootstrap values. Similarly, other members of the cruciferae and poaceae families resolve their members in the same cluster, which is indicative of evolutionary conservation among the members of different families.

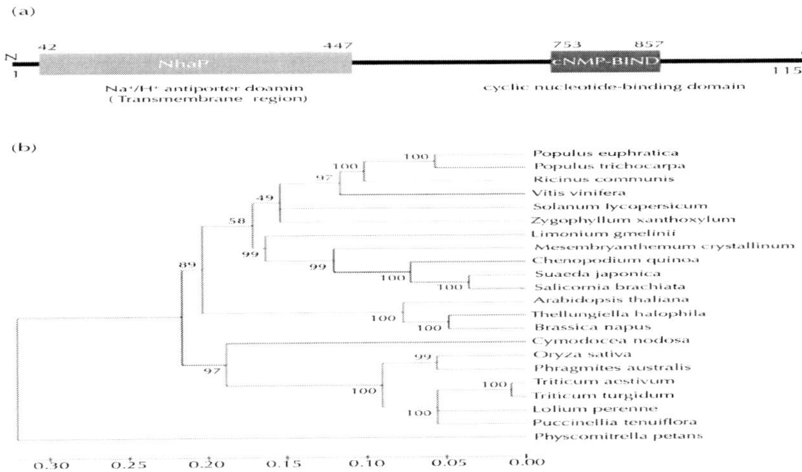

Figure 1: (a) Schematic diagram showing the transmembrane domains and cyclic nucleotide-binding domain. N and C indicates N-terminal and C-terminal end of the SbSOS1, respectively. (b) The phylogenetic relationship of SbSOS1 with SOS1 from other plant species. The phylogenetic tree was constructed using MEGA ver. 4.0 and the bootstrap values were calculated from 100 replicates and are shown next to branches. The higher bootstrap value signifies resilience in the phylogenetic position of the protein. The scale bar indicates substitutions per site. The protein sequences used for construction of the phylogenetic tree are as follows:Suaeda japonica (BAE95196.1), Chenopodium quinoa (ACN66494.1), Mesembryanthemum crystallinum (ABN04858.1), Limonium gmelinii (ACF05808.1), Ricinus communis(XP_002521897.1), Populus trichocarpa (XP_002315837.1), Vitis vinifera (ACY03274.1),Populus eu-

phratica (ABF60872.1), Zygophyllum xanthoxylum (ACZ57357.1), Solanum lycopersicum (BAL04564.1), Cymodocea nodosa (CAD20320.1), Phragmites australis(BAF41924.1), Oryza sativa Japonica (AAW33875.1), Triticum turgidum (ACB47885.1),Puccinellia tenuiflora (BAK23260.1), Triticum aestivum (CAX83738.1), Brassica napus(ACA50526.1), Lolium perenne (AAY42598.1), Arabidopsis thaliana (AF256224.1),Physcomitrella patens (CAM96566.1) and Thellungiella halophila (BAJ34642.1).

Differential Expression Analysis of Sbsos1 As a Result Of Exposure to Salt Stress

To study the expression of SbSOS1, real-time PCR was performed with cDNA from shoot and root tissues. Constitutively expressed SbSOS1 was 4.5 times higher in root tissue than in shoot tissue (Figure 2a). In root tissue, its expression was increased by 7-fold at 100 mM NaCl, relative to 0 mM NaCl; however, SbSOS1 expression remained constant at concentrations above 100 mM that were as high as 2.0 M (Figure 2b). In shoot tissue, SbSOS1 expression increased from 1.5- to 4-fold in response to increasing NaCl concentration (Figure 2c).

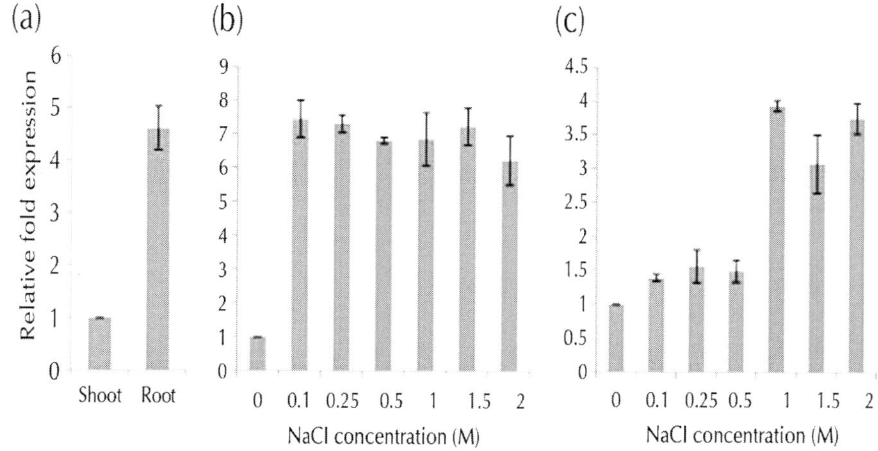

Figure 2: Real-time PCR analysis of the SbSOS1 gene in response to NaCl stress. (a) The relative expression of the SbSOS1 transcript in root tissue compared with shoot tissue at 0 mM NaCl, (b) the level ofSbSOS1 transcript in root tissue at different NaCl concentrations (M), (c) SbSOS1 transcript expression in shoot tissue at different NaCl concentrations (M). The relative expres-

sion in root tissue (a) was calculated using the C_T value of shoot tissue; in figures b and c, it was calculated using the values for 0 mM salt.

OVEREXPRESSION OF SBSOS1 INCREASES SALINITY TOLERANCE

Analysis of T_0 Transgenic Lines

The pCAMBIA2301-35S:SbSOS1 construct (Figure 3a) was introduced into tobacco plants for functional validation of SbSOS1 gene. Putative transgenic lines were selected on kanamycin-containing medium and were subsequently verified by GUS analysis (Figure 3f). Seventy individual transgenic lines derived from independent transgenic events were analysed by the GUS assay. GUS-positive plants were subsequently transferred to plastic pots containing garden soil and again to earthen pots after 15 days of hardening.

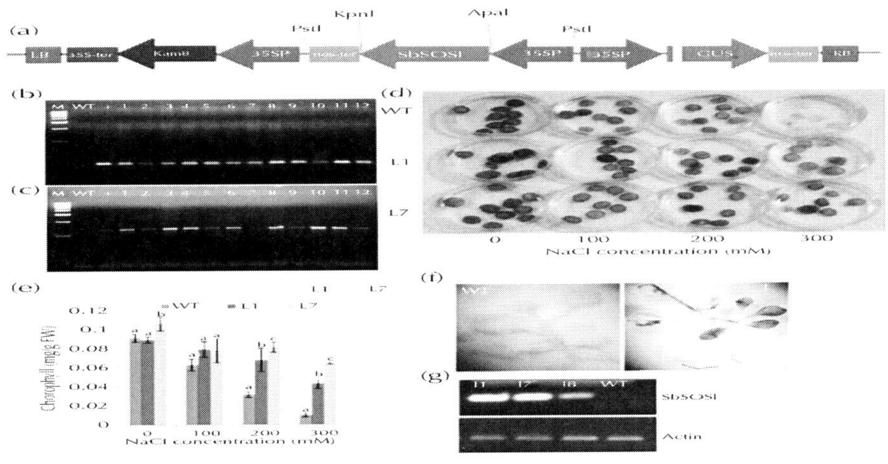

Figure 3: Characterisation of transgenic tobacco plants. (a) Schematic representation of the pCAMBIA2301 - 35S: SbSOS1 construct used to transform tobacco plants with the SbSOS1 gene. (b, c) Verification of transgenic lines via PCR analysis of the SbSOS1 gene with real-time PCR primers and gus-specific primers in T_0 plants. (d) Leaf disc assay of WT and transgenic lines (L1 and L7) at different NaCl concentrations in T_0 plants. (e) The graph represents

the mean and standard deviation (SD) of chlorophyll content in leaf discs of WT and transgenic lines (L1 and L7) at different NaCl concentrations. (f) GUS assay of T_1 plants showing positive GUS expression in the transgenic line and negative expression in WT plants. (g) Transcript levels of the SbSOS1 gene in transgenic lines (L1, L7 and L8) and WT T_1 plants via semiquantitative RT-PCR.

The GUS-positive transgenic lines were further verified by PCR with gene specific real-time PCR primers (RT R and RT F) and gus-specific primers (gusA F and gusA R) (Figure 3b, c). Leaf disc assays were performed to determine the salinity tolerance of T_0 transgenic plants. Leaf discs of uniform size from WT and transgenic plants (L1 and L7) were incubated in 0, 100, 200 and 300 mM NaCl. Leaf discs from WT plants began to turn yellow after 9 days and completely bleached after 17 days; in contrast, leaf discs from L1 and L7 plants remained relatively green (Figure 3d). To measure chlorophyll content in WT and transgenic plants, quantitative analyses were performed. The chlorophyll content in WT plants was significantly reduced as a function of increasing salt concentration. Of the two transgenic lines, L7 contained a larger amount of chlorophyll in the NaCl treatment and its chlorophyll content was not significantly affected in 200 and 300 mM NaCl relative to WT plants (Figure 3e).

Analysis of T_1 Transgenic Lines

T_0 seeds exhibited the expected 3:1 ratio of Kan^r/ Kan^s during germination in kanamycin-containing medium. Three independent T_1 transgenic lines were selected on the basis of GUS intensity and were further analysed for transgene expression via semi-quantitative RT-PCR (Figure3g). To study the effect of salt stress on germination, WT and T_0 seeds (L1, L7 and L8) were germinated in MS medium supplemented with 0, 50, 100, 150, 200 or 300 mM NaCl. Upon exposure to higher salt stress, the transgenic seeds exhibited better germination efficiency than WT seeds (Figure 4g). The efficiency of germination was reduced by increasing NaCl concentration for both WT and transgenic seeds (Figure 4a-g). In 300 mM NaCl, neither of the transgenic lines, nor the WT seeds, germinated until 15 days of seed inoculation (Figure 4f). In addition to seed germination assays, the growth of T_1 transgenic seedlings exposed to salt stress conditions were also examined (Figure 4h-j). T_0 seeds of WT and transgenic tobacco were allowed to germinate in MS medium for 8 days. Subsequently, T_1 seedlings were transferred to medium containing 0, 100 or 200 mM NaCl (Figure 4h-j). The T_1 transgenic lines exhibited significant enhancements in shoot and root length and leaf area relative to WT plants (Figure 5a-c). The transgenic lines also exhibited significant increases in both fresh and dry weight relative to WT (Figure 5d-e). The L7 transgenic line performed better in response to salt stress in terms leaf area, shoot and root length, relative water content and fresh and dry weight relative to both WT and other transgenic lines (Figure 4h-j, Figure 5a-f). The WT leaves exhibited dehydration in the presence of NaCl, whereas the transgenic lines were better hydrated and possessed higher relative water content (Figure 4h-j). Transgenic lines also exhibited better growth than their WT counterparts when subjected to salt stress in hydroponic culture (Figure 4k-m).

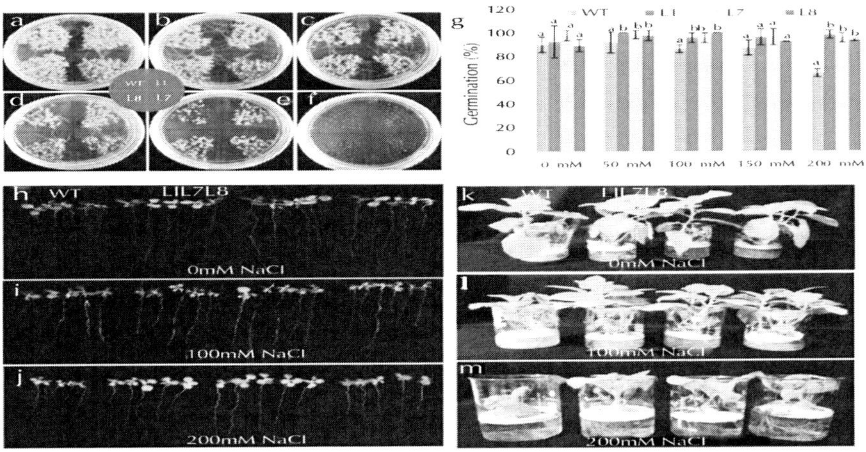

Figure 4: Phenotypic comparison of the growth of WT and T1 transgenic lines (L1, L7 and L8) overexpressing the SbSOS1 gene.(a-f) Germination of seeds from transgenic lines (L1, L7 and L8) and WT plants in (a) 0 mM, (b) 50 mM, (c) 100 mM, (d) 150 mM, (e) 200 mM, and (f) 300 mM NaCl. (g) The graph represents the per cent germination of transgenic lines (L1, L7 and L8) and WT plants in 0 mM, 50 mM, 100 mM, 150 mM, and 200 mM NaCl after 15 days. (H-j) Growth comparison of transgenic lines (L1, L7 and L8) and WT T_1 seedlings in (h) 0 mM, (i) 100 mM, and (j) 200 mM NaCl. (k-m) Growth of whole plants from transgenic lines (L1, L7 and L8) and WT plants at different NaCl concentrations: (k) 0 mM, (l) 100 mM and (m) 200 mM in hydroponic culture.

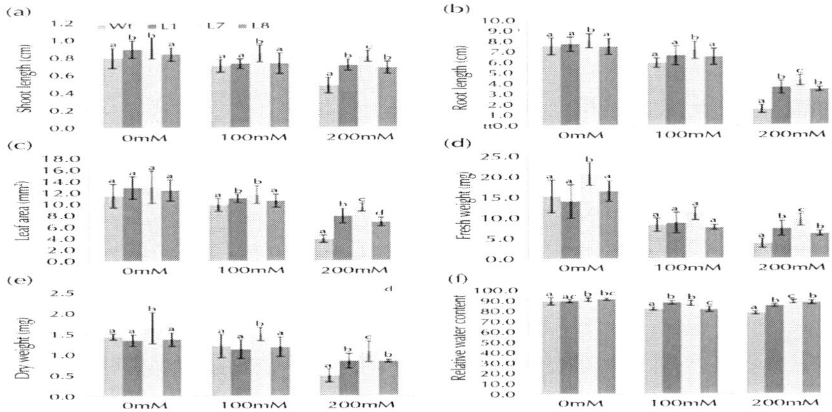

Figure 5: Comparison of growth parameters of 30-day-old seedlings from transgenic lines (L1, L7 and L8) and WT plants in 0 mM, 100 mM and 200 mM NaCl. (a) Shoot length, (b) root length, (c) leaf area, (d) fresh weight, (e) dry weight and (f) relative water content (RWC).

Electrolyte Leakage, Membrane Stability and MDA Content Analysis

Relative to WT plants, the transgenic lines exhibited significantly reduced electrolyte leakage during salt stress (Figure 6a). MSI analyses revealed that the cell membrane of lines L1 and L7 was more stable than WT plants (Figure 6b). Transgenic lines accumulated less MDA than WT in response to salt stress, which was indicative of reduced oxidative damage in plants overexpressingSbSOS1 (Figure 6c).

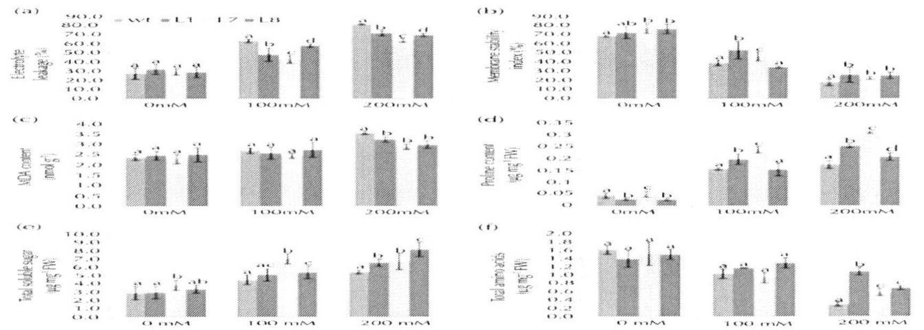

Figure 6: Electrolyte leakage (a), membrane stability index (b), MDA(c), proline (d), total soluble sugar (e), and amino acid (d) contents of transgenic lines (L1, L7 and L8) and WT plants grown in hydroponic culture with 0 mM, 100 mM, and 200 mM NaCl.

Proline, Total Soluble Sugar and Amino Acid Content Analysis

Proline, total soluble sugar and amino acid contents were measured in WT and transgenic plants to substantiate the function of the SbSOS1 gene. At 0 mM NaCl, proline content was nearly identical in WT and transgenic plants; however, in the presence of 100 and 200 mM NaCl, the transgenic plants had higher proline content relative to WT (Figure 6d). Transgenic lines and WT both exhibited a gradual accumulation of total soluble sugar by increasing NaCl concentration; however, transgenic lines exhibited accumulation in all treatments that was greater than that of WT (Figure6e). Whereas the total amino acid content decreased significantly in response to increased NaCl concentration for both WT and transgenic lines, the transgenic lines exhibited a more modest decrease in amino acid content relative to WT treated with 200 mM NaCl (Figure 6f).

In vivo Localisation and Quantification of O_2^- and H_2O_2 in T_1 Plant Leaves

Control, non-stressed leaves from WT and transgenic plants exhibited similar staining. In contrast, WT leaves exhibited more staining than transgenic lines in the presence of salt (Figure 7a, c). This result demonstrates that WT leaves accumulated more O_2^- and H_2O_2 than transgenic lines, confirming that SbSOS1 helps to minimise NaCl-induced oxidative stress in situ. Further these results were confirmed by quantification of O_2^- and H_2O_2 accumulation (Figure 7b, d). Both $O_2^-$and H_2O_2 contents were found higher in the WT plants compared to transgenic lines (Figure 7b, d).

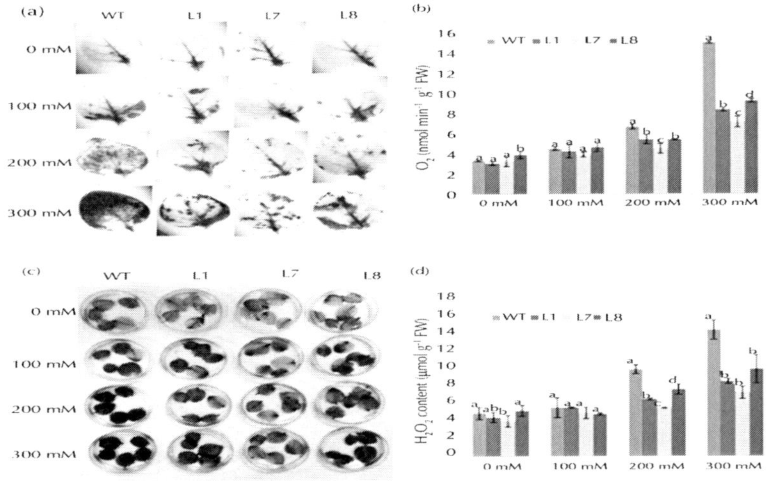

Figure 7: In vivo localisation and quantification of $O_2^-$and H_2O_2 in leaves of 35S-SbSOS1 and WT plants. (a) Localisation of O_2^- by NBT staining, (b) Quantification of O_2^- content, (c) Localisation of H_2O_2 by DAB staining and (d) Quantification of H_2O_2 content.

ION CONTENT ANALYSIS

Na^+ Content

At 0 mM NaCl (i.e., normal conditions) transgenic and WT plants exhibited approximately equal Na^+ content in root, stem and leaf tissues individually. However, in the presence of 100 and 200 mM NaCl; root and leaf tissues exhibited significant reductions in Na^+ ion content in all three transgenic lines (L1, L7 and L8) relative to WT plants (Figure 8). However, the accumulation of Na^+ions was found to be significantly higher in stem tissues in transgenic lines compared with WT. In general, stem tissues exhibited a 5-fold increase in Na^+ content compared with root and leaf tissues. All transgenic lines exhibited higher concentrations of Na^+ than WT in xylem sap under both normal and stress conditions (Figure 9).

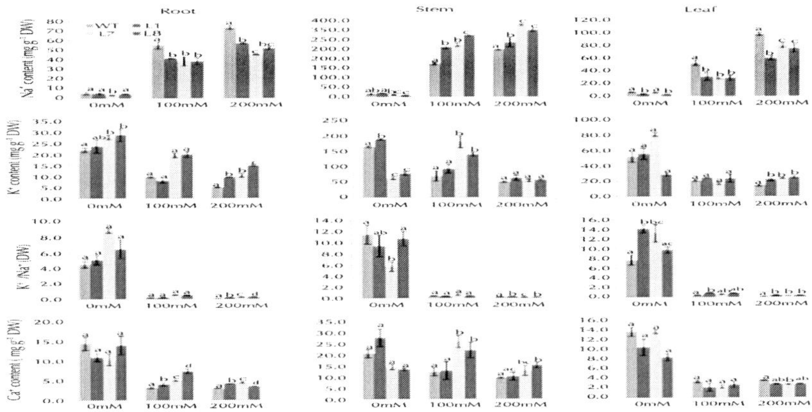

Figure 8: Na^+, K^+ and Ca^{2+} contents in root, shoot and leaf tissues of transgenic lines (L1, L7 and L8) and WT plants grown in hydroponic culture with 0 mM, 100 mM, and 200 mM NaCl.Individual K^+/Na^+ ratios of different plant organs are also shown in the graphs.

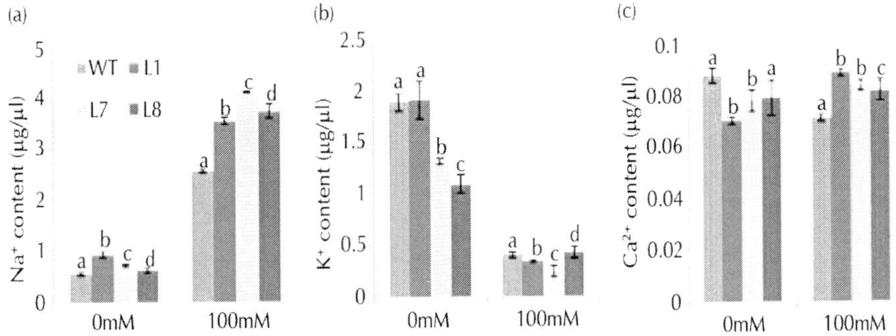

Figure 9: Na^+, K^+ and Ca^{2+} contents in xylem sap of transgenic lines (L1, L7 and L8) and WT plants grown in hydroponic culture with 0 and 100 mM NaCl.

K^+ Ion Content

K^+ content decreased as a function of increasing NaCl concentration in all tissues in both transgenic and WT plants (Figure 8). However, transgenic lines contained higher K^+ content than WT in root, stem and leaf tissues. The maximum accumulation observed in stem tissue was 6-fold greater than that in leaf and root tissues. Transgenic lines and WT plants accumulated equivalent levels of K^+ in xylem sap at 100 mM NaCl (Figure 9). Relative to WT, transgenic lines showed an increased K^+/Na^+ ratio in root and leaf tissues at all NaCl concentrations. In stem tissues, the transgenic lines and WT exhibited ratios that were nearly equal.

Ca^{2+} Ion Content

The transgenic lines showed greater Ca^{2+} content than WT in both root and stem tissues at both NaCl concentrations (Figure 8). Stem tissues exhibited a 2-fold increase in Ca^{2+} relative to root and leaf tissues. However, transgenic lines showed a slightly reduced Ca^{2+} content in leaf tissue relative to WT in stress conditions (Figure 8). Transgenic lines possessed a higher Ca^{2+} content in xylem sap compared with WT under stress conditions (Figure 9).

DISCUSSION

Salinity impedes plant growth and survival. Plants survive various stresses via mechanisms that operate on the cellular and molecular level. Salt stress can potentially perturb ion equilibrium in the plant system, and antiporters play an important role in adjusting cation concentrations to maintain cellular homeostasis. To avoid the toxic effects of salt, Na^+ ions are compartmentalised into vacuoles, or extruded to the apoplast, and remain at low concentrations in the cytosol [50-53]. In response to salt stress, plants maintain high concentrations of K^+ and low concentrations of Na^+in the cytosol by regulating the expression and activity of K^+ and Na^+ transporters and H^+ pumps[54]. Until now, few reports have described the in planta overexpression of the SOS1 gene fromArabidopsis [21, 29]; Thellungiella [30] and Puccinellia tenuiflora[31]. Subsequently, a small number of reports have described the SOS1 genes from different plant sources, which were characterised primarily in yeast systems [25, 26, 28, 55, and 56]. To elucidate its function in salt tolerance, we have cloned the SbSOS1 gene from the halophyte Salicornia brachiata. The SbSOS1 protein is predicted to contain 11 strong transmembrane regions (TM) at the amino-terminus, and a large cytoplasmic region at the carboxyl terminus. The N-terminal portion that forms the membrane pore exhibited the greatest sequence homology, whereas C-terminal domains were less similar. The TM3, TM5, TM8, TM10 and TM11 regions are highly conserved. Amino acid alignment of SbSOS1 revealed a high degree of similarity to Suaeda japonica SjSOS1, Chenopodium quinoa CqSOS1, Mesembryanthemum crystallinum McSOS1, Populus euphratica PeSOS1, Solanum lycopersicum SlSOS1, Oryza sativaOsSOS1, Thellungiella halophila ThSOS1, Brassica napus BnSOS1, Triticum aestivum TaSOS1 andArabidopsis thaliana AtSOS1 by 90%, 83%, 75%, 68%, 66%, 62%, 61%, 61%, 60% and 57%, respectively. A phylogenetic analysis revealed that SbSOS1 is closely related to Caryophyllales (M. crystallinum, S. japonica, C. quinoa and Limonium gmelinii Kuntze) (Figure 1b).

Real-time PCR revealed that the expression of SbSOS1 remained higher in root tissues than in shoot tissues (Figure 2a). Similarly, Kant et al. [57] observed a 7-fold increase in ThSOS1transcript expression in roots relative to shoots. In shoot tissue, the expression was slightly increased in response to 0.5 M NaCl and increased significantly at higher concentrations. In contrast, even at 100 mM, expression increased by 7-fold in the root and remained constant thereafter (Figure 2b, c). Maughan et al. [58] also observed that SOS1 expression was high at low salt concentration in the root tissue of C. quinoa, which indicates that the SOS1 gene is hyper-inducible in the roots of halophytic plants at even low salt concentrations. Similarly, the SOS1 gene from other plants was expressed more highly in the root in response to salt stress [17, 27]. P. euphratica PeSOS1 mRNA remained consistently high in shoot tissue when treated with 200 mM for 0, 24 or 48 h [28].

Transgenic plants overexpressing SbSOS1 exhibited increased salt tolerance and higher rates of seed germination. Furthermore, in response to salt stress, transgenic plants exhibited increases in root length, shoot length, leaf area, fresh weight, dry weight, relative water content (RWC) and chlorophyll content relative to WT (Figures 3d-e: 4a-j: 5a-f). NaCl also causes the dehydration stress in plants because of the physiological drought stress. Munns and Tester [59] have mentioned that the salinity stress first causes the rapid osmotic stress followed by slow Na^+ stress. The rapid osmotic stress in plants consecutively follows rapid dehydration. Oh et al. [30] studied that Thellungiella Thsos1-RNAi lines were highly salt sensitive and faster Na+ accumulating and showed severe water loss in shoots under salt stress and slower removal of Na+ from the roots compared with the WT. In the present study, the water retention was found higher in transgenic lines, confirming that SbSOS1 help in water retention during Na+ stress to prevent from physiological desiccation. The chlorophyll content was found higher in Arabidopsis transgenic plants than WT in the presence of salt stress. The increase in chlorophyll content indicates that SOS1transgenic plants have better light harvesting and photosynthetic capacities than WT during salt stress [21]. The morphological features of the transgenic plants were similar to those of WT plants under normal growth conditions.

In the present study, transgenic plants experiencing salt stress exhibited low electrolyte leakage and greater cell membrane stability (Figure 6a-b). This observation was corroborated by the reduction in MDA. Analysis of the in vivo localisation of ROS via NBT and DAB revealed thatSbSOS1-overexpressing lines maintained less ROS in response to salt treatment, which indicates a reduced potential for oxidative damage resulting from salt stress. Katiyar-Agarwal et al. [60] has reported that SOS1 interacts with RCD 1 (regulator of oxidative stress responses) via its predicted cytoplasmic tail to regulate the expression of ROS-scavenging genes. The relative abundance of proline, total soluble sugar and amino acids are important biochemical indicators of salinity in plants [61]. Proline protects the plants in response to salt stress by regulating the accumulation of usable nitrogen, which contributes to membrane stability and mitigates the disruptive effect of NaCl. Transgenic plants overexpressing SbSOS1 accumulate higher proline than WT plants during salt stress (Figure 6d). Total soluble sugar and amino acids are known to accumulate in plants during salt stress. SbSOS1 transgenic lines accumulated higher levels of total soluble sugar and amino acids relative to WT plants (Figure 6e-f). Soluble sugars help transgenic plants to survive salt stress by regulating osmoregulation and by acting as molecular chaperones that stabilise protein structure [62]. Similarly, the increased accumulation of amino acids suggests a higher rate of protein synthesis in plants experiencing salt stress.

Studying ion partitioning in different plant tissues is important to fully understand the function of the transporter genes in that system. Previously, Shi et al. [11] observed that, at 100 mM NaCl, sos1 mutant plants accumulated more Na^+ in the shoot and xylem sap than WT, which suggests that SOS1 functions in retrieving Na^+ from the xylem stream in response to severe salt stress. Additionally, they overexpressed AtSOS1 in Arabidopsis, which led to a reduced accumulation of Na^+ in the shoot and xylem sap at 100 mM NaCl [21]. However, both of these studies were performed in Arabidopsis plants. As described by Olias et al. [27], Arabidopsis plants limit the precise dissection of the relative content of Na^+ in stem vs. leaf; however, the ability to make this distinction is critical in assessing the role of SOS1 in xylem loading/unloading, as well as Na^+export in roots and retention in stems. Therefore, the present study was performed to examine the ion partition in different plant organs by overexpressing the SbSOS1 gene in tobacco, which has more convenient anatomy [29]. Transgenic tobacco plants accumulated lower Na^+ content in root and leaf at higher salt stress; in contrast, Na^+ content was higher in stem and xylem sap relative to WT, which suggests a role in Na^+ loading to xylem from root and leaf tissue. Whereas, Shi et al.[11] suggested that SOS1 function in retrieving Na^+ from the xylem stream under severe salt stress, our study supports the hypothesis of Olias et al. [27], which suggests that the SOS1 functions in Na^+ loading to xylem from root and leaf tissues.

Potassium is important to the plant system because it plays a role during plant growth and development. In response to NaCl stress, the SbSOS1 transgenic lines showed higher K^+ content in root and leaf tissue compared to WT plants. Transgenic lines showed higher ratios of K^+/Na^+relative to WT plants in root and leaf tissues at all NaCl concentrations. Even under normal conditions, the transgenic lines exhibited high K^+/Na^+ ratios. However, in stem tissues, there was no significant difference between transgenic lines and WT plants (Figure 8). Wu et al. [13] showed that sos1 mutants are hypersensitive to Na^+ stress and defective in high-affinity K^+ uptake; furthermore, these two phenotypes always co-segregated. Shi et al. [11] and Quintero et al. [24] demonstrated that a yeast strain (AXT3K) expressing AtSOS1 accumulated less Na^+ and exhibited improved K^+ status compared with empty vector controls. It has also been reported that AtSOS1does not suppress defects of the yeast K^+ transport mutant (trk1 trk2) [11, 24, 25] and aids indirect uptake and transport of K^+ ions. Accordingly, sos1 mutant plants displaced defective K^+ uptake at low external concentrations, and their transport capacity was not altered by overexpression of the SOS1 protein at reduced K^+ concentrations. However, Cymodocea SOS1 promoted uptake of K^+ in bacteria [25]. Similarly, Wu et al. [28] revealed that PeSOS1-expressing bacteria maintained lower Na^+ and higher K^+ in the presence of 200 mM NaCl relative to vector alone, which resulted in an increase in the K^+/Na^+ ratio.

Stress and other extracellular stimuli influence intracellular Ca^{2+} concentration [63, 64]. Ca^{2+}-mediated signal transduction is considered one of the earliest events in salt signalling and plays an essential role in ion homeostasis that permits salt tolerance in plants [54]. The SbSOS1 transgenic lines showed higher Ca^{2+} content than WT in root and stem tissues at both NaCl concentrations (Figure 8). However, transgenic lines showed slightly lower Ca^{2+} content in leaf tissues relative to WT plants exposed to salt stress (Figure 8). Transgenic lines exhibited increases in Ca^{2+} content relative to WT plants in xylem sap at stress conditions (Figure 9). Similar to our studies, Guo et al. [65] proposed a link between Ca^{2+} transport and SOS1 activity, and also demonstrated that thesos mutant alters the activity of Ca^{2+} transport systems in both normal and NaCl stress conditions. In WT plants, the Ca^{2+} influx rate increased after a 5 min treatment with NaCl, resulting in an increase in cytoplasmic Ca^{2+}. During the same time interval, Ca^{2+} efflux was observed in sos1mutants, which resulted in a decrease in cytoplasmic Ca^{2+}. The variability of Ca^{2+} content in theSbSOS1 transgenic lines, compared with WT, can be attributed to changes in Ca^{2+} transporter activity. Therefore, it is possible that SbSOS1 affects the Ca^{2+} transporter.

CONCLUSIONS

In conclusion, the SbSOS1 gene from the extreme halophyte S. brachiata was cloned which showed expression during salt stress. SbSOS1 expression was consistently higher in the root tissue than shoot tissue and was up regulated by salt stress. Overexpression of SbSOS1 in tobacco conferred a high degree of salt tolerance, enhanced plant growth and altered physiological and biochemical parameters in response to salt stress. Furthermore, the transgenic lines exhibited lower levels of MDA and ROS accumulation as a result of reduced cytosolic Na^+ content and oxidative damage. Transgenic tobacco plants accumulated reduced Na^+ content in root and leaf tissues, but higher Na^+ content in stem and xylem sap, relative to WT, which resulted from enhanced Na^+ loading in xylem from root and leaf tissues. This finding demonstrates that, in addition to the Na^+ efflux outside the plasma membrane, SbSOS1 transporter also helps to maintain differing concentrations of Na^+ in various organs. These results support a broader role ofSbSOS1 in planta and suggest that this gene can be utilised to develop salt-tolerant transgenic crops in the future.

AUTHORS' CONTRIBUTIONS

NSY has carried out gene cloning, raised transgenic plants and analyzed them for functional validation. PSS and AJ carried out the biochemical assay of transgenic lines. PKA and BJ designed, coordinated the experiments and finalized MS. All authors read and approved the final manuscript.

ACKNOWLEDGEMENTS

The authors are thankful to DST, New Delhi and Council of Scientific & Industrial Research, New Delhi (NWP-020), for financial assistance. NSY is thankful to CSIR for Senior Research Fellowship.

REFERENCES

1. Epstein E, Bloom AJ: Mineral nutrition of plants: principles and perspectives. 2nd edition. Sunderland, Massachusetts: Sinauer Associations; 2005:342-351.
2. FAO: FAO land and plant nutrition management service. 2008. http://www.fao.org/ag/agl/agll/spush/
3. Hasegava M, Bressan R, Pardo JM: The dawn of plant salt tolerance genetics. Trends Plant Sci 2000, 5:317-319.
4. Wyn Jones RG, Pollard A: Encyclopedia of plant physiology. Berlin: Springer- Verlag; 1983.

5. Hernández JA, Olmos E, Corpas FJ, Sevilla F, Del Río LA: Salt-induced oxidative stress in chloroplasts of pea plants. Plant Sci 1995, 105:151-167.
6. Huang GT, Ma SL, Bai LP, Zhang L, Ma H, Jia P, Liu J, Zhong M, Guo ZF: Signal transduction during cold, salt, and drought stresses in plants. Mol Biol Rep 2011, 39:969-987.
7. Agarwal PK, Jha B: Transcription factors in plants and ABA dependent and independent abiotic stress signalling. Biol Plant 2010, 54:201-212.
8. Niu X, Bressan RA, Hasegawa PM, Pardo JM: Ion homeostasis in NaCl stress environments. Plant Physiol 1995, 109(3):735-742.
9. Rajendran KA, Tester M, Roy SJ: Quantifying the three main components of salinity tolerance in cereals. Plant Cell Environ 2009, 32:237-249.
10. Gaxiola RA, Rao R, and Sherman A, Grifasi P, Alpier SL, Fink GR: The Arabidopsis thaliana proton transporters, AtNHX1 and Avp1, can function in cation detoxification in yeast. Proc Natl Acad Sci USA 1999, 96:1480-1485.
11. Shi H, Quintero FJ, Pardo JM, Zhu JK: The putative plasma membrane Na+/H+ antiporter SOS1 controls long-distance Na+ transport in plants. Plant Cell 2002, 14:465-477.
12. Blumwald E, Aharon GS, Apse MP: Sodium transport in plant cells. Biochim Biophys Acta 2000, 1465:140-151.
13. Wu SJ, Ding L, Zhu JK: SOS1, A genetic locus essential for salt tolerance and potassium acquisition. Plant Cell 1996, 8:617-627.
14. Zhu JK, Liu J, Xiong L: Genetic analysis of salt tolerance in Arabidopsis: evidence for a critical role of potassium nutrition. Plant Cell 1998, 10(7):1181-1191.
15. Liu J, Zhu JK: A calcium sensor homolog required for plant salt tolerance. Science 1998, 280:1943-1945.
16. Liu JP, Ishitani M, Halfter U, Kim CS, and Zhu JK: The Arabidopsis thaliana SOS2 gene encodes a protein kinase that is required for salt tolerance. Proc Natl Acad Sci USA 2000, 97:3730-3734.
17. Shi H, Ishitani M, Kim C, and Zhu JK: The Arabidopsis thaliana salt tolerance gene SOS1 encodes a putative Na+/H+ antiporter. Proc Natl Acad Sci USA 2000, 97:6896-6901.
18. Pardo JM, Reddy MP, Yang S, Maggio A, Huh GH, Matsumoto T,

Coca MA, Paino-D'Urzo M, Koiwa H, and Yun DJ: Stress signaling through Ca2+/calmodulin-dependent protein phosphatase calcineurin mediates salt adaptation in plants. Proc Natl Acad Sci USA 1998, 96:4718-4723.

19. Halfter U, Ishitani M, and Zhu JK: The Arabidopsis SOS2 protein kinase physically interacts with and is activated by the calcium-binding protein SOS3. Proc Natl Acad Sci USA 2000, 97:3735-3740.
20. Quan RD, Lin H, Mendoza I, Zhang Y, Cao W, Yang Y, Shang M, Chen S, Pardo JM, Guo Y:SCABP8/CBL10, a putative calcium sensor, interacts with the protein kinase SOS2 to protect Arabidopsis shoots from salt stress. Plant Cell 2007, 19:1415-1431.
21. Shi H, Lee BH, Wu SJ, Zhu JK: Overexpression of a plasma membrane Na+/H+ antiporter gene improves salt tolerance in Arabidopsis thaliana. Nat Biotechnol 2003, 21:81-85.
22. Gao XZ, Zhao RY, Zhang H: Overexpression of SOD2 increases salt tolerance of Arabidopsis. Plant Physiol 2003, 133:1873-1881.
23. Wu L, Fan Z, Guo L, Li Y, Chen ZL, Qu LJ: Overexpression of the bacterial nhaA gene in rice enhances salt and drought tolerance. Plant Sci 2005, 168:297-302.
24. Quintero FJ, Ohta M, Shi H, Zhu JK, Pardo JM: Reconstitution in yeast of the Arabidopsis SOS signaling pathway for Na+ homeostasis. Proc Natl Acad Sci USA 2002, 99:9061-9066.
25. Garciadeblás B, Haro R, Benito B: Cloning of two SOS1 transporters from the seagrass cymodocea nodosa. SOS1 Transporters from Cymodocea and Arabidopsis mediate potassium uptake in bacteria. Plant Mol Biol 2007, 63:479-490.
26. Martinez-Atienza J, Jiang X, Garciadeblas B, Mendoza I, Zhu JK, Pardo JM, Quintero FJ:Conservation of the salt overly sensitive pathway in rice. Plant Physiol 2007, 143:1001-1012.
27. Olías R, Eljakaoui Z, Li J, De Morales PA, Marín-Manzano MC, Pardo JM, Belver A: The plasma membrane Na+/H+ antiporter SOS1 is essential for salt tolerance in tomato and affects the partitioning of Na+ between plant organs. Plant Cell Environ 2009, 32:904-916.

28. Wu Y, Ding N, Zhao X, Zhao M, Chang Z, Liu J, Zhang L: Molecular characterization of PeSOS1, the putative Na+/H+ antiporter of populus euphratica. Plant Mol Biol 2007, 65:1-11.
29. Yue Y, Zhang M, Zhang J, Duan L, Li Z: SOS1 gene overexpression increased salt tolerance in transgenic tobacco by maintaining a higher K+/ Na+ ratio. J Plant Physiol 2012, 169:255-261.
30. Oh DH, Leidi E, Zhang Q, Hwang SM, Li Y, Quintero FJ, Jiang X, D'Urzo MP, Lee SY, Zhao Y, Bahk JD, Bressan RA, Yun DJ, Pardo JM, Bohnert HJ: Loss of halophytism by interference with SOS1 expression. Plant Physiol 2009, 151:210-222.
31. Wang X, Yang R, Wang B, Liu G, Yang C, Cheng Y: Functional characterization of a plasma membrane Na+/H+ antiporter from alkali grass (Puccinellia tenuiflora). Mol Biol Rep 2011, 38:4813-4822.
32. Gong Q, Li P, Ma S, Indu Rupassara S, Bohnert HJ: Salinity stress adaptation competence in the extremophile Thellungiella halophila in comparison with its relative Arabidopsis thaliana. Plant J 2005, 44:826-839.
33. Livak KJ, Schmittgen TD: Analysis of relative gene expression data using real-time quantitative PCR and the 2(−Delta Delta C(T)) method. Methods 2001, 25:402-408.
34. Töpfer R, Matzeit V, Gronenborn B, Schell J, Steinbiss HH: A set of plant expression vectors for transcriptional and translational fusions. Nucleic Acids Res 1987, 15:5890.
35. Horsch RB, Fry JE, Hoffmann NL, Eichholtz D, Rogers SG, Farley RT: A simple and general method for transferring genes into plants. Science 1985, 227:1229-1231.
36. Murashige T, Skoog F: A revised medium for rapid growth and bioassay with tobacco tissue cultures. Physiol Plant 1962, 15:473-497.
37. Jefferson RA: Assaying chimeric genes in plants: the GUS fusion system. Plant Mol Biol Rep 1987, 1987(5):387-405.
38. Lichtenthaler HK: Chlorophylls and carotenoids: pigments of photosynthetic biomembranes. Methods Enzym 1987, 148:350-382.
39. Hayat S, Hasan SA, Yusuf M, Hayat Q, Ahmad A: Effect of 28-homobrassinolide on photosynthesis, fluorescence and

antioxidant system in the presence or absence of salinity and temperature in Vigna radiate. Environ Exp Bot 2010, 69:105-112.
40. Sairam RK: Effect of homobrassinolide application on metabolism and grain yield under irrigated and moisture stress conditions of two wheat varieties. Plant Growth Regul 1994, 14:173-181.
41. Lutts S, Kinet JM, Bouharmont J: NaCl-induced senescence in leaves of rice (Oryza sativa L.) cultivars differing in salinity resistance. Ann Bot 1996, 78:389-398.
42. Hodges DM, DeLong JM, Forney CF, Prange RK: Improving the thiobarbituric acid-reactive-substances assay for estimating lipid peroxidation in plant tissues containing anthocyanin and other interfering compounds. Planta 1999, 207:604-611.
43. Shi J, Fu XZ, Peng T, Huang XS, Fan QJ, Liu JH: Spermine pretreatment confers dehydration tolerance of citrus in vitro plants via modulation of antioxidative capacity and stomatal response. Tree Physiol 2010, 30:914-922.
44. Liu F, Pang SJ: Stress tolerance and antioxidant enzymatic activities in the metabolism of the reactive oxygen species in two intertidal red algae grateloupia turuturu and palmaria palmate. J Exp Mar Bio Eco 2010, 328:82-87.
45. Mukherjee SP, Choudhuri MA: Implications of water stress-induced changes in the levels of endogenous ascorbic acid and hydrogen peroxide in vigna seedlings. Physiol Plant 1983, 58:166-170.
46. Bates LS, Waldern R, Teare ID: Rapid determination of free proline for water stress studies. Plant Soil 1973, 39:205-207.
47. Irigoyen JJ, Emerich DW, Sánchez-Díaz M: Water stress induced changes in concentrations of proline and total soluble sugars in nodulated alfalfa (medicago sativa) plants. Physiol Plant 1992, 84:67-72.
48. Shukla PS, Agarwal PK, Jha B: Improved salinity tolerance of Arachis hypogaea (L.) by the interaction of halotolerant plant-growth-promoting rhizobacteria. J Plant Growth Regul 2012, 31:195-206.
49. Chen K, Kurgan L, Rahbari M: Prediction of protein crystallization using collocation of amino acid pairs. Biochem Bioph Res Co 2007, 355:764-769.

50. Zhu JK: Salt and drought stress signal transduction in plants. Annu Rev Plant Biol 2002, 4:401-406.

51. Fukuda A, Chiba K, Maeda M, Nakamura A, Maeshima M, Tanaka Y: Effect of salt and osmotic stresses on the expression of genes for the vacuolar H+-pyrophosphatase, H+-ATPase subunit A, and Na+/H+ antiporter from barley. J Exp Bot 2004, 55:585-594.

52. Horie T, Schroeder JI: Sodium transporters in plants: Diverse genes and physiological functions. Plant Physiol 2004, 136:2457-2462.

53. Nakayama H, Horie T, Yonamine I, Shinmyo A, Yoshida K: Improving salt tolerance in plant cells. Plant Biotechnol 2005, 22:477-487.

54. Zhu JK: Regulation of ion homeostasis under salt stress. Curr Opin Plant Biol 2003, 6:441-445.

55. Oh DH, Gong Q, Ulanov A, Zhang Q, Li Y, Ma W, Yun DJ, Bressan RA, Bohnert HJ: Sodium stress in the halophyte Thellungiella halophyla and transcriptional changes in a thsos1-RNA interference line. J Integr Plant Biol 2007, 49:1484-1496.

56. Xu H, Jiang X, Zhan K, Cheng X, Chen X, Pardo JM, Cui D: Functional characterization of a wheat plasma membrane Na+/H+ antiporter in yeast. Arch Biochem Biophys 2008, 473:8-15.

57. Kant S, Kant P, Raveh E, Simon B: Evidence that differential gene expression between the halophyte, Thellungiella halophila, and Arabidopsis thaliana is responsible for higher levels of the compatible osmolyte proline and tight control of Na+ uptake in T. Halophila. Plant Cell Environ 2006, 29:1220-1234.

58. Maughan PJ, Turner TB, Coleman CE, Elzinga DB, Jellen EN, Morales JA, Udall JA, Fairbanks DJ, Bonifacio A: Characterization of salt overly sensitive 1 (SOS1) gene homoeologs in quinoa (chenopodium quinoa willd.). Genome 2009, 52:647-657.

59. Munns R, Tester M: Mechanisms of salinity tolerance. Annu Rev Plant Biol 2008, 59:651-681.

60. Katiyar-Agarwal S, Zhu JH, Kim KM, Agarwal M, Fu XM, Haung A, Zhu JK: The plasma membrane Na+/H+ antiporter SOS1 interacts with RCD1 and function in oxidative stress tolerance in Arabidopsis. Proc Natl Acad Sci USA 2006, 103:18816-18821.

61. Ashraf M, Harris PJ: Potential biochemical indicators of salinity tolerance in plants. Plant Sci 2004, 166:3-16.
62. Song SY, Chen Y, Chen J, Dai XY, Zhang WH: Physiological mechanisms underlying OsNAC5-dependent tolerance of rice plants to abiotic stress. Planta 2011, 234:331-345.
63. Knight H, Brandt S, Knight MR: A history of stress alters drought calcium signaling pathways in Arabidopsis. Plant J 1998, 16:681-687.
64. ReddyVS, Reddy AS: Proteomics of calcium-signaling components in plants. Phytochemistry 2004, 65:1745-1776.
65. Guo KM, Babourina O, Rengel Z: Na+/K+ antiporter activity of the SOS 1 gene: analysis and electrophysiological studies on Arabidopsis seedlings. Physiol Plant 2009, 137:155-165.

Chapter 10

Gender-Specific Selection on Codon Usage in Plant Genomes

Carrie-Ann Whittle*, Meghna R Malik, and Joan E Krochko

Plant Biotechnology Institute, National Research Council of Canada, 110 Gymnasium Place, Saskatoon, SK, S7N 0W9, Canada

ABSTRACT

Background

Currently, there is little data available regarding the role of gender-specific gene expression on synonymous codon usage (translational selection) in most organisms, and particularly plants. Using gender-specific EST libraries (with > 4000 ESTs) from *Zea mays* and *Triticum aestivum*, we assessed whether gender-specific gene expression *per se* and gender-specific gene expression level are associated with selection on codon usage.

Results

We found clear evidence of a greater bias in codon usage for genes expressed in female than in male organs and gametes, based on the variation in GC content at third codon positions and the frequency of species-preferred codons. This finding holds true for both highly and for lowly expressed genes. In addition, we found that highly expressed genes have greater codon bias than lowly expressed genes for both female- and male-specific genes. Moreover, in both species, genes with female-specific expression show a greater usage of species-specific preferred codons for each of the 18 amino acids having synonymous codons. A supplemental analysis of *Brassica napus*suggests that bias in codon usage could also be higher in genes expressed in male gametophytic tissues than in heterogeneous (flower) tissues.

Conclusion

This study reports gender-specific bias in codon usage in plants. The findings reported here, based on the analysis of 1 497 876 codons, are not caused either by differences in the biological functions of the genes or by differences in protein lengths, nor are they likely attributable to mutational bias. The data are best explained by gender-specific translational selection. Plausible explanations for these findings and the relevance to these and other organisms are discussed.

BACKGROUND

Although tissue-specific gene expression has been associated with bias in codon usage in certain multicellular organisms including humans, *Drosophila melanogaster*, and *Arabidopsis thaliana* [1-5], very little data currently exists for most organisms, particularly regarding the role of gender-specific tissues and gametes. One of the few studies addressing the effect of gender, in *Drosophila*, suggests that genes having a higher ratio of female to male expression have a greater bias in codon usage [4, 6]. A largely unstudied biological system where gender-specific gene expression could significantly alter codon usage is plants. Recent findings, in *A. thaliana*, have indicated that male gametes pass on a greater number of induced harmful mutations to their offspring, suggesting that mutations are subject to less selection in male tissues/gametes than in female tissues/gametes [7]. Such findings at the population level (short-term), suggest that gender could also impact the selective processes that alter molecular evolution in plants, including the usage of synonymous codons. Given that gender-specific selective pressures on codon usage could alter gene evolution and structure, and thereby influence population genetics, disease, and/or reproductive biology, and given the general lack of data to date, further investigation is warranted. Here, we focus on the evaluation of gender-specific codon usage in plants.

Nonrandom use of synonymous codons is a prevalent phenomenon observed in a diverse range of organisms [1, 8-17]. A bias in codon usage occurs when synonymous codons are not all used at the same frequency in coding DNA [14, 18, 19]. Such bias in codon usage could result from mutational pressure, as indicated by a positive correlation between the nucleotide content of third codon positions and adjacent introns[20-23], or from selective pressure. Selective pressure has been supported by two findings. Firstly, greater levels of bias in codon usage are generally associated with a greater frequency of "preferred" codons (those used most frequently in the most biased genes)[24], a trend that corresponds to the abundance and/or gene number of tRNA in bacteria, yeast, *C. elegans*, *Drosophila*,*Arabidopsis* and other organisms [1,8,9,12-15,25-27]. Secondly, bias in codon usage has been well correlated to the level of gene expression, with the greatest bias occurring in highly expressed genes [8, 10, 12, 28, and 29]. Each of these findings suggests that the use of preferred codons confers fitness benefits that enhance translational efficiency, a phenomenon particularly advantageous for the highly expressed genes [8, 12]. In this regard, gene expression level is an essential component of understanding gender-specific influences on codon usage.

The main challenges for comparing male and female codon usage relative to gene expression for plant species, where the availability of genomic DNA sequences is often limited, are obtaining sufficient coding DNA data to assess codon usage in those tissues and determining the level of gene expression. EST datasets provide an effective solution to both issues. In particular, EST data have proven to be an effective means of quantifying gene expression in a range of tissues as the extent of redundancy in ESTs reflects the abundance of mRNA in the tissue or cells from which the library was obtained [8, 10, 12, 30-32]. In addition, the increased availability of EST data, the long sequence length (200 to 800 bp) and the level of accuracy of the sequence data (from efficient sequence techniques and base-calling software, and the general use of only high quality reads) [33-35] makes it possible to study codon usage directly from EST sequences, even before the corresponding genomic sequences are available [36, 37]. In this regard, the recent availability of male- and female-specific EST libraries in plants provide an effective resource to better understand selection on codon usage and how it may be influenced by gender.

In the present study, the main goal was to assess whether gender-specific gene expression *per se*and gender-specific gene expression level are correlated with codon usage in *Zea mays* and*Triticum asestivum*. As a supplemental analysis, we compared the bias in codon usage for genes expressed only in gametophytic tissue (male germline cells, microspores) and in flower tissue (composed of the somatic and reproductive tissues) at both high and low expression levels in *Brassica napus*. Given that gene function and protein length [8, 10], have previously been found to influence codon usage, we also evaluated the role of those parameters within our analysis.

RESULTS

In order to compare codon usage relative to gender-specific expression, we collected data from sperm and egg EST libraries for *Z. mays* and anther and ovary libraries for *T. aestivum* (Table 1). Microspore and flower libraries were obtained for *B. napus*. In brief, we obtained gender/tissue-specific datasets in the following manner: 1) clustering and assembly of ESTs from each library using CAP3 [38], 2) identification of unigenes (from now forward referred to as "genes") having translation products that matched known or hypothetical proteins in *A. thaliana* (using BLASTX [39-41]), 3) extraction of genes with tissue-specific expression by comparisons between the two compared tissues for each species (using MEGABLAST) [41], and 4) determination of the expression level per gene (based on the number of ESTs, see Methods). The six tissue-specific sequence datasets obtained are: *Z. mays* sperm-specific genes (N = 955), *Z. mays* egg-specific genes (N = 946), *T. aestivum* anther-specific genes (N = 3326), *T. aestivum* ovary-specific genes (N = 1489), *B. napus* microspore-specific genes (N = 1675) and

B. napus flower-specific genes (*N*= 3181). Note that gender-specific genes represent those that are specific to a particular tissue or gamete (e.g. sperm) when compared to only one other tissue or gamete (e.g., egg,) and not relative to all tissues from the plant. Thus, these gene sets are larger than one would have found if the ESTs had been compared to all tissues of a plant. Subsequently, the GC content at third nucleotide positions (*GC3*) and the frequency of preferred codons (*Fpr*), each of which have been shown to be effective indicators of bias in codon usage [1,8,31], were determined for every gene from each of the tissue-specific datasets. Bias in codon usage was quantified using a single EST sequence to represent each gene (i.e., genes are represented by the longest EST in the contig or a singleton EST, see Methods). The data show, as described in detail below, that genes specific to female tissues and gametes have a greater bias in codon usage in both *Z. mays* and *T. aestivum*than genes expressed in male tissues and gametes. As well, male microspores have a greater bias than the heterogeneous tissues of the flower in *B. napus*.

Table 1: The cDNA libraries used in the present investigation, The number of ESTs from each library that match genes in *Arabidopsis thaliana*, and the associated number of contigs and singletons are provided

Species and tissue or cell type	Relevant properties of tissue or cells	Name of library in NCBI and contributor(s)	Total number of ESTs per library (March 2006)	Number of ESTs matching genes in Arabidopsis (e < 10-7)	Number of contigs identified using CAP3	Number of singletons identified using CAP3	Unigenes (contigs plus singletons)
Zea mays							
Sperm	Male	Zea mays sperm cell cDNA library (S. McCormick) Engel et al.[64]	5174	2457	384	781	1165
Egg	Female	Zea mays egg cell cDNA library (H. Yang and S. McCormick) Yang et al. [65]	4085	1957	264	863	1127
Triticum aestivum							
Anther-meiotic	Male	Wheat meiotic anther cDNA library (O. Anderson)	9139	6118	860	3058	3918
Ovary	Female	Ov (Genoplante)	5311	3699	813	1295	2108
Brassica napus							
Microspores (from 3 to 4 mm buds)	Male	65JKBNM0 (M. Malik and J. Krochko)	6539	5325	781	1620	2401
Buds and young flowers	Heterogeneous (male, female and somatic)	ADIS-MPIZ 021 (B. Weisshaar)	8747	6164	995	2665	3660

The GC content at third nucleotide positions and the frequency of preferred codons were each statistically significantly higher for genes expressed specifically in eggs as compared to sperm in *Z. Mays* and for genes expressed in ovary as compared to anther in *T. aestivum* (Figure 1). This result was statistically significant for genes expressed at both high and at low levels (high > 5 ESTs per 10 000; low ≤ 5 ESTs per 10 000) as well as across all genes. Statistically significant higher values for *GC3* were detected in microspore-specific genes as compared to the flower-specific genes in *B. napus* for genes expressed at low levels, but not for highly expressed genes (only *GC3* was determined for *B. napus* as preferred codons have not been described yet). Within the male-specific and the female-specific genes, *GC3* and *Fpr* were statistically significantly greater for the highly expressed as opposed to the lowly expressed genes in both *Z. mays* and *T. aestivum* and for the male-specific specific as compared to flower-specific genes in *B. napus*. Notably, lowly-expressed female-specific genes have a statistically significant higher *GC3* and *Fpr* than the highly expressed male-specific genes in *Z. mays* and no difference was detected between these two groups for *T. aestivum*. Most of the statistically significant comparisons (of 35 pairwise comparisons in total) remained significant after Bonferroni correction except for some contrasts among genes for high versus low expression within a gender (i.e., male-specific tissue/gametes in *Z. mays* and *T. aestivum* and within the flower in *B. napus*) and a single between gender comparison for *Fpr* at high expression levels in *T. aestivum* (*GC3* for this comparison remained statistically significant). All other between gender comparisons remained statistically significant.

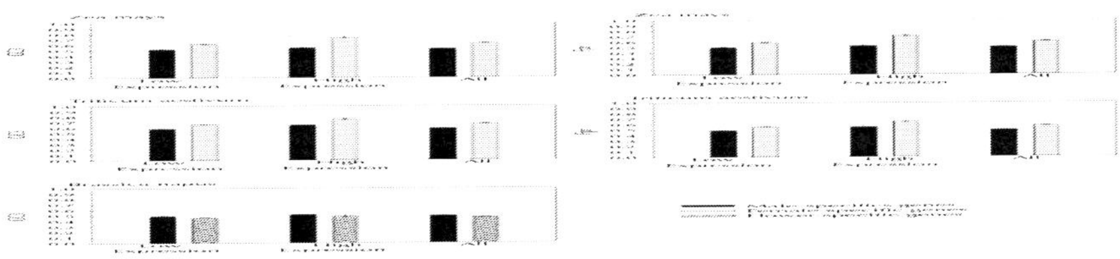

Figure 1: The GC content at third codon positions and the frequency of preferred codons for genes expressed specifically in male and female gametes or tissues in *Zea mays* and *Triticum aestivum* and the GC content of male-specific and flower-specific genes in *Brassica napus*relative to gene expression level. *P*-values between male and female (flower for *B. napus*) expressed

genes for *GC3* and for *Fpr* are as follows (respectively, in parenthesis):*Z. mays*: high ($< 1 \times 10^{-16}$**, $< 1 \times 10^{-16}$**), low ($< 1 \times 10^{-16}$ **, $< 1 \times 10^{-16}$**), all expression levels ($< 1 \times 10^{-16}$**, $< 1 \times 10^{-16}$**); *T. aestivum*: high ($< 1 \times 10^{-16}$*, 0.001*), low ($< 1 \times 10^{-16}$**, $< 1 \times 10^{-16}$**), all expression levels ($< 1 \times 10^{-16}$**, $< 1 \times 10^{-16}$**); *B. napus* (*GC3* only): high (0.141), low ($< 1 \times 10^{-16}$**) and all expression levels ($< 1 \times 10^{-16}$**). *P*-values for high versus low expression level within gender (or flower): *Z. mays*: Female ($< 1 \times 10^{-16}$**, $< 1 \times 10^{-16}$**), Male (0.012*, 0.010*); *T. aestivum*: Female ($< 1 \times 10^{-16}$**, $< 1 \times 10^{-16}$**), Male (0.036*, 0.025*); *B. napus*: Male (1×10^{-16}**), Flower (0.018*). *P*-values for lowly expressed female-specific genes versus highly expressed male-specific genes: *Z. mays* ($< 1 \times 10^{-16}$**, $< 1 \times 10^{-16}$**) and *T. aestivum* (0.867, 0.553). *P*-values for highly expressed female-specific genes versus lowly expressed male-specific genes: *Z. mays* ($< 1 \times 10^{-16}$**, $< 1 \times 10^{-16}$**) and *T. aestivum* ($< 1 \times 10^{-16}$**, $< 1 \times 10^{-16}$**). *P*-value for highly expressed male-specific genes versus lowly expressed genes in flower in *B. napus* (0.760). *P*-value for highly expressed male-specific genes versus lowly expressed genes in the flower in *B. napus*($< 1 \times 10^{-16}$).* Indicates statistical significance ($P < 0.05$), **Indicates that *P*-value remains statistically significant after the Bonferroni correction.

The relative synonymous codon usage (*RSCU*) represents the observed frequency of a codon divided by the expected frequency (i.e., if all synonymous codons were used equally) [19]. Values different from 1, thus indicate the presence of bias. Analysis of *RSCU* for genes with male-specific and female-specific expression (concatenated across all genes) indicated that the bias in codon usage towards species-specific preferred codons was consistently higher in the female than in male tissues/gametes for *Z. mays* and *T. aestivum* (Table 2). In particular, 26 of 27 of the species-specific preferred codons for 18 amino acids in *Z. mays* (i.e, 18 have synonymous codons, Table 2) [42] were more frequent in female-specific than in male-specific genes. This represents a higher usage of at least one species-specific preferred codon for every amino acid with synonymous codons (noting that some amino acids have more than one preferred codon). For *T. aestivum*, female-specific genes had a greater usage of all 23 of the species-specific preferred codons (for 22 of 23 comparisons, the difference in *RSCU* was greater than 0.1). In addition, hierarchical clustering was conducted using Pearson correlation coefficients between *RSCU* values for each combination of species and gender-specific tissues/gametes for *Z. mays* and *T. aestivum*[43]. The results indicate that these groups cluster by gender rather than by species, consistent with gender being a major parameter in shaping codon usage (Figure 2). The *RSCU* data also show that *B. napus* demonstrates a preference towards GC ending codons in male-specific as compared to flower-specific genes (Table 2). The entire dataset across all species consisted of 1 497 876 codons.

Table 2: Relative synonymous codon usage (*RSCU*) as determined from concatenated EST sequences of female- and male-specific genes in *Zea mays* and *Triticum aestivum* and male-specific and flower-specific sequences in *Brassica napus*

Amino acid	Codon	RSCUa										
		Zea mays				Triticum aestivum				Brassica napus		
		Female	Male	Difference (female minus male) b	Codon(s) previously identified as Preferredc	Female	Male	Difference (female minus male)	Codon(s) previously identified as Preferredc	Male	Flower	Difference (male minus flower)d
Phe	UUU	0.65	0.86	-0.21		0.60	0.82	-0.22		0.83	0.90	-0.07
	UUC	1.35	1.14	+0.21	*	1.40	1.18	+0.22	*	1.17	1.10	+0.07
Leu	UUA	0.26	0.43	-0.17		0.27	0.42	-0.15		0.49	0.61	-0.12
	UUG	0.87	1.07	-0.20	*	0.79	0.98	-0.19		1.26	1.27	-0.01
	CUU	1.13	1.30	-0.17		0.98	1.28	-0.30		1.44	1.46	-0.02
	CUC	1.61	1.23	+0.38	*	1.82	1.39	+0.43	*	1.46	1.29	+0.17
	CUA	0.47	0.56	-0.09		0.39	0.49	-0.10		0.52	0.60	-0.08
	CUG	1.66	1.40	+0.26	*	1.77	1.44	+0.33		0.82	0.76	+0.06
Ile	AUU	0.92	1.11	-0.19		0.84	1.07	-0.23		1.06	1.05	+0.01
	AUC	1.57	1.22	+0.35	*	1.65	1.27	+0.38	*	1.39	1.26	+0.13
	AUA	0.51	0.67	-0.16		0.52	0.65	-0.13		0.55	0.69	-0.14
Val	GUU	0.96	1.25	-0.29		0.89	1.19	-0.30		1.43	1.46	-0.03
	GUC	1.26	1.07	+0.19	*	1.30	1.10	+0.20	*	0.95	0.89	+0.06
	GUA	0.35	0.46	-0.11		0.33	0.43	-0.10		0.39	0.47	-0.08
	GUG	1.44	1.22	+0.22	*	1.49	1.28	+0.21	*	1.22	1.17	+0.05
Ser	UCU	0.98	1.20	-0.22		0.88	1.18	-0.30		1.52	1.54	-0.02
	UCC	1.23	1.02	+0.21	*	1.37	1.13	+0.24	*	1	0.88	+0.12
	UCA	0.93	1.26	-0.33		0.88	1.13	-0.25		1	1.11	-0.11

	UCG	0.75	0.61	+0.14	*	0.90	0.63	+0.27		0.75	0.68	+0.07
	AGU	0.61	0.82	-0.21		0.57	0.77	-0.2		0.77	0.84	-0.07
	AGC	1.51	1.09	+0.42	*	1.40	1.16	+0.24	*	0.96	0.95	+0.01
Pro	CCU	1.07	1.27	-0.20		0.93	1.18	-0.25		1.55	1.49	+0.06
	CCC	0.99	0.73	+0.26	*	1.07	0.84	+0.23	*	0.63	0.57	+0.06
	CCA	1.04	1.35	-0.31		0.92	1.21	-0.29		1.08	1.19	-0.11
	CCG	0.90	0.65	+0.25	*	1.08	0.78	+0.30	*	0.73	0.76	-0.03
Thr	ACU	0.92	1.13	-0.21		0.80	1.06	-0.26		1.26	1.25	+0.01
	ACC	1.41	1.05	+0.36	*	1.51	1.17	+0.34	*	1.07	0.96	+0.11
	ACA	0.91	1.21	-0.30		0.83	1.15	-0.32		0.93	1.09	-0.16
	ACG	0.77	0.60	+0.17	*	0.86	0.62	+0.24	*	0.74	0.71	+0.03
Ala	GCU	1.13	1.34	-0.21		0.87	1.19	-0.32		1.74	1.69	+0.05
	GCC	1.23	0.99	+0.24	*	1.44	1.12	+0.32	*	0.76	0.72	+0.04
	GCA	0.88	1.07	-0.19		0.74	1	-0.26		0.81	0.96	-0.15
	GCG	0.77	0.60	+0.17	*	0.96	0.68	+0.28		0.68	0.64	+0.04
Tyr	UAU	0.65	0.88	-0.23		0.63	0.82	-0.19		0.78	0.83	-0.05
	UAC	1.35	1.12	+0.23	*	1.37	1.18	+0.19	*	1.22	1.17	+0.05
His	CAU	0.85	1.08	-0.23		0.77	0.99	-0.22		0.99	1.07	-0.08
	CAC	1.15	0.92	+0.23	*	1.23	1.01	+0.22	*	1.01	0.93	+0.08
Gln	CAA	0.63	0.77	-0.14		0.57	0.73	-0.16		0.89	0.98	-0.09
	CAG	1.37	1.23	+0.14	*	1.43	1.27	+0.16	*	1.11	1.02	+0.09
Asn	AAU	0.68	0.97	-0.29		0.71	0.92	-0.21		0.76	0.79	-0.03
	AAC	1.32	1.03	+0.29	*	1.29	1.08	+0.21	*	1.24	1.21	+0.03
Lys	AAA	0.50	0.70	-0.20		0.49	0.64	-0.15		0.72	0.88	-0.16
	AAG	1.50	1.30	+0.20	*	1.51	1.36	+0.15	*	1.28	1.12	+0.16
Asp	GAU	0.87	1.09	-0.22		0.82	1.04	-0.22		1.18	1.21	-0.03
	GAC	1.13	0.91	+0.22	*	1.18	0.96	+0.22	*	0.82	0.79	+0.03

Glu	GAA	0.63	0.83	-0.20		0.60	0.75	-0.15		0.81	0.90	-0.09
	GAG	1.37	1.17	+0.20	*	1.40	1.25	+0.15	*	1.19	1.10	+0.09
Cys	UGU	0.58	0.77	-0.19		0.51	0.73	-0.22		0.98	1.01	-0.03
	UGC	1.42	1.23	+0.19	*	1.49	1.27	+0.22	*	1.02	0.99	+0.03
Arg	CGU	0.66	0.77	-0.11		0.60	0.72	-0.12		1.08	0.98	+0.10
	CGC	1.29	1.06	+0.23	*	1.57	1.21	+0.36	*	0.68	0.54	+0.14
	CGA	0.41	0.55	-0.14		0.34	0.46	-0.12		0.54	0.58	-0.04
	CGG	0.89	0.77	+0.12		1.06	0.91	+0.15		0.45	0.51	-0.06
	AGA	0.90	1.17	-0.27		0.84	1.13	-0.29		1.71	1.92	-0.21
	AGG	1.84	1.68	+0.16	*	1.58	1.56	+0.02	*	1.54	1.48	+0.06
Gly	GGU	0.86	1.03	-0.17		0.74	0.97	-0.23		1.27	1.22	+0.05
	GGC	1.41	1.13	+0.28	*	1.65	1.30	+0.35	*	0.66	0.64	+0.02
	GGA	0.86	1.02	-0.16		0.71	0.92	-0.21		1.34	1.41	-0.07
	GGG	0.88	0.83	-0.05		0.90	0.82	+0.08		0.72	0.73	-0.01

Positive differences in *RSCU* greater than 0.1 in the tissue comparisons (highlighting codon preferences in female tissues) are in bold. Start and termination codons and the single codon for tryptophan have been excluded. Preferred codon data is not available for *B napus*.

[a] The total number of codons used to estimate *RSCU*: *Zea mays* 116 919 (sperm) and 94 794 (egg), *Triticum aestivum* 453 286 (anther) and 199 397 (ovary) and *Brassica napus* 232 516 (microspore) and 400 964 (flower).*RSCU* indicates the difference from a 1:1 ratio for all available codons within synonymous codon groups.

[b] +, higher *RSCU* in female-specific genes; -, higher *RSCU* in male-specific genes.

[c] Previously described as preferred codons in *Z. mays* and *T. aestivum* as indicated by their frequencies in high-versus low-biased genes by Kawabe and Miyashita (2003) [42]. Asterisk (*) indicates a preferred codon. No values available for *B. napus*.

[d] +, higher *RSCU* in male-specific genes; -, higher *RSCU* in flower-specific genes.

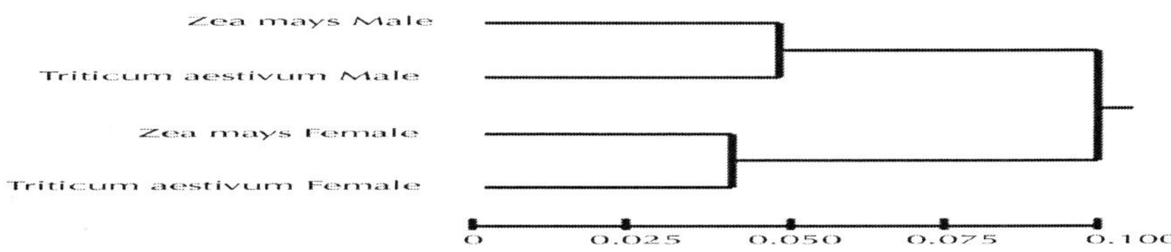

Figure 2: Hierarchical clustering based on Pearson correlation matrixes on relative synonymous codon usage values derived from concatenated EST sequences for each combination of species and gender (*Zea mays*and *Triticum aestivum* male-specific and female-specific genes).

Given that bias in codon usage has been inversely associated with protein length in certain eukaryotic organisms [e.g., [8]], we assessed whether it played a role in the observed gender-specific bias described above. For this, we identified the protein length for each gene under study (number of amino acids) in the *Arabidopsis thaliana* protein sequence database (i.e., genes previously identified as a match, see above). We used protein lengths from *Arabidopsis* because our DNA sequence data was derived from ESTs (that are partial sequences, not containing gene or protein length information) and because, unlike well-studied model organisms, the complete and/or annotated genomic DNA or protein sequences are not yet available for most of the genes in these plants. Protein lengths tend to be highly conserved among eukaryotes [44]. The data indicates that the mean protein length was greater for male-specific genes than female-specific genes in *Z. mays* and *T. aestivum* and for flower-specific genes as compared to male-specific genes in *B. napus* (mean protein length (± Standard Error): *Z. mays* egg 409.1 (± 13.2), *Z. mays*sperm 493.2 (± 13.3), *T. aestivum* ovary 426.3 (± 8.1), *T. aestivum* anther 571.3 (± 6.7), *B. napus* microspore 370 (± 6.7), *B. napus* flower 441.3 (± 5.6)). The higher mean lengths generally resulted from the presence of a relatively few genes that encoded very long proteins (between 1000 and 5000 amino acids). The number of genes encoding proteins with more than 1000 amino acids relative to the total number of genes are: *Zea mays* egg 38/946 = 4.0%,. *Z. mays* sperm 80/955 = 8.3%, *T. aestivum* ovary 69/1489 = 4.6%, *T. aestivum* anther 368/3326 = 11.1%, *B. napus* microspore 51/1675 = 3.0%, *B. napus* flower 176/3181 = 5.5%. We thus determined *RSCU*values for genes encoding proteins of similar lengths, in order to assess whether this protein length variation was related to our findings of gender-specific biases in codon usage. For this, male-specific and female-specific genes from each species under study (and *B. napus* flower-specific genes) were classified as being of either short (≤200 amino acids), medium (>200 and ≥400 amino acids) or of long length (>400 amino acids).

Comparisons were conducted for *RSCU* values relative to gender for *Z. mays* and *T. aestivum* (and male-versus flower-specific genes in *B. napus*) for the concatenated EST sequences (longest EST per contig) within each protein length category. The results show that female-specific genes consistently have higher values for *RSCU* for the preferred codons (as indicated by "+" sign in each of the columns) within each of these three protein length categories for *Z. mays* and *T. aestivum* (Table 3) [see Additional data file 1; Tables 1 to 3)] indicating that protein length variation does not explain the gender-specific bias observed for specific codons described in Table2 (see Table 3). The bias in codon usage for male-specific versus flower-specific genes in *B. napus*was most evident for genes encoding long proteins.

Table 3: Difference in the relative synonymous codon usage (*RSCU*) for short genes (less than or equal to 200 amino acids in the homologous *Arabidopsis thaliana* protein), medium length genes (more than 200 amino acids and less than or equal to 400) and long genes (more 400 amino acids) as determined from concatenated EST sequences of female- and male-specific sequences in *Zea mays* and *Triticum aestivum* and male-specific and flower-specific sequences in *Brassica napus*. 1

Amino acid	Codon	Difference in RSCU (Female-specific genes minus male-specific genes)a								Difference in RSCU (Male-specific genes minus flower-specific genes)b		
		Zea maysc				Triticum aestivumc				Brassica napusc		
		Short	Medium	Long	Codon(s) previously identified as preferredd	Short	Medium	Long	Codon(s) previously identified as preferredd	Short	Medium	Long
Phe	UUU	-0.15	-0.19	-0.23		-0.06	-0.12	-0.24		+0.05	-0.01	-0.21
	UUC	+0.15	+0.19	+0.23	*	+0.06	+0.12	+0.24	*	-0.05	+0.01	+0.21
Leu	UUA	-0.07	-0.15	-0.19		-0.06	-0.09	-0.14		-0.03	-0.01	-0.28
	UUG	-0.22	-0.20	-0.16	*	-0.01	-0.13	-0.15		+0.13	-0.05	-0.02
	CUU	-0.22	-0.10	-0.18		-0.17	-0.20	-0.30		+0.09	-0.09	-0.06
	CUC	+0.50	+0.34	+0.27	*	+0.24	+0.24	+0.32	*	-0.18	-0.05	+0.59
	CUA	-0.06	-0.10	-0.07		-0.03	-0.04	-0.10		-0.06	+0.02	-0.17
	CUG	+0.07	+0.23	+0.32	*	+0.04	+0.23	+0.37		+0.05	+0.18	-0.04
Ile	AUU	-0.14	-0.21	-0.18		-0.16	-0.14	-0.22		+0.08	+0.06	-0.08
	AUC	+0.24	+0.33	+0.35	*	+0.12	+0.25	+0.33	*	-0.09	-0.07	+0.46
	AUA	-0.09	-0.13	-0.17		+0.03	-0.12	-0.11		0	+0.01	-0.38
Val	GUU	-0.32	-0.22	-0.30		-0.20	-0.14	-0.30		+0.06	-0.07	-0.05
	GUC	+0.25	+0.22	+0.08	*	+0.06	+0.13	+0.15	*	-0.13	+0.01	+0.20
	GUA	-0.17	-0.08	-0.09		-0.04	-0.06	-0.10		-0.02	+0.03	-0.20
	GUG	+0.23	+0.08	+0.31	*	+0.18	+0.07	+0.25	*	+0.08	+0.03	+0.05
Ser	UCU	-0.19	-0.14	-0.28		-0.13	-0.17	-0.32		-0.01	-0.06	-0.01
	UCC	+0.19	+0.11	+0.22	*	+0.19	+0.19	+0.13	*	-0.06	+0.02	+0.30
	UCA	-0.33	-0.38	-0.25		-0.09	-0.28	-0.15		+0.04	+0.05	-0.34
	UCG	+0.11	+0.10	+0.15	*	+0.13	+0.23	+0.25		+0.02	-0.02	+0.15
	AGU	-0.19	-0.18	-0.20		-0.14	-0.13	-0.17		+0.09	-0.02	-0.21
	AGC	+0.41	+0.47	+0.36	*	+0.04	+0.16	+0.25	*	-0.09	+0.03	+0.10
Pro	CCU	-0.28	-0.19	-0.14		-0.10	-0.14	-0.24		+0.15	+0.03	+0.05
	CCC	+0.29	+0.26	+0.18	*	+0.22	+0.19	+0.13	*	-0.01	+0.05	+0.10
	CCA	-0.25	-0.23	-0.33		-0.22	-0.24	-0.21		-0.03	+0.03	-0.25
	CCG	+0.23	+0.17	+0.28	*	+0.10	+0.19	+0.31	*	-0.10	-0.12	+0.10
Thr	ACU	-0.21	-0.18	-0.21		-0.01	-0.26	-0.25		+0.09	+0.01	-0.01
	ACC	+0.29	+0.33	+0.31	*	+0.13	+0.25	+0.28	*	-0.10	-0.06	+0.37
	ACA	-0.17	-0.30	-0.30		-0.11	-0.19	-0.29		+0.01	+0.04	-0.43

	ACG	+0.10	+0.14	+0.20	*	-0.01	+0.19	+0.27	*	0	+0.01	+0.07
Ala	GCU	-0.11	-0.21	-0.22		-0.19	-0.27	-0.28		+0.18	0	+0.02
	GCC	+0.18	+0.19	+0.25	*	+0.22	+0.27	+0.22	*	-0.15	0	+0.16
	GCA	-0.10	-0.19	-0.19		-0.11	-0.18	-0.23		-0.04	+0.01	-0.33
	GCG	+0.03	+0.19	+0.17	*	+0.08	+0.18	+0.29		+0.02	-0.01	+0.15
Tyr	UAU	-0.13	-0.28	-0.19		-0.11	-0.08	-0.21		+0.06	+0.07	-0.24
	UAC	+0.13	+0.28	+0.19	*	+0.11	+0.08	+0.21	*	-0.06	-0.07	+0.24
His	CAU	-0.27	-0.22	-0.16		-0.03	-0.13	-0.20		+0.04	-0.01	-0.21
	CAC	+0.27	+0.22	+0.16	*	+0.03	+0.13	+0.20	*	-0.04	+0.01	+0.21
Gln	CAA	-0.11	-0.10	-0.15		-0.10	-0.10	-0.14		-0.09	-0.03	-0.17
	CAG	+0.11	+0.10	+0.15	*	+0.10	+0.10	+0.14	*	+0.09	+0.03	+0.17
Asn	AAU	-0.32	-0.32	-0.21		-0.01	-0.11	-0.19		+0.05	+0.06	-0.15
	AAC	+0.32	+0.32	+0.21	*	+0.01	+0.11	+0.19	*	-0.05	-0.06	+0.15
Lys	AAA	-0.14	-0.17	-0.20		-0.03	-0.10	-0.12		-0.09	-0.08	-0.27
	AAG	+0.14	+0.17	+0.20	*	+0.03	+0.10	+0.12	*	+0.09	+0.08	+0.27
Asp	GAU	-0.06	-0.21	-0.24		-0.09	-0.13	-0.20		+0.04	-0.01	-0.10
	GAC	+0.06	+0.21	+0.24	*	+0.09	+0.13	+0.20	*	-0.04	+0.01	+0.10
Glu	GAA	-0.08	-0.20	-0.20		-0.05	-0.07	-0.13		-0.08	-0.02	-0.16
	GAG	+0.08	+0.20	+0.20	*	+0.05	+0.07	+0.13	*	+0.08	+0.02	+0.16
Cys	UGU	-0.08	-0.14	-0.20		-0.10	-0.18	-0.17		+0.04	+0.02	-0.14
	UGC	+0.08	+0.14	+0.20	*	+0.10	+0.18	+0.17	*	-0.04	-0.02	+0.14
Arg	CGU	-0.01	-0.16	-0.07		-0.14	-0.12	-0.07		+0.12	-0.05	+0.22
	CGC	+0.14	+0.17	+0.14	*	+0.06	+0.33	+0.20	*	+0.02	-0.01	+0.22
	CGA	+0.01	-0.18	-0.12		-0.03	-0.08	-0.12		-0.11	+0.03	-0.03
	CGG	+0.03	+0.08	+0.18		+0.18	+0.10	+0.18		-0.05	-0.01	-0.10
	AGA	-0.21	-0.18	-0.26		-0.03	-0.15	-0.27		+0.04	+0.03	-0.57
	AGG	+0.04	+0.28	+0.12	*	-0.05	-0.07	+0.07	*	-0.02	-0.01	+0.20
Gly	GGU	-0.11	-0.11	-0.21		-0.16	-0.11	-0.23		+0.08	+0.05	+0.03
	GGC	+0.34	+0.16	+0.31	*	+0.20	+0.29	+0.29	*	-0.15	+0.01	+0.12
	GGA	-0.22	-0.08	-0.16		-0.07	-0.19	-0.18		-0.04	-0.06	-0.06
	GGG	0	+0.04	+0.07		+0.04	+0.01	+0.12		+0.11	0	-0.09

[a] +, higher *RSCU* in female-specific genes; -, higher *RSCU* in male-specific genes.

[b] +, higher *RSCU* in male-specific genes; -, higher *RSCU* in flower-specific genes.

[c] The total number of short genes and the average gene length (± Standard error), respectively, used to estimate *RSCU*: *Z. mays* sperm 149, 137.9 (± 3.2), *Z. mays* egg 204, 135.3 (± 2.9), *T. aestivum* anther 340, 142.3 (± 2.0), *T. aestivum* ovary 309, 138.8 (± 2.3), *B. napus* microspore 485, 144.8 (± 1.5), *B. napus* flower 598, 140.6 (± 1.6). The total number of medium length genes and the average gene length, respectively, used to estimate *RSCU*: *Z. mays* sperm 330, 295.2 (± 3.1), *Z. mays* egg 352, 296.2 (± 3.2), *T. aestivum* anther 893, 309.0 (± 1.9), *T. aestivum* ovary 520, 298.4 (± 2.6), *B. napus* microspore 608, 294.3 (± 2.4), *B. napus* flower 1127, 297.5 (± 1.7). The total number of long genes and the average gene length, respectively, used to estimate *RSCU*: *Z. mays* sperm 476, 741.6 (± 20.7), *Z. mays* egg 390, 655.4 (± 18.9), *T. aestivum* anther 2093, 753.4 (± 8.3), *T. aestivum* ovary 660, 661.7 (± 13.0), *B. napus* microspore 582, 639.5 (± 12.6), *B. napus* flower 1456, 676.1 (± 8.6).

[d] Previously described as preferred codons in *Z. mays* and *T. aestivum* as indicated by their frequencies in high-versus low-biased genes by Kawabe and Miyashita (2003) [42]. Asterisk (*) indicates a preferred codon. No values available for *B. napus*.

Given that the codons showing bias generally ended in G or C for each of the species examined (Table 2), we also compared the *GC3* content between each of the two tissue/gamete types per species for each of the protein length categories (Table 4). The results of pairwise comparisons show that the *GC3* values are statistically significantly higher for female-specific than male-specific genes in *Z. mays* and *T. aestivum* and for male-specific than flower-specific genes in *B. napus*-within each of the protein length categories, consistent with a gender-bias on codon usage. Examination of only those genes encoding very long proteins (equal to or more than 1000 amino acids) showed similar trends for each of these species. Notably, *GC3* values were inversely correlated with protein length within each of the six species and tissue-specific datasets (i.e., for the male-specific, female-specific and for flower-specific datasets, Table 4) [see also Additional data file 1; Table 4], consistent with the relationship between codon usage and protein length reported in other species [8].

Table 4: Mean *GC3* values for short genes (less than or equal to 200 amino acids in the homologous *Arabidopsis thaliana* protein), medium length genes (more than 200 amino acids and less than or equal to 400) and long genes (more 400 amino acids), and for genes encoding very long proteins only (equal to or more than 1000 amino acids) for female-specific and male-specific genes in *Zea mays* and *Triticum aestivum* and for male-specific and flower-specific genes in *Brassica napus*. 1

	Short			Medium		Long				Very long only[a]		
	Female	Male	P-value	Female	Male	P-value	Female	Male	P-value	Female	Male	P-value
Zea mays	0.702	0.632	$<1.0 \times 10^{-16}$	0.635	0.542	$<1.0 \times 10^{-16}$	0.593	0.483	$<1.0 \times 10^{-16}$	0.505	0.656	0.044
Triticum aestivum	0.768	0.729	0.015	0.677	0.617	$<1.0 \times 10^{-16}$	0.612	0.526	$<1.0 \times 10^{-16}$	0.518	0.436	$<1.0 \times 10^{-16}$
Brassica napus	Male	Flower		Male	Flow-er		Male	Flower		Male	Flower	
	0.523	0.501	0.002	0.489	0.472	0.0002	0.469	0.448	$<1.0 \times 10^{-16}$	0.441	0.423	0.340

P-values are those for the Mann-Whitney Rank Sum Test (t-tests yielded similar P-values). P-values less than 0.05 are in bold. P-values not remaining statistically significant after Bonferroni correction among these contrasts are bold italics. For P-values of comparisons between protein length categories.

[a], Sample sizes are relatively small for this group, influencing *P*-values for *Z. mays* and *B. napus*; the direction and magnitude of the differences are maintained for each comparison (see text). Mean protein lengths (± Standard Error) for this category are as follows. *Z. mays* sperm 1556.0 (± 63), *Z. mays* egg 1485.2 (± 112), *T. aestivum* anther 1377.0 (± 24.7), *T. aestivum* ovary 1334.7 (± 72.8), *B. napus* microspore 1370.5 (± 70.6), *B. napus* flower 1361.4 (± 34.0). Other mean protein length values are presented in Table 3.

Biological function has been proposed as a potential factor altering certain molecular evolutionary processes [10]; therefore we examined the gene profiles for each of the contrasting tissues for each species under study. The profile of biological functions of genes expressed only in the female tissues/gametes and male tissues/gametes were nearly identical for *Z. mays* and for *T. aestivum* (Figure 3). Similarly, the biological functions for genes specifically expressed in male microspores and those expressed in flowers were strongly associated in *B. napus*.

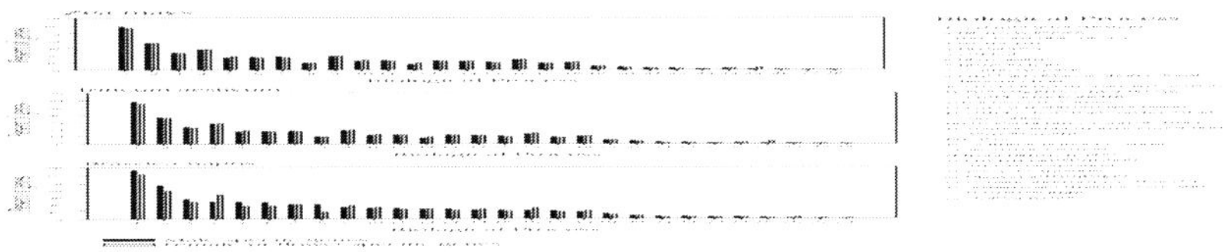

Figure 3: Biological functions of genes expressed only in the sperm and eggs in *Zea mays*, in the anther and ovary in *Triticum aestivum* and in the microspore and flower in *Brassica napus*. Sum of percentages may exceed 100 as some genes belong to more than one category.

DISCUSSION

Gender Differences in Codon Usage

The statistically significant higher GC content at third codon positions and the greater frequency of preferred codons for genes expressed specifically in eggs as compared to sperm in *Z. mays* and for genes expressed in the ovary as compared to anther in *T. aestivum*, provide strong evidence that there is a greater bias in codon usage for genes expressed in female tissues than in male tissues and/or gametes (Figure 1). These findings were similar when genes were classified as either high expressing or low expressing, suggesting that gender has a substantial impact on synonymous codon use. We can infer that these differences are likely due to selective pressure because the bias is associated with gene expression (tissue-specific gene expression or high versus low expression level, Figure 1). Furthermore, the data indicate that the gender effect cannot be attributed to variation in protein lengths or to gene function (Table 3, Table 4, Figure 3). Overall, these results, across a broad range of genes, provide evidence that codon usage is altered by gender-specific pressures in plants.

Genes expressed in eggs have a higher relative synonymous codon usage value than those expressed in sperm for 26 of the 27 previously identified preferred codons for *Z. mays*, and indicates, remarkably, that selective pressure in eggs specifically acts to enhance the frequency of preferred codons for each of the 18 amino acids that have synonymous codons (i.e., egg-specific genes have a greater frequency of at least one of the preferred codons per amino acid, Table 2). Similar findings for *T. aestivum*, showing that all 23 of the preferred codons for this species are enhanced in genes expressed in the ovary as compared to the anther, also demonstrate that a selective pressure inherent to these female organs and gametes is acting to enhance the incidence of preferred codons across all synonymous codon groups. In addition, the fact that the female-bias was detected for each gene length category (i.e., among short genes, among medium length genes, and among long genes, Tables 3 and 4) and that the gender-specific gene expression was the major determinant of hierarchical clustering (relative to *RSCU* values), rather than species (Figure 2), supports the notion that the codon usage bias demonstrated here is greatly influenced by gender-specific factors. It is notable for *T. aestivum* that in three cases the *RSCU* was greater for female-than male-specific genes (bold values, no asterisk, Table 2) for G or C ending codons that had not been previously identified as preferred, but had been described as preferred in *Z. mays* and other plant species [42]. The overall results presented here, showing greater use of preferred codons in genes expressed in female organs and gametes, suggests that these codons are probably also preferred in *T. aestivum*, at least for genes expressed in the reproductive tissues and gametes.

The greater bias in codon usage among genes expressed in female organs and gametes as compared to male organs and gametes, reflecting an increased propensity for translational selection, could be caused by several factors. In particular, it is possible that protein products of genes expressed in female organs and gametes experience a more diverse biochemical environment than their male counterparts, a phenomenon that could lead to greater selective constraint on proteins [45,46], and thus, on their translation. It is also possible that mutations at third codon positions in genes expressed in female organs and gametes may on average have greater effects on fitness, as has been proposed for genes expressed across a broad array of tissues (higher selection coefficients) [8,9]. This could occur, for example, if translational inefficiency in female organs and gametes alters the cellular energy resources or interferes with essential biological processes in a manner not prevalent in male regions. Mutations affecting female regions could also have greater fitness effects because of the general uncertainty in the pollination process, which makes it highly advantageous for each ovary, ovule and/or egg to be fully functional (thereby mutations in female regions may affect fitness overall more than for anthers, pollen or sperm) [47] and because maternal traits can have a much greater impact on seed production (seed number, size, and dispersal) and survival (and thus, on overall fitness)[48]. Another possible explanation for the observed results is that there are differences in gene function between female and male organs and gametes, a theory that has been proposed as a potential factor altering amino acid substitution rates [10]. As shown in Figure 3, however, this is not the likely explanation in this study, as there is remarkable similarity in the biological functions represented by the male-specific and female-specific genes in both *Z. mays* and *T. aestivum*. Nonetheless, subtle differences in gene function (e.g., specific genes that influence codon usage) or other, unidentified, functional differences between the male and female tissues/gametes could play a role [10]. An additional potential contributing factor worth consideration is that genes that have greater breadth of expression throughout the entire plant (that can have greater bias in codon usage) [1], are coincidentally also more commonly expressed in female organs and gametes than in males. Although this possibility cannot be definitively excluded, it seems unlikely given the similarity between the functional profiles of female- and male-specific genes. Altogether, it seems that the best explanations are differences in the amount of selective pressure for effective translation due to different cellular environments and/or a greater impact of mutations on female tissues and gametes. Further studies will nonetheless be needed to ascertain the mechanisms underlying the greater bias in codon usage in female organs and gametes in these plant species.

The relationship between gender-specific gene expression and codon usage in *Z. mays* and *T. aestivum* is consistent with the very limited data currently available for other organisms. It has been shown in humans, for example, that genes expressed in ovaries have likely been under slightly greater selective constraint than testes for codon usage following the divergence of humans and mice [3]. The trend notably corresponds to the generally high rates of protein evolution (and thus reduced selective pressure) reported in genes involved in spermatogenesis in primates [49]. In *Drosophila*, it has been found that the relative expression of genes in females versus males (the female: male ratio of gene expression) is well correlated to bias in codon usage[4]. In addition, in *Arabidopsis*, previous findings have indicated that more induced harmful mutations are passed to progeny by the sperm than by the eggs, consistent with the relatively lower selective pressure on mutations in male than in female tissues and gametes [7]. The present results extend these findings to include gender-specific selection on codon usage. Each of these gender-specific trends, in humans, *Drosophila*, and *Arabidopsis* are consistent with the findings we report here, and suggest that the higher bias in codon usage for genes expressed in female tissues could be inherent to a range of organisms. Further studies will be needed to better understand the full range of organisms for which gender-specific gene expression is associated with a bias in codon usage.

Gene Expression Level

Gene expression level has been shown to be positively correlated with bias in codon usage in many organisms [1,8-15,24-27,31]. Selection is the best explanation for this finding because higher levels of gene expression lead to greater opportunity for selection to alter codon usage[8,10,12] and because mutational bias has only rarely been associated with gene expression level (in certain microorganisms) [8,50,51]. In *Drosophila*, a positive relationship between gene expression and bias in codon usage has been reported for female tissues, but a relatively weak negative correlation was detected for male tissues [4]. Our findings of greater values for *GC3* and*Fpr* for highly expressed genes than for lowly expressed genes for both male-specific and female-specific genes from *Z. mays* and *T. aestivum* suggests that gene expression level is positively correlated to bias in codon usage for genes expressed in male and in female regions for these plants.

It is notable nonetheless that we found that the differences in the bias in codon usage between highly and lowly expressed genes were not as marked the male as in female tissues and gametes, as evidenced by the fact that the Bonferroni correction excluded the statistical significance of this comparison for both *Z. mays* and *T. aestivum*. In fact, the lowly expressed female-specific genes had statistically significantly higher bias in codon usage than the highly expressed male-specific genes in *Z. mays* and no difference was detected between these two groups for *T. aestivum*(Figure 1). It thus seems that female tissues/gametes maintain substantial selective pressure on codon usage even for genes with reduced expression, in a manner not characteristic of male tissues/gametes.

Selection and Gender-Bias

The greater bias in codon usage for female-specific than for male-specific genes is currently best explained by selection. This is for the following reason. The male-specific and female-specific gene sets examined here were determined based on calculations that these genes were solely or primarily expressed in one tissue type and not in the other (i.e., the lack of ESTs in the contrasting tissue indicates that the mRNA was very rare or absent). Thus, the observed effects are associated with gender-specific gene expression (gene expression allows opportunity for selection, and is not usually associated with mutational bias [8, 10, 12]). In addition, our data indicate that gene function and protein length variation between male and female tissues/gametes do not explain the observed bias in codon usage between the gender-specific gene sets. Gene expression level differences cannot be implicated because the gender bias was detected for genes expressed at similar levels (high versus high expression and low versus low expression, regardless of protein length, Figure 1,) [see Additional data file 1, Table 5]. Notably, because we examined ESTs for the present analysis, which rarely contain introns, and studied plant species where annotated genomic DNA (containing the introns) is not yet available (NCBI, personal communication), we do not include an analysis of the GC (or AT) content of introns versus third codon positions in our genes, an approach sometimes used to exclude mutational bias [8, 12, 24, 42, 45, 52]. Nevertheless, each of these trends, taken in their entirety, suggest that the bias in codon usage associated with gender-specific expression is best explained by differential selective pressure on genes expressed in male-specific tissues/gametes versus female-specific tissues/gametes.

Male-Specific versus Flower-Specific Genes in *Brassica napus*

Although the differences in bias in codon usage between the microspore and flower in *B. napus*were generally lower in magnitude than the previous between gender comparisons for *Z. mays*and *T. aestivum* (Figure 1, Table 2), the data overall indicates that genes expressed in the two *B. napus* tissues have specific patterns of codon usage. Specifically, the higher GC content in *B. napus* microspore-specific genes than in flower-specific genes suggests that the male portion of the flower may be under more selective pressure for codon usage than the flower as a whole. In particular, given that the flower and flower bud EST library should represent genes from the male, female and vegetative (somatic) tissue, one can infer that the combined vegetative and female tissue is under less selective pressure than the microspore. Given that the vegetative tissue usually represents the greatest fraction of the flower tissue (petals, sepals) [53], then it could, in turn, be inferred that the somatic region is likely under reduced selective pressure for codon usage than the microspore (with no inference regarding the pressure in female tissues). The fact that the GC content at third codon positions of the genes specifically expressed in the microspores varied markedly among synonymous codon groups (Table 2), and was found to be positively correlated to gene expression level (Figure 1), further supports the notion that translational selection is enhanced in the microspore component of the flower. Moreover, from examination of Table 2, it is evident that for six of seven comparisons where the differences in *RSCU* between tissues were greater than 0.1, the microspore had enhanced usage of G or C ending codons (which were notably also the preferred codons in *Z. mays* and *T. aestivum*), a trend consistent with greater selective pressure. Notably, analysis of *RSCU* values relative to protein length suggests that differences between male-specific and flower-specific genes are greatest for genes encoding longer proteins (>400 amino acids, Table 3) in the *B. napus* tissue comparisons as these genes have substantially greater usage of G or C ending codons in this category ("+" signs in Table 3). This effect could be partially caused by the greater percentage of genes encoding very long proteins in the flower-specific dataset or by greater male-specific effects on codon usage for genes encoding longer proteins. Nonetheless, all three of the gene length categories demonstrate higher*GC3* values for male-specific genes (Table 4). One possible interpretation of all of these findings in*B. napus*, when combined with the data from *Z. mays* and *T. aestivum*, is that the translational selection increases in the following order: flower-specific (heterogeneous) genes, male-specific genes, female-specific genes. Because these analyses are in different species, however, further evaluation of this possible relationship will be needed. Altogether, the totality of the findings here suggest that genes expressed in reproductive tissues may be under greater translational level selection than those expressed in vegetative (somatic) tissues, a factor consistent with the key role of reproductive success in fitness.

Protein Length and Gene Expression

The analysis of protein lengths indicates that genes encoding shorter proteins tend to generally have greater bias in codon usage, as indicated by *GC3*, for the species examined here (Table 4) [see also Additional data file 1; Table 4]. This is consistent with the trends reported in other organisms to date such as *Arabidopsis*, *Drosophila*, *C. elegans* and yeast [e.g., [8, 54]]. We also found marked evidence that the gene expression level in the species studied here is inversely correlated to protein length for each of the six datasets examined (across all genes per dataset), a result consistent with trends reported in humans, *Drosophila* and *Populus tremula* [2,55,56]. In particular, the Pearson correlation coefficients were: *Z. mays* male (R = -0.135, P = 2.6 × 10^{-5}),*Z. mays* female (R = -0.080, P = 0.010), *T. aestivum* male (R = -0.049, 4.3 × 10^{-3}), *T. aestivum*female (R = -0.18, P = 1.1 × 10^{-12}), *B. napus* male (R = -0.149, P = 1.9 × 10^{-9}), and *B. napus*flower (R = -0.081, P = 4.6 × 10^{-6}). This suggests that the tendency of shorter genes to have greater bias in codon usage (Table 4), at least for the genes examined here, may be due to greater levels of gene expression and an associated selective pressure [54]. Notably, in a complementary analysis to Figure 1, we found that the gender-bias in codon usage was evident among highly and among lowly expressed genes within each the three different protein length categories (short, medium, and long) [see Additional data file 1; Table 5].

Thus, the gender-specific biases in *Z. mays* and *T. aestivum* (and flower-specific differences in *B. napus*) at high and low levels of gene expression observed in Figure 1 cannot be explained by differences in protein lengths. In addition to the inverse association between protein length and bias in codon usage, it is also evident from Table 4 that the three species examined here tend to have different values for *GC3*, with decreasingly lower values occurring from *T. aestivum*, to *Z. mays* and to *B. napus*. Altogether, it is evident from our entire analysis that the gender-specific effects on codon usage can be detected across a range of protein lengths, gene expression levels, and for different plant species, thereby demonstrating that gender-specific factors play a significant role in genome evolution.

Notable Issues

It should be noted that the *B. napus* material used for the microspore cDNA library was grown at low temperatures (10°C/5°C) that could potentially alter some fraction of the gene expression in the microspore if stress-mediation genes were enhanced. In fact, we found less than 1% of the total genes in *B. napus* microspores were stress-related (data not shown). Another issue worth considering is the implication of previous findings of gender-specific mutation rates in plants, a trend that was based on the detection of higher evolutionary rates at silent sites, including third codon positions, in male gametes [57]. Higher mutation rates in sperm, however, should act to enhance per generational mutation rates across the entire genome, including those genes expressed in females and in males, and thus, not impact the observed bias in codon usage.

Nonetheless, it should be noted that the differential male/female inheritance of organelles (and the underlying mechanisms; pre- or post-zygotic), could influence whether these genes are expressed in male or female tissues/gametes, and potentially contribute to the codon usage for organellar genes and their substitution rates [57,58]). It is also notable that the abundance of tRNA of the preferred codons could be greater for female than the male tissues/gametes and contribute to the gender differences in codon usage [3,4]. This seems relatively unlikely given that the abundance of tRNAs would have to be higher in female-tissues/gametes for every single preferred codon and that such differences have been shown not to explain the gender-specific codon usage in*Drosophila* [4]. It should also be noted that the reproducibility of the results for *GC3* and *Fp*robserved in this study is consistent with the notion that *GC3* content alone could be an effective indicator of codon bias in some species [31].

CONCLUSIONS

This study reports findings of female-specific bias in codon usage in plants. The remarkable consistency of the increased GC content of third codon positions and the increased frequency of preferred codons for both *Z. mays* and *T. aestivum*, even across different gene expression levels and protein lengths, combined with the enhanced usage of species-specific preferred codons for each of the 18 amino acids having synonymous codons, strongly indicate that gender plays a key role in codon usage. The findings in *B. napus* suggest that the tissues of the reproductive system, including both male and female organs, have a greater impact on codon usage than somatic regions. Overall, it is apparent that gender needs to be a key player in furthering our understanding of translational level selection. Further study will be needed to ascertain whether this is a generalized phenomenon, inherent to other organisms, as it could play a key role in DNA and protein sequence changes relevant to epidemiology, population genetics and to molecular evolution.

METHODS

Sequence Data

Sperm and egg EST libraries from *Z. mays* and anther and ovary libraries for *T. aestivum* were extracted from Genbank using Entrez Nucleotide available at the National Center for Biotechnology Information [41] (Table 1). We chose these data because of the availability of large gender-specific EST libraries (>4000) in Genbank. When more than one library was available we chose the one most likely to reflect gamete expression (e.g., an ovary library was selected over a pistil library in *T. aestivum*). In addition to these libraries, we also collected *B. napus* sequences from an in-house cDNA library representing isolated late-uninucleate and early-binucleate microspores (male germline cells) and from a publicly available flower library (Table 1). The EST datasets used represent those that were available to us as of March 2006.

Expression Profiles and Preparation of Sequence Data

Each of the EST sequences from each of the six libraries was compared against the *A. thaliana*protein sequence database using BLASTX [39-41]. Only EST sequences having an e-value of less than 1×10^{-7} to known or hypothetical proteins of *A. thaliana* were kept for further analysis. Notably, this process automatically excludes all rRNAs from the analysis as they would not be in the protein database. Using these datasets, we clustered and assembled the ESTs for each library into contigs and singletons using the software program CAP3 [38] (Table 1). The expression profile for each gene was determined from the number of ESTs per contig, and this value was 1 for singleton ESTs. For each gene in each of the six datasets, we standardized the expression level by dividing these values by the total number of ESTs in the original and complete EST library (Table1), an approach that has been previously demonstrated to be an effective measure of expression level [8,12,31]. These values were multiplied by 10 000 (to obtain ESTs per 10 000) and the expression level for each gene was categorized as either high (>5 ESTs per 10 000) or low (≤5 ESTs per 10 000). Although many of the putative unigenes used here have not yet been definitively described as genes for those species, we nonetheless refer to them as "genes" here and in the text.

We identified genes that were expressed only in the male library and only in the female library (or flower for *B. napus*) as follows. Beginning with the male EST library from each species, the longest EST sequence per contig was identified and chosen as the representative for that gene. Each singleton represents its own gene. Each of these ESTs was then submitted to MEGABLAST [41] as a query against the original and redundant female-specific EST dataset (or flower library for *B. napus*, Table 1). The original female EST dataset was used in order to be conservative in the identification of male-specific ESTs. ESTs having more than 95% similarity were considered a match, which represents a level of similarity rigorous enough to distinguish among genes in conserved gene families [32]. The genes not having matches were categorized as male-specific. The process was then repeated for the female set of genes (or flower-specific for *B. napus*), after removing the sequences that were identified as matches to the male EST library. Specifically, the longest EST per contig and each singleton was queried against the original and redundant male-specific EST library in MEGABLAST. Genes that did not have matches were considered female-specific. The final datasets were: *Z. mays* sperm-specific genes (N = 955), *Z. mays* egg-specific genes (*N* = 946), *T. aestivum* anther-specific genes (*N* = 3326), *T. aestivum* ovary-specific genes (*N* = 1489), *B. napus* microspore-specific genes (*N* = 1675) and *B. napus* flower-specific genes (*N*= 1637). Genes with high expression (>5 ESTs per 10 000) represented approximately one quarter (or less) of each of these datasets. Notably, these gender-specific genes represent those that are specific to a particular tissue or gamete (e.g. sperm) when compared to only one other tissue or gamete (e.g., egg), and thus the tissue/gamete-specific genes identified here for each species are more numerous than one would observe if these had been compared to all the available libraries for that species (as is the standard analysis for ESTs). As well, the description of a gene as tissue-specific (e.g., female-specific) does not necessarily indicate that the gene is not expressed in the opposing tissue (e.g., male-specific) but rather that there were no ESTs present in the publicly available EST dataset used for comparison. It should be noted that although the genomic DNA sequences were available for a small portion of the ESTs for each species, we only used EST sequences for this entire analysis for consistency.

The open reading frame for each singleton and an EST representing every contig (longest EST per contig) was identified for every gene from each of the six reduced-sized gender-specific sequence datasets (see above) using alignments from BLASTX against the *A. thaliana* protein database. The BLASTX amino acid based algorithm provides alignments of the six-frame translated EST relative to the protein database and thereby is more sensitive to elements of functionality and homology than DNA alignments and accurately reveals reading frames [41,59]. Using the amino acid alignments between the translated ESTs and the *A. thaliana* homologues, we identified and extracted that portion of each EST sequence representing the reading frame (these generally did not include the start or termination codons). Most edited sequences were between 200 and 700 bp in length. Gaps in the alignments were rare, but when identified these regions were excluded from the EST, as were any occasional missense codons (resulting from the less than 1% sequencing errors in most large scale EST sequencing projects; in-house ESTs had PHRED scores [33] of greater than 20, representing less than a 1% error). All DNA sequence editing was conducted using BioEdit 7.0.5.3 [60].

Data Analysis

The GC content at third codon positions (*GC3*) and the frequency of preferred codons (*Fpr*) for each gene was determined using CodonW [61]. *GC3* content has been shown to be well-correlated with the degree of biased codon usage for *A. thaliana* and other plant species [1, 31]. For the determination of *Fpr,* we used the preferred codons (sometimes called favoured codons) for *Z. mays* and *T. aestivum* previously identified by Kawabe and Miyashita (2003) [41]. Preferred codons are those that are most frequently used in the most highly-biased genes (as compared to lowly-biased genes) per degenerate codon group, and have been well-correlated to the optimal codons for many species (those codons used in the most highly expressed genes) [8]. *Fpr* was not determined for *B. napus* as the preferred codon data have not been described yet (and cannot be determined here as then the same data would then have to be used to determine both the preferred codons and *Fpr*).

A series of pairwise comparisons were conducted with respect to each combination of gender and gene expression level using *GC3* and *Fpr* (including the flower for *B. napus*). Additional pairwise contrasts were conducted to assess the impact of gene expression within each gender-specific dataset for each species (and the flower in *B. napus*) for a total of 35 contrasts. Tests were conducted using the non-parametric Mann-Whitney Rank Sum test (as normality was not detected in some contrasts; *t*-tests nonetheless yielded similar results). Statistical significance required $P < 0.05$. A Bonferroni correction was applied across all contrasts. All statistical analyses were conducted in SigmaPlot 10.0 and SigmaStat 3.5 for Windows (Systat© software 2006). In addition to these pairwise tests, we also determined the relative synonymous codon usage (*RSCU*) for the concatenated reading frames for the entire dataset of male-specific and for female-specific (and the flower for *B.*

napus) genes for each species using CodonW [61]. See main text for the description of gender-specific *RSCU*, *GC3*, and gene expression relative to protein length. Hierarchical clustering was conducted based on the Pearson correlation coefficients between *RSCU*values for each combination of species and gender-specific tissues/gametes for *Z. mays* and *T. aestivum* (Systat, 2004)[43].

As a means to assess whether the profile of gene functions conducted by male-specific and female-specific genes differed for *Z. mays* and for *T. aestivum*, each of these gender-specific sequence datasets (open reading frames) were submitted to MIPs [62]. A similar analysis was conducted for male-specific and flower-specific genes in *B. napus*. The gene functions were determined by comparison to proteins characterized in the *A. thaliana* database (MATDB)[62] with annotation from The Arabidopsis Information Resource (TAIR) and as implemented by the software Classification Superviewer [63].

AUTHORS' CONTRIBUTIONS

CAW conceived the project, conducted the majority of the bioinformatics and comparative analysis, and prepared the manuscript. MRM conducted laboratory work for the *B. napus*microspore EST library and participated in the analysis and the editing of the manuscript. JEK is the research group supervisor and contributed to the conception of the study, managed the data from the *B. napus* microspore library, and was involved in the data analysis and the editing of the manuscript. All authors read and approved the final manuscript.

ACKNOWLEDGEMENTS

The authors thank three anonymous reviewers for their valuable comments on the manuscript. We also thank the researchers listed in Table 1 for making their EST data publicly available. This work was funded by the National Research Council Canada – Genomics and Health Initiative III. This paper is National Research Council Canada publication No. 48420.

REFERENCES

1. Chiapello H, Lisacek F, Caboche M, Henaut A: Codon usage and gene function are related in sequences of *Arabidopsis thaliana*. *Gene* 1998, 209:GC1-GC38.
2. Urrutia AO, Hurst LD: Codon usage bias covaries with expression breadth and rate of synonymous evolution in humans, but this is not evidence for selection. *Genetics* 2001, 159:1191-1199.
3. Plotkin JB, Robins H, Levine AJ: Tissue-specific codon usage and the expression of humans genes. *Proc Natl Acad Sci USA* 2004, 101:12588-12591.
4. Hambuch TM, Parsch J: Patterns of synonymous codon usage in *Drosophila melanogaster* genes with sex-biased expression. *Genetics* 2005, 170:1691-1700.
5. Kotlar D, Lavner Y: The action of selection on codon bias in humans is related to frequency, complexity and chronology of amino acids. *BMC Genomics* 2006, 7:67.
6. Zhang Z, Hambuch TM, Parsh J: Molecular evolution of sex-biased genes in*Drosophila*. *Mol Biol Evol* 2004, 21:2130-2139.
7. Whittle CA, Johnston MO: Male-biased transmission of deleterious mutations to the progeny in *Arabidopsis thaliana*. *Proc Natl Acad Sci USA* 2003, 100:4055-4059.
8. Duret L, Mouchiroud D: Expression pattern and, surprisingly, gene length shade codon usage in *Caenorhabditis*, *Drosophila*, and *Arabidopsis*. *Proc Natl Acad Sci USA* 1999, 96:4482-4487.
9. Duret L: tRNA gene number and codon usage in the *C. elegans* genome are coadapted for optimal translation of highly expressed genes. *Trends Genet* 2000, 16:287-289.
10. Akashi H: Gene expression and molecular evolution. *Curr Opin Genet Dev* 2001, 11:660-666.
11. Akashi H: Translational selection and yeast proteome evolution. *Genetics* 2003, 164:1291-1303.
12. Wright SI, Yau CB, Loosely M, Myers C: Effects of gene expression on molecular evolution in *Arabidopsis thaliana* and *Arabidopsis lyrata*. *Mol Biol Evol* 2004, 21:1719-1726.
13. Guoy M, Gautier C: Codon usage in bacteria: correlation with gene expressivity. *Nucleic Acids Res* 1982, 10:7055-7074.

14. Sharp PM, Li W-H: An evolutionary perspective on synonymous codon usage in unicellular organisms. *J Mol Evol* 1986, 24:28-38.
15. Stenico M, Lloyd T, Sharp PM: Codon usage in *Caenorhabditis elegans*: delineation of translational selection and mutational biases. *Nucleic Acids Res* 1994, 22:2437-2446.
16. Ermolaeva MD: Synonymous codon usage in bacteria. *Curr Issues Mol Biol* 2001, 3:91-97.
17. Chen SL, Lee W, Hottes AK, Shapiro L, Harley H: Codon usage between genomes is constrained by genome-wide mutational processes. *Proc Natl Acad Sci USA* 2004, 101:3480-3485.
18. Miyata T, Hayashida H: Extraordinarily high evolutionary rate of pseudogenes: evidence for the presence of selective pressure against changes between synonymous codons. *Proc Natl Acad Sci* 1981, 78:5739-5743.
19. Sharp PM, Li WH: The codon adaptation index – a measure of directional synonymous codon usage bias, and its potential applications. *Nucleic Acids Res* 1987, 15:1281-1295.
20. Osawa S, Ohama T, Yamao F, Muto , Jukes TH, Ozeki H, Umesono K: Directional mutation pressure and transfer RNA in choice of the third nucleotide of synonymous two-codon sets. *Proc Natl Acad Sci USA* 1988, 85:1124-1128.
21. Sueoka N: Directional mutation pressure and neutral molecular evolution. *Proc Natl Acad Sci USA* 1988, 85:2653-2657.
22. Kano A, Andachi Y, Ohama T, Osawa S: Novel anticodon composition of transfer RNAs in *Micrococcus luteus*, a bacterium with a high genomic G + C content: Correlation with codon usage. *J Mol Biol* 1991, 221:387-401.
23. Sharp PM, Averof M, Lloyd AT, Matassi G, Peden JF: DNA sequence evolution: The sounds of silence. *Philos Trans R Soc Lond B Biol Sci* 1995, 349:241-247.
24. Wright SI: The "effective number of codons" used in a gene. *Gene* 1990, 87:23-29.
25. Ikemura T: Codon usage and tRNA content in unicellular and multicellular organisms. *Mol Biol Evol* 1985, 2:13-34.
26. 25.Kanaya S, Yamada Y, Kinouchi M, Kudo Y, Ikemura T: Codon usage and tRNA genes in eukaryotes: correlation of codon usage

diversity with translation efficiency and with CG-dinucleotide usage as assessed by multivariate analysis. *J Mol Evol* 2002, 53:290-298.

27. Moriyama EN, Powell JR: Codon usage bias and tRNA abundance in *Drosophila*. *J Mol Evol* 1997, 45:514-523.
28. Grantham R, Gautier C, Guoy M, Jacobzone M, Mercier M: Codon catalog usage is a genome strategy modulated for gene expressivity. *Nucleic Acids Res* 1981, 9:r43-r49.
29. Murray EE, Lotzer J, Eberle M: Codon usage in plant genes. *Nucleic Acids Res* 1989, 17:477-498.
30. Duret L, Mouchiroud D: Determinants of substitution rates in mammalian genes: expression patterns affects selection intensity but not mutation rate. *Mol Biol Evol* 2000, 17:68-74.
31. Wright SI, Lauga B, Charlesworth D: Rates and patterns of molecular evolution in inbred and outbred *Arabidopsis*. *Mol Biol Evol* 2002, 19:1407-1420.
32. Subramanian S, Kumar S: Gene expression intensity shapes evolutionary rates of the proteins encoded by the vertebrate genome. *Genetics* 2004, 168:373-381.
33. Ewing B, Hillier L, Wendl MC, Green P: Base-calling of automated sequencer traces using PHRED. I Accuracy Assessment. *Genome Res* 1998, 8:175-185.
34. Staden R, Beal KF, Bonfield JK: The Staden Package, 1998. In *Bioinformatics Methods and Protocols*. Edited by Misener S, Krawetz SA. Totowa, NJ: The Humana Press; 1999:115-130.
35. Adzhubei AA, Laerdahl JK, Vlasova AV: preAssemble: A tool for automatic sequencer trace data processing. *BMC Bioinformatics* 2006, 17:17-22.
36. Tiffin P, Hahn MW: Coding sequence divergence between two closely related plant species: *Arabidopsis thailiana* and *Brassica rapa* ssp. pekinensis. *J Mol Evol* 2002, 54:746-753.
37. Mitreva M, Wendl MC, Martin J, Wylie T, Yin Y, Larson A, Parkinson J, Waterston RH, McCarter JP: Codon usage patterns in Nematoda: analysis based on over 25 million codons in thirty-two species. *Genome Biol* 2006, 7:R75.
38. Huang X, Madan A: CAP3: A DNA sequence assembly program. *Genome Res* 1999, 9:868-877.

39. Altschul SF, Gish W, Miller W, Myers WE, Lipman DJ: Basic local alignment search tool. *J Mol Biol* 1990, 215:403-410.
40. Altschul SF, Madden TL, Schaffer AA, Zhang J, Zhang Z, Miller W, Lipman DJ: Gapped BLAST and PSI-BLAST: a new generation of protein database search programs. *Nucleic Acids Res* 1997, 25:3389-3402.
41. National Center for Biotechnology Information [http://www.ncbi.nlm.nih.gov]
42. Kawabe A, Miyashita N: Patterns of codon usage bias in three dicot and four monocot plant species. *Genes Genet Syst* 2003, 78:343-352.
43. *Systat 11 Statistics, Vol. I* Systat Software, Inc., Richmond, CA; 2004.
44. Wang D, Hsieh M, Li W-H: A general tendency for conservation of protein lengths across eukaryotic Kingdoms. *Mol Biol Evol* 2004, 22:142-147.
45. Maside X, Lee AW, Charlesworth B: Selection on codon usage in *Drosophila americana*. *Curr Biol* 2004, 14:150-154.
46. Hastings KE: Strong evolutionary conservation of broadly expressed protein isoforms in the troponin I gene family and other vertebrate gene families. *J Mol Evol* 1996, 42:631-640.
47. Charlesworth D: Why do plants produce so many more ovules than seeds? *Nature* 1989, 338:21-22.
48. Mazer SJ, Snow AA, Stanton ML: Fertilization dynamics and parental effects upon fruit development in *Raphanus raphanistrum*: consequences for seed size variation. *Am J Bot* 1986, 73:500-511.
49. Wyckoff G, Wang W, Wu CI: Rapid evolution of male reproductive genes in the descent of man. *Nature* 2000, 403:304-309.
50. Datta A, Jink-Roberton S: Association of increased spontaneous mutation rates with high levels of transcription in yeast. *Science* 1995, 268:1616-1619.
51. Beletskii A, Bhagwat AS: Transcription-induced mutations: increase in C to T mutations in the nontranscribed strand during transcription in *Escherichia coli*. *Proc Natl Acad Sci USA* 1996, 93:13919-13924.

52. Kliman RM, Hey J: The effects of mutation and natural selection on codon bias in the genes of *Drosophila*. *Genetics* 1994, 137:1049-1056.

53. Polowick PL, Sawhney VK: A scanning electron microscopic study on the initiation and development of floral organs of *Brassica napus* cv. Westar. *Am J Bot* 1986, 73:254-263.

54. Moriyama EN, Powell JR: Gene length and codon usage bias in *Drosophila melanogaster, Saccharomyces cerevisiae* and *Escherichia coli*. *Nucleic Acids Res* 1998, 26:3188-3193.

55. Lemos B, Bettencourt BR, Meiklejohn CD, Hartl DL: Evolution of proteins and gene expression levels are coupled in *Drosophila* and are independently associated with mRNA abundance, protein length, and number of protein-protein interactions. *Mol Biol Evol* 2005, 22:1345-1354.

56. Ingvarsson PK: Gene expression and protein length influence codon usage and rates of sequence evolution in *Populus tremula*. *Mol Biol Evol* 2007, 24:836-844.

57. Whittle CA, Johnston MO: Male-driven evolution of mitochondrial and chloroplastidial DNA sequences in plants. *Mol Biol Evol* 2002, 19:938-949.

58. Birky CW: Uniparental inheritance of mitochondrial and chloroplast genes: mechanisms and evolution. *Proc Natl Acad Sci USA* 1995, 92:11331-11338.

59. Gish W, States DJ: Identification of protein coding regions by database similarity search. *Nat Genet* 1993, 3:266-272.

60. Hall T: BioEdit: A user-friendly biological sequence alignment program for Windows 95/98/NT. *Nucleic Acids Symp Ser* 1999, 41:95-98.

61. Peden J: Analysis of codon usage. In *PhD Thesis*. Department of Genetics, University of Nottingham, UK; 1999.

62. Mewes HW, Frishman D, Guldener U, Mannhaupt G, Mayer K, Mokrejs M, Morgenster B, Munsterkkotter M, Rudd S, Weil B: MIPS: A database for genomes and protein sequences. *Nucleic Acids Res* 2002, 30:31-34.

63. Provart N: Classification Superviewer. [http://bbc.botany.utoronto.ca] 2006.

64. Engel ML, Charboud A, Dumas C, McCormick S: Sperm cells of *Zea mays* have a complex complement of mRNAs. *Plant J* 2003, 34:697-707.

65. Yang H, Kaur N, Kiriakopolos S, McCormick S: EST generation and analyses towards identifying female gametophyte-specific genes in *Zea mays* L. *Planta* 2006, 224:1004-1014.

CHAPTER 1

Adrian Reyes-Prieto, and Ahmed Moustafa, Plastid-Localized Amino Acid Biosynthetic Pathways of Plantae are Predominantly Composed of Non-Cyanobacterial Enzymes, doi: 10.1038/srep00955.

CHAPTER 2

RLM Guedes, F Prosdocimi, GR Fernandes, LK Moura, HAL Ribeiro, and JM Ortega, Amino acids biosynthesis and nitrogen assimilation pathways: a great genomic deletion during eukaryotes evolution, doi: 10.1186/1471-2164-12-S4-S2.

CHAPTER 3

Takashi Yuasa, Yuri Nagasawa, Katsumasa Osanai, et al., "Induction of a bZIP Type Transcription Factor and Amino Acid Catabolism-Related Genes in Soybean Seedling in Response to Starvation Stress," Journal of Botany, vol. 2013, Article ID 935479, 8 pages, 2013. doi:10.1155/2013/935479.

CHAPTER 4

Mohammad Anwar Hossain, Pukclai Piyatida, Jaime A. Teixeira da Silva, and Masayuki Fujita, "Molecular Mechanism of Heavy Metal Toxicity and Tolerance in Plants: Central Role of Glutathione in Detoxification of Reactive Oxygen Species and Methylglyoxal and in Heavy Metal Chelation," Journal of Botany, vol. 2012, Article ID 872875, 37 pages, 2012. doi:10.1155/2012/872875.

CHAPTER 5

Christian Hettenhausen, Maria Heinrich, Ian T Baldwin and Jianqiang Wu*,Fatty acid-amino acid conjugates are essential for systemic activation of salicylic acid-induced protein kinase and accumulation of jasmonic acid in Nicotiana attenuate , doi:10.1186/s12870-014-0326-z.

CHAPTER 6

Judith Katharina Paulus, Daniel Schlieper & Georg Groth; Greater efficiency of photosynthetic carbon fixation due to single amino-acid substitution, doi: 10.1038/ncomms2504.

CHAPTER 7

Zita Kovács, Livia Simon-Sarkadi, Ildikó Vashegyi, and Gábor Kocsy, "Different Accumulation of Free Amino Acids during Short- and Long-Term Osmotic Stress in Wheat," The Scientific World Journal, vol. 2012, Article ID 216521, 10 pages, 2012. doi:10.1100/2012/216521.

CHAPTER 8

Bhaskar Gupta and Bingru Huang, "Mechanism of Salinity Tolerance in Plants: Physiological, Biochemical, and Molecular Characterization," International Journal of Genomics, vol. 2014, Article ID 701596, 18 pages, 2014. doi:10.1155/2014/701596.

CHAPTER 9

Narendra Singh Yadav, Pushp Sheel Shukla, Anupama Jha, Pradeep K Agarwal, and Bhavanath Jha, he SbSOS1 gene from the extreme halophyte Salicornia brachiata enhances Na+ loading in xylem and confers salt tolerance in transgenic tobacco, doi:10.1186/1471-2229-12-188.

CHAPTER 10

Carrie-Ann Whittle, Meghna R Malik, and Joan E Krochko, Gender-Specific Selection on Codon Usage in Plant Genomes, doi:10.1186/1471-2164-8-169.

Index

A

B

V

W

Y

Z